Experimental Techniques in Magnetism and Magnetic Materials

The field of magnetism and magnetic materials has diverse applications in areas including information technology, wireless communication, microelectronics, information storage, and biotechnology. This text discusses experimental techniques for high-resolution measurements of magnetic properties of materials at both macroscopic and microscopic levels. It is meant to be a comprehensive introduction to graduate students and academicians in the fields of physics and materials science wishing to conduct research in magnetism and magnetic materials.

Starting with a brief history of the subject and insights into its contemporary applications, the book provides a concise exposition of the phenomenology of magnetism. It then covers a wide spectrum of experimental methodologies to study both the microscopic and macroscopic aspects of various kinds of magnetic phenomena and materials. The physical principles behind each kind of experimental technique with a broad-based introduction to instrumentation are also presented. The main text is augmented by appendices on the generation of the magnetic field, units in magnetism, and demagnetization fields.

Sindhunil Barman Roy is Emeritus Professor and Raja Ramanna Fellow at the UGC-DAE Consortium for Scientific Research, Indore, India. His research includes experimental and theoretical aspects of magnetism and superconductivity. He has published more than 200 research papers in international peer-reviewed journals and holds two US patents.

Experimental Techniques in Magnetism and Magnetic Materials

Sindhunil Barman Roy

CAMBRIDGE
UNIVERSITY PRESS

University Printing House, Cambridge CB2 8BS, United Kingdom

One Liberty Plaza, 20th Floor, New York, NY 10006, USA

477 Williamstown Road, Port Melbourne, VIC 3207, Australia

314 to 321, 3rd Floor, Plot No. 3, Splendor Forum, Jasola District Centre, New Delhi 110025, India

103 Penang Road, #0506/07, Visioncrest Commercial, Singapore 238467

Cambridge University Press is part of the University of Cambridge.

It furthers the University's mission by disseminating knowledge in the pursuit of education, learning and research at the highest international levels of excellence.

www.cambridge.org
Information on this title: www.cambridge.org/9781108489980

© Sindhunil Barman Roy 2022

This publication is in copyright. Subject to statutory exception and to the provisions of relevant collective licensing agreements, no reproduction of any part may take place without the written permission of Cambridge University Press.

First published 2022

Printed in India by Avantika Printers Pvt. Ltd.

A catalogue record for this publication is available from the British Library

ISBN 978-1-108-48998-0 Hardback

Cambridge University Press has no responsibility for the persistence or accuracy of URLs for external or third-party internet websites referred to in this publication, and does not guarantee that any content on such websites is, or will remain, accurate or appropriate.

To all my teachers, the first of the lot being my late parents

Contents

Preface   xiii

I   Introduction to Magnetism and Magnetic Materials   1

1   A Short History of Magnetism and Magnetic Materials   3

2   Role of Magnetism and Magnetic Materials in Modern Society   13

II   Basic Phenomenology of Magnetism   19

3   Magnetic Moment and the Effect of Crystal Environment   23
     3.1   Magnetic Moment and Magnetization 23
         3.1.1   Magnetic Moment 23
         3.1.2   Magnetization 25
     3.2   Classical Electromagnetism, Magnetic Moment, and Angular Momentum 26
     3.3   Precession of Magnetic Moment 30
     3.4   Magnetization, Magnetic Field, and Magnetic Susceptibility . . . 31
     3.5   Orbital and Spin Angular Momentum of Electron and Magnetic Moment 32
     3.6   Magnetism of Isolated Atoms and Ions 33
     3.7   Diamagnetism 34

- 3.8 Paramagnetism 37
  - 3.8.1 Classical theory of paramagnetism 37
  - 3.8.2 Quantum theory of paramagnetism 40
  - 3.8.3 van Vleck paramagnetism 41
- 3.9 Ground State of Ions and Hund's Rules 42
  - 3.9.1 Fine structure 42
  - 3.9.2 Hund's rules 43
  - 3.9.3 Russel–Saunders coupling versus $\vec{j}$–$\vec{j}$ coupling 43
- 3.10 Effect of Crystal Environment on Magnetic Ions 44
  - 3.10.1 Crystal fields and their origin 44
  - 3.10.2 Quenching of orbitals 47
  - 3.10.3 Jahn–Teller effect 48

4 Exchange Interactions and Magnetism in Solids 49
- 4.1 Coupling between Spins 49
- 4.2 Origin of Exchange Interactions 51
- 4.3 Physical Meaning of Exchange Energy 54
- 4.4 Direct Exchange 55
- 4.5 Indirect Exchange in Insulating Solids 55
  - 4.5.1 Superexchange 55
  - 4.5.2 Double exchange 59
  - 4.5.3 Anisotropic exchange interaction: Dzyaloshinski–Moriya interaction 61
- 4.6 Indirect Exchange in Metals 62

5 Magnetically Ordered States in Solids 63
- 5.1 Ferromagnetism 63
  - 5.1.1 Magnetic susceptibility in a ferromagnet 67
- 5.2 Antiferromagnetism 68
  - 5.2.1 Magnetic susceptibility in an antiferromagnet 69
- 5.3 Ferrimagnetism 71
- 5.4 Helical Magnetic Order 71
- 5.5 Spin-Glass Order 72
- 5.6 Spin Waves in Ferromagnets 74
- 5.7 Domains and Domain Wall 79
- 5.8 Magnetic Skyrmion 82
- 5.9 Magnetic Anisotropy 83
- 5.10 Magnetization Process in Ferromagnets 85

III Experimental Techniques in Magnetism   89

6 Conventional Magnetometry   93
- 6.1 Force Method 94
  - 6.1.1 Gouy and Faraday balance 94
  - 6.1.2 Alternating gradient magnetometer 96
  - 6.1.3 Cantilever beam magnetometer 99
- 6.2 Induction Method 102
  - 6.2.1 Vibrating sample magnetometer 104
  - 6.2.2 Superconducting quantum interference device magnetometer 108
  - 6.2.3 SQUID-VSM 115
  - 6.2.4 Extraction magnetometer 116
- 6.3 AC Susceptibility 117
- 6.4 Summary 122

7 Magnetic Resonance and Relaxation   125
- 7.1 Nuclear Magnetic Resonance 126
- 7.2 Electron Paramagnetic Resonance 136
- 7.3 Ferromagnetic Resonance 145
- 7.4 Muon Spin Rotation 151
- 7.5 Mössbauer Spectroscopy 162
- 7.6 Summary 166

8 Optical Methods   169
- 8.1 Magneto-optical Effects 169
  - 8.1.1 Principles of magneto-optical effects ... 170
  - 8.1.2 Experimental methods 172
- 8.2 Scanning Near-field Optical Microscopy 178
  - 8.2.1 Principle of scanning near-field optical microscope 178
  - 8.2.2 Magneto-optical measurement using scanning near-field optical microscope 180
- 8.3 Brillouin Light Scattering 182
  - 8.3.1 Principles of Brillouin light scattering .. 183
  - 8.3.2 Experimental method for Brillouin light scattering 184
- 8.4 Summary 189

9 Neutron Scattering   191
- 9.1 Neutron Sources and Neutron Scattering Facilities 193
- 9.2 Basics of Neutron Scattering 196
  - 9.2.1 Neutron cross sections 197
  - 9.2.2 Conservation of energy and momentum ... 198

		9.2.3 Master formula for neutron scattering	198
		9.2.4 Nuclear scattering	201
		9.2.5 Magnetic scattering	204
		9.2.6 Classification of magnetic structures	206
	9.3	Neutron Scattering Experiments	211
		9.3.1 Neutron powder diffraction	214
	9.4	Single-Crystal Experiments	217
	9.5	Polarized Neutron Scattering	218
	9.6	Magnetic Small-Angle Neutron Scattering	219
	9.7	Inelastic Neutron Scattering	223
	9.8	Polarized Neutron Reflectometry	229
	9.9	Summary	233

10 X-ray Scattering   237

    10.1 Magnetic and Resonant X-ray Diffraction 238
        10.1.1 Classical formalism of magnetic X-ray scattering 239
        10.1.2 Quantum mechanical theory of magnetic and resonant X-ray scattering 244
        10.1.3 Resonant X-ray scattering 247
        10.1.4 X-ray magnetic circular dichroism and X-ray magnetic linear dichroism 250
    10.2 Summary 255

11 Microscopic Magnetic Imaging Techniques   259

    11.1 Electron-Optical Methods 259
        11.1.1 Scanning electron microscopy 260
        11.1.2 Transmission electron microscopy 268
    11.2 Imaging with Scanning Probes 279
        11.2.1 Magnetic force microscopy 280
        11.2.2 Spin-polarized scanning tunneling microscopy 284
        11.2.3 Scanning Hall probe and scanning SQUID microscopy . . . 287
    11.3 Magnetic Imaging Using Synchrotron Radiation Sources 290
        11.3.1 Scanning X-ray microscopy 291
        11.3.2 Transmission X-ray microscopy 292
        11.3.3 X-ray photoelectron microscopy 294
    11.4 Summary 297

12 Nano-Scale Magnetometry with Nitrogen Vacancy Centre   301

    12.1 Physics of the Nitrogen-Vacancy (NV) Centre in Diamond 302
    12.2 A Brief Introduction to the Principle of NV Magnetometry 305
    12.3 Diamond Materials and Microscopy 306

12.4 Optically Detected Magnetic Resonance		308
12.5 NV Magnetometers		309
12.5.1 Samples for NV Magnetometry		309
12.5.2 DC magnetometer		310
12.5.3 AC Magnetometer		312
12.5.4 Sensitivity of NV magnetometers		313
12.5.5 Some experimental results		316
12.6 Summary		317

Appendix A   Magnetic Fields and Their Generation   **319**

    A.1 Steady Field 319

    A.2 Pulsed-Field 321

Appendix B   Units in Magnetism   **325**

Appendix C   Demagnetization Field and Demagnetization Factor   **329**

    C.1 Phenomenology 329

    C.2 Experimental aspects 335

Index   339

Preface

Magnetism and magnetic materials have been subjects of considerable interest for more than 3000 years, going back to ancient times. In modern times, myriad are the applications of magnetism and magnetic materials ranging from generation of electrical power to communications and information storage. Magnetic materials are absolutely indispensable in modern technology, and the intensity and importance of their applications are reflected in the multi-billion dollar market for magnetic materials in three broad areas: permanent magnets, soft magnets, and the magnetic recording medium. Continuous evolution in the field of magnetic materials has not, however, remained confined only to these well-identified areas. It is now well recognized that fresh applications are possible through the coupling of magnetism with other physical properties of materials such as magneto-thermal, magneto-elastic, magneto-optic, and magneto-electrical couplings. Newer classes of magnetic materials are being discovered with interesting new functions stimulating further growth of newer technology in various areas including information technology, wireless communication, microelectronics, biotechnology, and such like.

Against this backdrop, it is natural that the subject of magnetism finds a prominent place in solid state/condensed matter physics textbooks taught in the advanced undergraduate and postgraduate courses at universities all over the world. However, this coverage is mostly confined to a basic understanding of the phenomenon of magnetism as one of the physical properties of solid materials within the general framework of quantum mechanics. Detailed theoretical exposition of the subject is left to the specialized books on magnetism, and there are not too many of such books. Unlike high energy and particle physics, magnetism with its huge potential in technological applications is mostly an experimental science, where experimental techniques and related instruments play a very crucial role. The quantum many-body theories of magnetism (and for that matter condensed matter physics in general) are continuously evolving to explain the classes of emerging phenomena being discovered through experimental work on magnetic materials (and other classes of condensed matter). It is experimentation which is leading the field in the case of condensed matter/magnetism rather than theory as in the case of high energy and particle physics.

In the area of high energy and particle physics, the students are at least aware of the necessary theoretical techniques and the relevant experimental methods before they enter the research field at the Ph.D. level. The same, however, may not apply to the experimental sciences of magnetism and magnetic materials. The specialized textbooks on magnetism are mostly focused on the theoretical aspects of magnetism. Experimental aspects and techniques are mostly ignored in such textbooks or discussed in a rather cursory manner. The only dedicated book (to the best of this author's knowledge) available on the experimental methods in magnetism, which is accessible to the undergraduate and postgraduate level students and young researchers in the field, is the rather old book *Experimental Methods in Magnetism* by H. Zijlstra (North-Holland) published in 1967. Since that time there has been a huge development in the area of experimental techniques and entirely new classes of sensitive instrumentation have evolved for the high-resolution measurements of magnetic properties of materials at both the macroscopic and the microscopic levels. Such sophisticated instruments are now quite common in research laboratories all over the world. In addition, mega-science facilities of neutron and synchrotron radiation sources have become more accessible to researchers in both academic institutions and industries. Except for a few monographs and lecture series, there is, however, no book with comprehensive coverage of experimental techniques in magnetism and magnetic materials. Even these monographs are often a collection of specialized articles focusing on a particular class of experimental methods/characterization techniques. Such literature is useful to just a section of experienced specialist researchers.

There is also a general observation that the sophisticated laboratory instruments available today for experimental studies on magnetism and magnetic materials come with a computerized data acquisition and control system. With easy access to such state-of-the-art instruments, there is a tendency among the entry-level young researchers to use these instruments as a kind of black box without a proper understanding of their working principle. In this situation, it is not uncommon that the instruments may sometimes generate misleading experimental data. With the ever-increasing importance of the field of magnetism, there is now a need for a book that will introduce experimental magnetism to the advanced undergraduate and postgraduate level students in a pedagogical way, while also addressing young researchers in the areas of physics, materials science, and solid-state chemistry.

The present book aims to make the student aware of the experimental aspects of magnetism, and the complementarity between various experimental techniques to understand and appreciate various aspects of magnetism and magnetic materials. In addition, this book aims to become a good reference source to the researchers involved in research and development work in industrial laboratories.

The book consists of three parts. Part I has two short chapters. The first one outlines a brief history of magnetism and magnetic materials. The second one discusses the role of magnetism and magnetic materials in modern society with a short exposition of the current technological applications of magnetic materials and devices. Part II of the book provides a concise exposition of the phenomenology of magnetism. The contents of Part II are more than that usually available in the general solid-state physics textbooks, but much less than that in a few excellent textbooks available in the market on the physics of magnetism. This part was included to make the book self-contained and to provide a ready reference to the reader in the context of the next part of the book. Part III of the book has seven chapters covering a wide spectrum of experimental methodologies to study both the microscopic and macroscopic aspects of various kinds of magnetic phenomena and materials. The physical principles behind each kind of experimental technique with a broad-based introduction to instrumentation are also presented. Further, there are three appendices on the generation of the magnetic field, units in magnetism, and demagnetization fields for augmenting the main body of the text.

There will be plenty of technical references for the interested readers for more detailed information. The technical references chosen in the book are of two types: (i) review papers, monographs on a particular class of experimental techniques, and edited books with a collection of specialized review articles; (ii) technical references with experimental results on actual magnetic materials obtained using various experimental techniques. The second kind of references is chosen randomly from the vast available literatures to give some flavour of the information that can be obtained from various kinds of experiments performed on magnetic materials.

I am grateful to Sudip Pal, D. T. Adroja, M. K. Chattopadhyay, Dileep Gupta, and B. Raghavendra Reddy who took the trouble to read some of the chapters in the book and let me have their criticism and suggestions, and correct errors that had crept in. For the remaining errors (if any) in the book I remain solely responsible. I am also grateful to Sudip Pal for helping me to draw some of the illustrations in the book. I acknowledge the Department of Atomic Energy, Government of India, for providing financial support in the form of the Raja Ramanna Fellowship. My wife Gopa is a source of constant support in my academic pursuit, and her inspiration to complete the writing of this book is deeply appreciated. I acknowledge my nephew Soumyanil for his support and encouragement while this book was being written. I also acknowledge the help and support received from Vaishali Thapliyal, Vikash Tiwari and Aniruddha De of Cambridge University Press.

Part I
Introduction to Magnetism and Magnetic Materials

The book begins with an exposition of the interesting history of magnetism and magnetic materials. This is followed by a short chapter discussing the role of magnetism and magnetic materials in modern society and current technological applications of magnetic materials and devices. This second chapter highlights why magnetism is considered to be more of an applied or experiemntal science rather than a theoretical one.

1

A Short History of Magnetism and Magnetic Materials

The magnetite iron ore FeO - Fe_2O_3 (or Fe_3O_4), famously known as lodestone, is the first known natural magnet. Folklore is that roughly around 2500 BC a Greek shepherd was tending his sheep in a region of ancient Greece called Magnesia (now in modern Turkey), and the nails that held his shoe together were stuck to the rock he was standing on. There were more such ancient stories about iron parts being pulled out from hulls of the ships sailing past the islands in the south Pacific and ones about the disarming and immobilizing of knights in their iron armor. Depending on the time and places where Fe_3O_4 or magnetite ore was found, it was variously known as the *Magnesia stone, lodestone,* the *stone of Lydia, l'aimant* in France, *chumbak* in India, or *ts'u she* in China. The modern name *magnet* is possibly derived from early lodestones found in the ancient Greek region of Magnesia.

The chronicled history of magnetism dates back to 600 BC. Lodestone's magnetic properties were studied and documented by the famous Greek philosopher Thales of Miletus (Fig. 1.1) in 600 BC [1]. Around the same period, the magnetic properties of lodestone were known in India, and the well-known ancient physician sage Sushruta (see Fig. 1.1) applied it to draw out metal splinters from bodies of injured soldiers [2]. However, Chinese writings dating back to 4000 BC mention magnetite, and indicate the possibility that original discoveries of magnetism might have taken place in China [3]. The Chinese were the first to notice that lodestone would orient itself to point north if not hindered by gravity and friction. The early Chinese compasses, however, were used in fortune-telling through the interpretation of lines and geographic alignments as symbols of the divine. These were also used for creating harmony in a room or building with the alignment of various features to different

FIGURE 1.1 Greek philosopher Thales of Miletus and Indian sage physician Sushruta. (Source: https://commons.wikimedia.org/)

FIGURE 1.2 Ladle shaped Chinese *south pointer* compass. (Source: CC BY-SA 3.0, https://commons.wikimedia.org/w/index.php?curid=553022)

compass points. The first navigational lodestone compasses emerged from China. They had a unique design with the lodestone being shaped as a ladle (Fig. 1.2). The lodestone ladle sat in the center of a bronze or copper plate/disc. These compasses rotated freely when pushed and usually came to rest with the handle part of the ladle pointing south and so were known as *south pointers*. The copper/bronze base would be inscribed with cardinal direction points and other important symbols.

An important discovery (attributed to Zheng Gongliang in 1064) was that iron acquired a thermoremanent magnetization when quenched from high temperature

FIGURE 1.3 Medieval navigational compass. (Source: https://www.freepnglogos.com/pics/compass)

red hot conditions [4]. The first artificial permanent magnets were the steel needles thus magnetized in the Earth's field. These needles aligned themselves with the Earth's field when suitably suspended or floated; that eventually led to their use as navigational compasses especially for sea voyages. Such compasses were reinvented in Europe a century later and played a very important role in the great voyages of discovery. Fig. 1.3 shows a medieval navigational compass.

A systematic study of magnetism was started in the sixteenth century by an English physicist, William Gilbert of Colchester (Fig. 1.4), who also happened to be the personal physician of Queen Elizabeth I of England. Gilbert performed experiments with a magnetic needle placed on the surface of a lodestone sphere and conjectured that the Earth is a large spherical magnet. Within this framework, the north-south pointing property of a compass needle could be conceptualized with the attraction of unlike poles of magnets. He also investigated the temperature-dependent magnetic properties of iron and observed that iron was no more attracted to a magnet when it became red hot. Gilbert also could distinguish clearly between static electricity and magnetism and documented all his investigations of magnetic and static electricity phenomena in a monograph, *De Magnete*, first published in London in 1600. This monograph is possibly the first printed textbook in any branch of modern physics [3].

Significant progress was made during the 1700s in making artificial magnets from iron bars and steel. Another important event was the invention of the horseshoe magnet in 1743 by Daniel Bernoulli (Fig. 1.4). This was to become the most classic shape of the magnet, and remains as an icon of magnetism even today. However, the

FIGURE 1.4 Clockwise from the top, William Gilbert (1544–1603), Daniel Bernoulli (1700–1782), Simeon Denis Poisson (1781–1840), and Charles-Augustin de Coulomb (1736–1806). (Source: https://www.famousscientists.org and https://en.wikipedia.org/w/index.php?title=Sim%C3%A9on_Denis_Poisson&oldid=1072295007)

next major development in the understanding of the subject took place only in the eighteenth century when Charles-Augustin de Coulomb (Fig. 1.4) experimentally found the inverse square force law, namely the Coulomb law, between electrical charges as well as magnetic poles. Subsequently, Simeon Denis Poisson (Fig. 1.4) and Carl Friedrich Gauss (Fig. 1.5) formulated an elegant mathematical theory for Coulomb's experimental findings. In 1820 Danish physicist Hans Christian Oersted (Fig. 1.5) discovered that a wire carrying an electrical current can deflect a magnetic needle placed nearby. An explanation for this discovery was put forward by the French scientist Andre-Marie Ampere (Fig. 1.5) in terms of a *magnetic force* around a current-carrying wire. This idea led to a law, which is now known as the Ampere law that was further formulated in a mathematical form by James Clerk Maxwell (Fig. 1.6). Ampere also carried out further experiments on the magnetic effects of electric currents, including measurements of forces between current-carrying wires.

The next important development in the field of magnetism was due to Michael Faraday (Fig. 1.5), when in 1831 he discovered the law of electromagnetic induction. He put two coils of insulated copper wire around a thick iron ring on two opposite sides, and connected one of the coils to a galvanometer and the other to a battery through a switch. Faraday observed sudden deflection in the galvanometer when the switch was closed or opened. This observation marked the discovery of the phenomenon of electromagnetic induction. Subsequently, Faraday also found that a current could be induced in a coil with the movement of a permanent magnet

FIGURE 1.5 Clockwise from the top, Carl Friedrich Gauss (1777–1855), Hans Christian Oersted (1777–1851), Michael Faraday (1791–1867), and Andre-Marie Ampere (1775–1836). (Source: https://www.famousscientists.org)

near the coil. He carried forward the studies of Oersted and Ampere and introduced the revolutionary concept of magnetic and electric fields. Faraday also discovered diamagnetism and paramagnetism in substances like bismuth and oxygen, and observed that diamagnetic substances were repelled from stronger magnetic fields while paramagnetic substances were attracted towards these.

James Clerk Maxwell (Fig. 1.6) expressed experimental findings of Faraday in elegant mathematical language in the form of the now-famous Maxwell equations. Not only Faraday's but the findings of Coulomb, Oersted, Ampere, and Poisson too could now be expressed by Maxwell equations. That subsequently led to further developments in magnetism and ultimately a technological revolution. Lorentz introduced an additional equation for the force on a particle with charge q moving with velocity v and subjected to a combined electric and magnetic field.

The experimental findings of Michael Faraday on paramagnetic and diamagnetic materials were advanced further by Pierre Curie (Fig. 1.6). He investigated the possibility of phase transitions between various kinds of magnetism in a given material by performing a thorough study of magnetic properties of some twenty substances [5]. His studies led to three important discoveries: (1) diamagnetism is generally approximately temperature-independent; (2) paramagnetism in various magnetic materials is temperature-dependent and magnetic susceptibility is inversely proportional to temperature; this is now famously known as the Curie law; (3) ferromagnetism disappears when the temperature is raised above a critical point called the Curie temperature. Paul Langevin (Fig. 1.6) provided the first microscopic mathematical theory of the behavior of diamagnetic and paramagnetic substances

FIGURE 1.6 Clockwise from the top, James Clerk Maxwell (1831–1879), Pierre Curie (1859–1906), Pierre Weiss (1865–1940), and Paul Langevin (1872–1946). (Source: https://commons.wikimedia.org)

using the statistical mechanics of Boltzmann and Gibbs. Langevin deduced the correct temperature dependence of magnetization of a paramagnetic sample with an ad-hoc assumption of a tiny magnetic moment associated with each atom of the sample and the inclusion of the thermal agitation of magnetic moments. Langevin's theory was generalized further by Pierre Weiss (Fig. 1.6) to explain ferromagnetism by introducing the important concept of *mean molecular field*. Explanation of ferromagnetism by the theory of Weiss, however, was only partial and a complete understanding had to wait for a quantum mechanical explanation.

George Uhlenbeck and Samuel Goudsmit (Fig. 1.7) in 1925 discovered the intrinsic spin of the electron, which is quantized in such a manner that it can have just two possible orientations, up and down, in an applied magnetic field. The intrinsic magnetic moment of the electron the Bohr magneton: $\mu_B = 9.274 \times 10^{-24} \mathrm{Am}^2$ or $9.274 \times 10^{-24} \mathrm{JT}^{-1}$ originates from the spin, and the magnetic properties of solids originate from the magnetic moments of their atomic electrons. Werner Heisenberg (Fig. 1.7) in 1929 showed that the Weiss molecular field in

Werner Heisenberg

Wolfgang Pauli

George Uhlenbeck and Samuel Goudsmit

Louis Neel

FIGURE 1.7 Clockwise from the top, Werner Heisenberg (1901–1976), Wolfgang Pauli (1900–1958), Louis Neel (1904–2000), and George Uhlenbeck (1900–1988) and Samuel Goudsmit (1902–1978). (Source: https://commons.wikimedia.org)

ferromagnets arose from two interconnected effects, namely the quantum mechanics of the exclusion principle introduced by Wolfgang Pauli (Fig. 1.7) and electrostatic repulsion between two electrons. The core idea is that a quantum mechanical exchange interaction causes the ferromagnetic alignment of tiny atomic magnetic moments. Exchange interaction in ferromagnets has positive values, but it can also have negative values leading to antiparallel rather than parallel alignment of magnetic moments. Louis Néel (Fig. 1.7) in 1936 and 1948 pointed out the possibility of such antiferromagnetism and ferrimagnetism, depending on the crystal structure of the material. The archetypal natural magnetic material magnetite Fe_3O_4 is a ferrimagnet.

On the materials and technology front, in 1750 an English physicist and astronomer, John Michell, developed a method of making magnets from soft iron and hardened steel, and the latter retained magnetization for a considerably longer time. Another technical landmark was William Sturgeon's invention of the iron-cored electromagnet in 1824. The electromagnet had a horseshoe-shaped core, which was magnetized by the magnetic field produced with the flow of electric current in the magnet windings [4]. The era of modern technology, however, started in the 1930s when it was found that the carbon-free iron-rich ternary alloy with 25% nickel and 10% aluminum was a ferromagnet with a relatively high magnetic coercivity, and which did not require hardening. The properties of these magnets were improved further during the next 50 years by substituting part of the iron with cobalt and adding copper and titanium. The magnets were also prepared by sintering powdered alloys, and also by powder bonding with plastic and then subjecting to pressure. These magnets are known as Alnico magnets, and they have been widely used in military as well as civil electronic applications such as automotive and aircraft sensor applications.

In the 1950s ceramic materials barium ferrite oxides ($BaFe_{12}O_{19}$) and strontium ferrite oxides ($SrFe_{12}O_{19}$), better known as ferrite magnets, were introduced. These are ferrimagnets with both ferromagnetic and antiferromagnetic coupling between atomic moments with magnetic interaction depending on the specific crystallographic position of iron ions. Despite their poorer service parameters (brittleness, low magnetization value at room temperature) compared to Alnico magnets, ferrite magnets have the commercial and industrial advantage due to their low cost and chemical inertness and also because they are easy to process.

In the mid-1960s new permanent magnets were prepared with the alloying of rare earth elements with $3d$-transition metals (Fe, Ni, Co). The examples of such permanent magnets are $SmCo_5$, Sm_2Co_{17} and Sm-Fe-N magnets, especially $Sm_2Fe_{17}N_3$. Research in this field eventually led to the development of Nd-Fe-B magnets in the mid-1980s. It may be noted that forming perfect $Nd_2Fe_{14}B$ and $Sm_2Fe_{17}N_3$ phases is a difficult if not impossible task. Thus the focus of research moved towards improving the microstructure of the materials. Magnets of nanocrystalline structure, nanocomposite, and anisotropic diphase nanocrystalline magnets were produced. Fig. 1.8 shows the evolution of permanent magnetic materials since the early twentieth century.

It may be noted here that while fundamental understanding of magnet science was definitely of help, it was not necessarily a prerequisite for metallurgical and initial technological progress. The progression from the poorly distinguished soft and hard magnetic steels in the early twentieth century to the wealth of different magnetic materials available today are more due to metallurgy and solid-state and structural chemistry. Quantum mechanics, however, started contributing significantly to the development of magnetic materials involving rare earth and transition metals [4]. The interplay between science and technology over the last century, especially from the early 1950s, led to an immense expansion in the applications of magnetic materials, more so during the last five decades. Much of

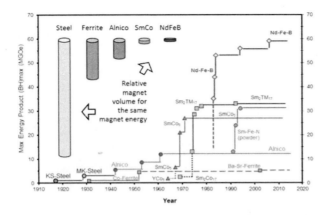

FIGURE 1.8 Evolution of permanent magnetic materials from early twentieth century. (Source: https://www.intechopen.com/chapters/63847; DOI: http://dx.doi.org/10.5772/intechopen.81386; Creative Commons Attribution 3.0 License)

the progress that has taken place in the field of computers, telecommunications equipment and various consumer goods is linked to the advances in permanent magnets, magnetic recording materials and high-frequency magnetic materials. Permanent magnets replaced electromagnets in billions of tiny motors manufactured every year, whereas magnetic recording sustains the revolution in information technology and the Internet. There have also been tremendous advances in earth science, medical imaging, and the theory of phase transitions in materials that can be linked to magnetism.

A new horizon is now opening up in the form of spin electronics or spintronics [4]. It is now well recognized that current complementary metal-oxide-semiconductor (CMOS) technology is fast approaching its fundamental limits, and there is driving interest in the newer kinds of materials with newer concepts in physics leading to devices for information processing and storage. CMOS technology involves only the charge of the electrons and ignores the spin on the electron. The research effort is now on to learn how to make good use of the spin by controlling and manipulating spin currents [6]. Then there is another class of transition metal-based insulating magnetic oxide materials, the so-called Mott insulators, where the insulating state arises out of strong electron–electron interaction. Although Mott insulators were known since the late 1930s, they started drawing attention only after the 1990s. The Mott insulators with their newer physics concepts can lead to devices for information processing and storage including metal-oxide-semiconductor field-effect transistor (MOSFET) and memory devices [7]. All these possibilities have now been recognized and recorded in the International Technology Roadmap for Semiconductors [8]. Then there are new classes of magnetic materials, namely two-dimensional (2D), curved, and topological magnetic materials, multiferroic and magnetocaloric materials, and also materials exhibiting unconventional magnetic phase transitions. While research and applications in the twentieth century involving magnetic substances and phenomena were mainly focused on collinear magnetization configurations, such efforts in the twenty-first century have been increasingly dominated by noncollinear magnetism, for example, spin spirals and skyrmions [9]. These developments in turn give rise to the possibility of newer and novel technological applications [10].

The three core pillars of modern-day magnetism, (i) magnetic materials, (ii) magnetic phenomena and associated characterization techniques, and (iii) applications of magnetism, are now well identified. The future progress in the field of magnetism will essentially be based on these three pillars and their interconnectivity [11, 12].

Bibliography

[1] Shoenberg, D. (1949). *Magnetism.* London: SIGMA Books Limited.
[2] Singh, N., and Jayannavar, A. M. (2019). *A Brief History of Magnetism.* Condmat ArXiv 1903.07031.
[3] Mattis, D. C. (2006). *The Theory of Magnetism Made Simple.* World Scientific.
[4] Coey, J. M. D. (2009). *Magnetism and Magnetic Materials.* Cambridge: Cambridge University Press.
[5] *Biography of Pierre Curie* in Encyclopedia Britannica (www.britannica.com/biography/Pierre-Curie).
[6] Zutit, I., Reyren, N., and Cros, V. (2017). *Nature Reviews Materials,* 2: 17031.
[7] Roy, S. B. (2019). *Mott Insulators: Physics and Applications.* IOP Publishing.
[8] Emerging Research Devices. (2015). *International Technology Roadmap for Semi Conductors.* https://irds.ieee.org/.
[9] Fert, A., Reyren, N., and Cros, V. (2017). Magnetic Skyrmions: Advances in Physics and Potential Applications. *Nature Reviews Materials,* 2 (7): 17301.
[10] Tokura, Y., and Kanazawa, N. (2020). Magnetic Skyrmion Materials. *Chemical Reviews* 121 (5): 2857–2897. https://dx.doi.org/10.1021/ acs.chemrev.0c00297.
[11] Sander, D., Valenzuela, S. O., Makarov, D., et al. (2017). The 2017 Magnetism Roadmap. *Journal of Physics D: Applied Physics,* 50 (36): 363001.
[12] Vedmedenko, E. Y., Kawakami, R. K., Sheka, D. D., et al. (2020). The 2020 Magnetism Roadmap. *Journal of Physics D: Applied Physics,* 53 (45): 453001.

2

Role of Magnetism and Magnetic Materials in Modern Society

Magnets have been used in society for centuries. In ancient times they were considered paranormal or mysterious substances. Nobody knew how or why the magnets attracted certain but not all materials. As we have seen in Chapter 1, it was not until the seventeenth century that there was considerable understanding of electromagnetism and a progressive increase in the use of magnetic materials as useful functional materials. Nowadays magnets are all-pervasive in modern society, starting from home to medical applications, to transport, and industrial sectors (Fig. 2.1).

We utilize magnetism all over our homes although it may not be very obvious. Electric motors create force by using electricity and magnetic fields. So nearly all household appliances such as fans, washing machines, vacuum cleaners, and blenders that use electricity to create motion invariably have magnets. Many households have small magnets holding paper notes or small items to the metal refrigerator door. While some magnets are visible, the others are hidden inside various items and appliances such as computers, cellphones, DVDs, iPods, cameras, sensors, doorbells, and toys of children. The dark stripe on the backside of credit cards is a magnetic strip storing the relevant data of the cardholder.

Computers use hard disk drives to store information. Hard disks are memory devices where magnets alter the direction of magnetic material on disk segments. Information is processed in computers in binary language, the base-2 units of which correspond to a magnetic field aligned to either the north or the south. These fields are spun in a hard disk, and a magnetic sensor is used to read these. Inside the small

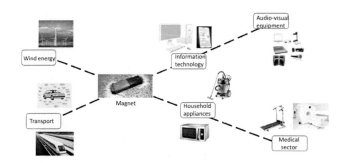

FIGURE 2.1 Some applications of magnets in modern society.

speaker found in computers, televisions and radios, electrical signals are converted into sound vibrations by wire coil and magnet.

Magnets are used profusely in the industrial world. Mechanical energy is converted into electricity with the use of magnets in electric generators. On the other hand, motors use magnets to convert electricity back into mechanical work. Sorting machines using magnets are deployed in mines to separate useful metallic ores from crushed rock. In the food processing industry magnets are used for removing small metallic particles from grains and other food. Electrically powered magnets in cranes are used in construction sites as well as in the recycling process to sort out and move large pieces of metal, some weighing thousands of kilograms.

In the medical sector, Magnetic Resonance Imaging (MRI) machines are quite common nowadays. In an MRI scan, powerful magnets are used to align hydrogen atoms in a patient's body, which enables the creation of detailed three-dimensional images of areas in the patient's body. Biocompatible magnetic nanoparticles offer a wide variety of applications in biomedicine: contrast agents in MRI, site-specific magnetic targeting, magnetic hyperthermia treatment, multimodal imaging, magnetic field-dependent controlled drug delivery, magnetofection, biomedical separation, tissue repair, and so many others [1]. In magnetic hyperthermia treatment, the magnetic energy absorption of magnetic nanoparticle-containing tissues in a radio frequency field induces localized heating, and that at a critical temperature range above 42–45 ^{0}C kills targeted cancer cells [2]. Fig. 2.2 presents a schematic illustration of various biomedical applications of magnetic nanomaterials.

More importantly, modern society is now at a crossroads with human development causing a rapid depletion of natural energy resources and climate changes with unpredictable consequences. There is now an urgent requirement of new concepts for energy generation and utilization in the progressive replacement of oil-based fuels in transportation by electric motors. Improvement in the efficiency of electricity transmission and end use is also very much on the cards. To this end, functional magnetic materials, such as advanced hard and soft magnets, magnetic

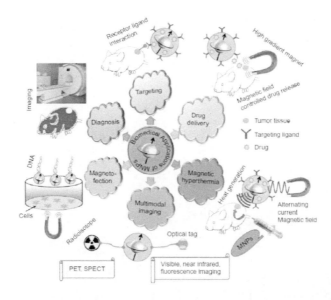

FIGURE 2.2 A schematic representation of various biomedical applications of magnetic nanomaterials. (From [1] with permission from Elsevier)

refrigerants, magnetic microelectromechanical systems, magnetic shape memory alloys, and magnetorheological fluids and elastomers, can have a big influence on the progress of the new emerging technologies [3].

Electric motors, generators, transformers, and actuators are all-pervasive in our technological society [3]. Motors ranging from a few watts to several hundred kilowatts have been widely employed in office and household appliances, the transportation sector, and industrial drives (Fig. 2.3). In developed and developing countries a huge percentage of the electricity generated is consumed by electric motors alone. Large economical and environmental savings are possible with an improvement in energy efficiency for electric motors, and even small progresses matter because there are so many of them. To this end, magnetic materials are essential components of motors, generators, transformers, actuators, and suchlike, and improvement in magnetic materials will have a significant positive impact on society. For example, a newer improved magnet can lead to a smaller, lighter-weight electric motor with the necessary power and torque required for transportation and actuation applications [3].

There is now an increasing need for environment-friendly electric transportation including hybrid and electric vehicles. This will increase the consumption of electricity significantly above its current rates of annual increase. Innovative designing and development of high-efficiency motors, generators, and power converters will not be enough to offset this increased consumption of electricity. In the intense research and development effort on renewable or sustainable energy

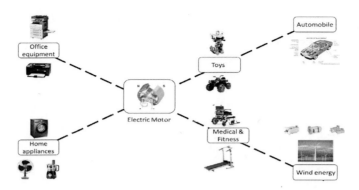

FIGURE 2.3   Electric motors in office and household appliances, transport and medical sectors.

sources like solar, wind, ocean, hydropower, nuclear, and geothermal energy that can be converted to electrical power, it should be noted that magnetic materials will be critical for the very efficiencies of many such renewable energy technologies [4], and we provide a few examples below.

The bidirectional flow of energy between sources, storage, and the electrical grid involves the conversion of electrical power [5]. This is accomplished through the use of power electronics. Power conversion electronics and motor controllers can operate at much higher frequencies with the use of wide bandgap semiconductors and promise more energy-efficient and high-rotational speed electrical machines. Soft magnetic materials are key to the efficient operation of such power electronics and next-generation electrical motors and generators [5].

Another example is refrigeration for food and air conditioning, which contribute significantly to the total electrical power consumption in the average household of developed and developing countries. To this end, vapor-compression refrigerators are bulky, heavy, and energy inefficient and at the same time contribute to greenhouse gas emissions. Here the use of adiabatic demagnetization effect in magneto-caloric materials can provide energy-efficient and environment-friendly solid-state refrigeration technologies.

Finally, the application of spin of the electron for electronic devices, information processing and storage, charge transport, etc., referred to collectively as spintronics, has generated much interest in the scientific community during the last two decades. This is exemplified by the 2020 magnetism roadmap [6] and the 2021 quantum materials roadmap [7]. Realization of such spin-based phenomena at nanoscale mainly stimulates the search for low-dimensional magnetic materials. Reduction of dimensionality gives rise to a radical change that occurs in the material characteristics (electrical, magnetic, structural, etc.) with newer and exciting prospects for future technological applications.

Bibliography

[1] Guleria, A., Priyatharchini, K., and Kumar, D. (2018). Biomedical Applications of Magnetic Nanomaterials. *Applications of Nanomaterials*. Elsevier, pp. 318. https://doi.org/10.1016/B978-0-08-101971-9.00013-2.

[2] Rivas, J., Bañobre-López, M., Piñeiro, Y., et al. (2012). Magnetic Nanoparticles for Application in Cancer Therapy. *Journal of Magnetism and Magnetic Materials*, 324 (21): 3499–3502.

[3] Gutfleisch, O., Willard, M. A., Brck, E., et al. (2011). Magnetic materials and devices for the 21st century: stronger, lighter, and more energy efficient. *Advanced Materials*, 23: 821

[4] Matizamhuka, W. (2018). The Impact of Magnetic Materials in Renewable Energy-Related Technologies in the 21st Century Industrial Revolution: The Case of South Africa. *Advances in Materials Science and Engineering*. https://doi.org/10.1155/2018/3149412.

[5] Silveyra, J. M., Ferrara, E., Huber, D. L., and Monson, T. C. (2018). Soft Magnetic Materials for a Sustainable and Electrified World. *Science*, 362–418.

[6] Vedmedenko, E. Y., Kawakami, R. K. Sheka, D. D., et al. (2020). The 2020 Magnetism Road Map. *Journal of Physics: Materials*, 3 042006. https://doi.org/10.1088/1361-6463/ab9d98. IOP Publishing.

[7] Giustino, F., Lee, J. H., Trier, F., et al. (2020). The 2021 Magnetism Road Map. *Journal of Physics: Materials*, 3 042006. https://doi.org/10.1088/2515-7639/abb74e. IOP Publishing.

Part II
Basic Phenomenology of Magnetism

The magnetic properties in solids originate mainly from the magnetic moments associated with electrons. The nuclei in solids also carry a magnetic moment. That, however, varies from isotope to isotope of an element. The nuclear magnetic moment is zero for a nucleus with even numbers of protons and neutrons in its ground state. The nuclei can have a non-zero magnetic moment if there are odd numbers of either or both neutrons and protons. However, the magnetic moment of a nucleus is three orders of magnitude less than that of the electron.

The microscopic theory of magnetism is based on the quantum mechanics of electronic angular momentum, which has two distinct sources: orbital motion and the intrinsic property of electron spin [1]. The spin and orbital motion of electrons are coupled by the spin–orbit interaction. The magnetism observed in various materials can be fundamentally different depending on whether the electrons are free to move within the material (such as conduction electrons in metals) or are localized on the ion cores. In a magnetic field, bound electrons undergo Larmor precession, whereas free electrons follow cyclotron orbits. The free-electron model is usually a starting point for the discussion of magnetism in metals. This leads to temperature-independent Pauli paramagnetism and Landau diamagnetism. This is the case with noble metals and alkali metals. On the other hand, localized non-interacting electrons in $3d$-transition metals, $4f$-rare earth elements, $5f$-actinide elements, and their alloys and intermetallic compounds with incompletely filled inner shells exhibit Curie paramagnetism. Many transition metal-based insulating oxide and sulfide compounds also show Curie paramagnetism. In the presence of magnetic interactions, many such systems eventually develop long-range magnetic order if the magnetic interaction can overcome thermal fluctuations in some temperature regimes.

Against the above backdrop, in the next three chapters, we will introduce the readers to the basic phenomenology of magnetism, concentrating mainly on solid materials with some electrons localized on the ion cores. There are some excellent textbooks available on the subject, including those by J. M. D. Coey [1], B. D. Cullity and C. D. Graham [2], D. Jiles [3], S. J. Blundell [4], and N. W. Ashcroft and N. D. Mermin [5]. The aim here, however, is to make the present book on experimental techniques a self-contained one. Hence, the subject matters presented will be rather introductory and mostly in line with the books of Blundell [4], and Ashcroft and Mermin [5].

The study of the subject of magnetism is complicated by the existence of two different systems of units: the SI (International System) or MKS, and the CGS (electromagnetic or emu) system. For example, a magnetic field can mean either $\vec{B}$-field or $\vec{H}$-field. The SI units for these fields are tesla (T) and amperes per meter (Am^{-1}), respectively, whereas in CGS units these are gauss (G) and oersted (Oe), all of which are currently in use. It is possible to use either the B-field or the H-field when one is dealing with the magnetic field. They can be distinguished by their behavior at a boundary between media having different relative permeabilities (μ_r). Across the boundary of the media, the normal component of the B-field and the tangential component of the H-field will be continuous. In free space $\vec{B} = \mu_0 \vec{H}$. In SI units $\mu_0 = 4\pi \times 10^{-7}$ henreys per meter (Hm^{-1}), hence $\vec{B}$-field and $\vec{H}$-field have different numerical values. On the other hand, in CGS system $\mu_0 = 1$, hence $\vec{B}$-field or the $\vec{H}$-fields have identical numerical values. This causes some confusion in the nomenclature of gauss for the $\vec{B}$-field and oersted for the $\vec{H}$-field. Both SI and CGS systems are found in scientific literature, research papers, materials, and instrument specifications; so we will use both sets of units in this book. Different magnetic units and their conversion factors are summarized in tabular forms in Appendix B. Readers should refer to Appendix B for a more detailed discussion on the magnetic units.

3

Magnetic Moment and the Effect of Crystal Environment

3.1 Magnetic Moment and Magnetization

When we examine a magnetic material, it is first essential to identify parameters that characterize the response of the magnetic material to an applied magnetic field. We will see that these parameters are magnetic moment and magnetization.

3.1.1 Magnetic Moment

All of us at some point in our lives have come across magnets and experienced the strange forces of attraction and repulsion between them. These magnetic forces appear to originate in regions called poles, which are located near the ends of, say, a bar magnet. In magnets, poles always occur in pairs, but it is impossible to separate them. A magnetic field is created by a magnetic pole, which pervades the region around the pole [2]. This magnetic field causes a force on a second pole nearby. This magnetic force is directly proportional to the product of the pole strength p and field strength or field intensity $\vec{H}$, which can be verified experimentally:

$$\vec{F} = kp\vec{H} \tag{3.1}$$

This equation defines $\vec{H}$ if the proportionality constant k is put equal to 1. A magnetic field of unit strength causes a force of 1 dyne on a unit pole [2]. In CGS units, a field of unit strength has an intensity of 1 oersted (Oe).

Let us now consider a bar magnet with poles of strength p located near each end and separated by a distance l, which is placed at an angle θ to a uniform field

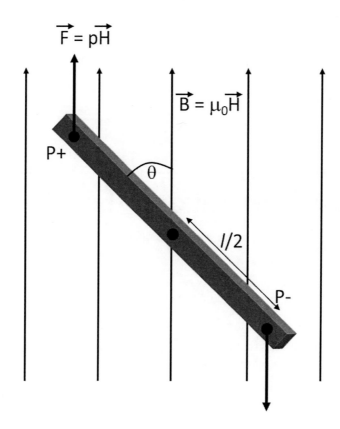

FIGURE 3.1 Schematic presentation of a bar magnet in a uniform magnetic field.

$\vec{B} = \mu_0 \vec{H}$ (Fig. 3.1). The magnet will experience a torque, which will tend to turn the magnet parallel to the magnetic field. The moment μ of this torque is expressed as [2]:

$$\mu = pHl\sin\theta \qquad (3.2)$$

When $\vec{H} = 1$ Oe and $\theta = 90^0$, the moment is given by $\mu = pl$. The *magnetic moment* of the magnet is defined as the moment of the torque experienced by the magnet when it is at right angles to a uniform field of 1 Oe. In a non-uniform magnetic field, the magnet will also feel a translational force acting on it. We will see in the subsequent sections that *magnetic moment* is a fundamental quantity for magnetism in materials.

We note from Fig. 3.1 that a magnet not aligned parallely to the magnetic field will have a potential energy E_p relative to the parallel position. Thus, the work done against this magnetic field in turning the magnet through an angle $d\theta$ is [2]:

$$dE_p = 2(pH\sin\theta)\left(\frac{l}{2}\right)d\theta = \mu H \sin\theta d\theta \qquad (3.3)$$

Conventionally the zero of energy is fixed at the $\theta = 90^0$ position. Therefore:

$$E_p = \int_{90^0}^{\theta} \mu H \sin\theta d\theta = -\mu H \cos\theta \qquad (3.4)$$

The magnetic moment μ is a vector, and in vector notation, the energy equation is written as:

$$E_p = -\vec{\mu}.\vec{H} \qquad (3.5)$$

If the energy E_p is expressed in ergs, then the unit of magnetic moment μ is erg/oersted. This quantity is the electromagnetic unit of magnetic moment and is generally termed simply as *emu*.

3.1.2 Magnetization

A piece of iron becomes magnetized if it is placed in a magnetic field. A quantity is needed to describe the degree of magnetism, which depends on the strength of the magnetic field. Suppose two bar magnets of the same size and shape, and each having the same pole strength p and interpolar distance l are placed side by side, as in Fig. 3.2(a). The poles add up and the magnetic moment $\mu = 2pl$ is double the moment of each magnet. In Fig. 3.2(b) the adjacent poles cancel if two magnets are placed end to end, and the magnetic moment becomes doubled to $\mu = 2pl$. These examples show that the total magnetic moment is the sum of the magnetic moments of the individual magnets. The magnetic moment is thus doubled by doubling the volume. However, the *magnetic moment per unit volume* is not changed, and this is, therefore, a quantity that describes the degree to which the magnets are magnetized [2]. This is termed as the *intensity of magnetization*, or simply *magnetization*, and is expressed as $\vec{M}$.

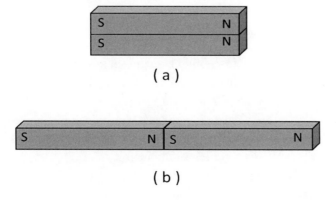

FIGURE 3.2 (a) Two bar magnets placed side by side; (b) Two bar magnets placed end to end.

Since the magnetization $\vec{M}$ is magnetic moment $\vec{\mu}$ per unit volume, the unit of $\vec{M}$ in the CGS system is erg/oersted cm^3. However, it is more often written simply as emu/cm^3 [2]. The mass of a small sample can be measured more accurately than its volume, hence it is sometimes convenient to express the value of magnetization in terms of unit mass instead of unit volume. We will also see below that in small volumes like the unit cell of magnetic solids, the magnetic moment is often expressed in a unit called Bohr magnetons.

3.2 Classical Electromagnetism, Magnetic Moment, and Angular Momentum

In classical electromagnetism the magnetic moment ($\vec{\mu}$) can be equated with a current loop. The magnetic moment $\vec{d\mu}$ associated with a current I flowing in an infinitesimal current loop of area $|\vec{dS}|$ can be expressed in the units of Am^2 as:

$$\vec{d\mu} = I\vec{dS} \quad (3.6)$$

The direction of the current I around the loop determines the direction of the vector $\vec{dS}$, which is normal to the plane of the loop (Fig. 3.3(a)). The magnetic moment $\vec{\mu}$ associated with a current-carrying loop of finite size can be estimated by assuming that this finite loop consists of many equal-sized infinitesimal current loops distributed throughout the area of the bigger loop, and then adding the contributions of the magnetic moments from each infinitesimal loop. However, it can be seen easily that the currents from the neighboring infinitesimal loops get cancelled, which leaves only the current flowing through the perimeter of the big finite-sized loop. This leads to:

$$\vec{\mu} = \int \vec{d\mu} = I \int \vec{dS} \quad (3.7)$$

Fig. 3.3(b) shows the magnetic moment $\vec{d\mu}$ associated with a current I flowing in an elementary finite current loop. The current in a loop arises due to the motion of electrical charge(s). Hence, there will also be an angular momentum vector associated with the electrical charge(s) circulating the loop (Fig. 3.3(c)). The magnetic moment $\vec{d\mu}$ associated with the elementary current loop can be either parallel or antiparallel to this angular momentum.

Let us now study the effect of an applied magnetic field $\vec{B}$ on a single charge and then use the result to estimate the magnetic response of a system of charges within the framework of classical electromagnetism. The force on a particle with mass m, charge q, and moving with a velocity $\vec{v}$ in an electric field $\vec{E}$ and magnetic field $\vec{B}$

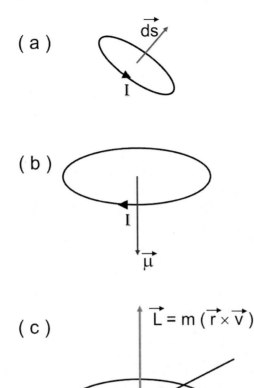

FIGURE 3.3 (a) An infinitesimal loop of area $\vec{ds}$ carrying a current I; (b) magnetic moment $\vec{\mu}$ associated with a finite current carrying loop; (c) angular momentum vector associated with the electrical charge circulating around the loop.

is given by the Lorentz force:

$$\vec{F} = q(\vec{E} + \vec{v} \times \vec{B}) \tag{3.8}$$

Now using the well-known relation between $\vec{E}$, $\vec{B}$, electric potential V, and magnetic vector potential $\vec{A}$:

$$\vec{E} = -\nabla V - \frac{\partial \vec{A}}{\partial t} \tag{3.9}$$

$$\vec{B} = \nabla \times \vec{A} \tag{3.10}$$

the Eqn. 3.8 can be expressed as:

$$F = m\frac{d\vec{v}}{dt} = -q\nabla V - q\frac{\partial \vec{A}}{\partial t} + q\vec{v} \times (\nabla \times \vec{A}) \tag{3.11}$$

Eqn. 3.11 can be rewritten further by using the vector identity:

$$\vec{v} \times (\nabla \times \vec{A}) = \nabla(\vec{v}.\vec{A}) - (\vec{v}.\nabla)\vec{A} \qquad (3.12)$$

thus leading to:

$$m\frac{d\vec{v}}{dt} + q\left(\frac{\partial \vec{A}}{\partial t} + (\vec{v}.\nabla)\vec{A}\right) = -q\nabla(V - \vec{v}.\vec{A}) \qquad (3.13)$$

Here the force $md\vec{v}/dt$ on the charge particle q is measured in a coordinate system that moves with the particle, whereas the partial derivative $\partial \vec{A}/\partial t$ is a measure of the rate of change of $\vec{A}$ at a fixed point in space. Now Eqn. 3.13 can be rewritten as:

$$\frac{d}{dt}(m\vec{v} + q\vec{A}) = -q\nabla(V - \vec{v}.\vec{A}) \qquad (3.14)$$

where

$$\frac{d\vec{A}}{dt} = \frac{\partial \vec{A}}{\partial t} + (\vec{v}.\nabla)\vec{A} \qquad (3.15)$$

Eqn. 3.15 measures the rate of change of the magnetic vector potential $\vec{A}$ at the location of the moving charge q, and Eqn. 3.14 takes the form of Newton's second law. This, in turn, leads to the definition of canonical momentum of the charged particle:

$$\vec{p} = m\vec{v} + q\vec{A} \qquad (3.16)$$

and a velocity-dependent effective potential energy $q(V - \vec{v}.\vec{A})$ experienced by the charged particle. In the absence of any magnetic field $\vec{A} = 0$, and the canonical momentum takes the familiar form of $\vec{p} = m\vec{v}$. The kinetic energy can be written in terms of the canonical momentum as $(\vec{p} - q\vec{A})^2/2m$.

The energy of a magnetic moment $\vec{\mu}$ in a magnetic field $\vec{B}$ is given by:

$$U = -\vec{\mu}.\vec{B} \qquad (3.17)$$

To estimate the net magnetic moment arising from a system comprising of a collection of charges in a solid, we need to find the magnetization (i.e. magnetic moment per unit volume) $\vec{M}$ induced by the magnetic field $\vec{B}$. Now from Eqn. 3.17, magnetic moment $\vec{\mu}$ is proportional to the rate of change of energy of the system with the applied magnetic field $\vec{B}$. However, from Eqn. 3.8 we can see that the magnetic field always produces a force on moving charged particles in the direction perpendicular to their velocities. Hence, no work is done and the energy of the system does not depend on $\vec{B}$. This, in turn, implies that there can be no magnetization in the system. This idea is enclosed in the Bohr–van Leeuwen theorem, which states that in a classical system there is no thermal equilibrium magnetization [4]. In quantum mechanics, however, an electron in a state $\psi(\vec{r})$ can be correlated to a current density:

$$\vec{j}(r) = \frac{e\hbar}{2m_e i} \left(\psi^*(\vec{r}) \nabla \psi(\vec{r}) - \psi(\vec{r}) \nabla \psi^*(\vec{r}) \right) \tag{3.18}$$

This current density can be non-vanishing for a complex wave function ψ and gives rise to a magnetic moment. In atoms, a magnetic moment $\vec{\mu}$ can be associated with an orbiting electron. Let us now consider a hydrogen atom with an electron of charge $-e$ and mass m_e moving with a velocity $\vec{v}$ in a circular orbit of radius r around the nucleus. The current due to the orbiting electron is $I = -e/\tau$, where $\tau = 2\pi r/v$ is the orbital period. Here $v = |\vec{v}|$ is the speed of the electron. Hence, the magnetic moment associated with the orbiting electron is given by:

$$\vec{\mu} = \pi r^2 I = -\frac{evr}{2} \tag{3.19}$$

Now the angular momentum $\vec{L}$ associated with the orbiting electron is $m_e v r$, and that in the ground state must be equal to $\hbar$. Hence, Eqn. 3.19 can be written as:

$$\vec{\mu} = -\frac{e\hbar}{2m_e} \equiv -\vec{\mu}_B \tag{3.20}$$

Here $\vec{\mu}_B$ is called Bohr magneton, and is defined by:

$$\mu_B = \frac{e\hbar}{2m_e} \tag{3.21}$$

Bohr magneton is a suitable unit for the estimation of the size of atomic magnetic moment and it takes the value of 9.274×10^{-24} Am^2. We can now write the formal relation between atomic magnetic moment and the orbital angular momentum as:

$$\vec{\mu} = \gamma \vec{L} \tag{3.22}$$

Here γ is a constant known as the gyromagnetic ratio. The sign of the magnetic moment of the electron as expressed in Eqn. 3.20 is negative. Hence, from Eqn. 3.22 one can see that the magnetic moment of the electron is antiparallel to its angular momentum. This arises due to the negative charge of the electron. The gyromagnetic ratio for the electron is given by $\gamma = -e/2m_e$. The relation between the magnetic moment and angular momentum can be demonstrated through the Einstein-de Haas effect, in which an application of magnetic field applied along the length of a ferromagnetic rod suspended vertically from a torsion fibre causes a rotation of the rod [4]. The applied magnetic field causes an alignment of the atomic magnetic moments in the rod and that corresponds to a net angular momentum. The rod then turns about its long axis in the opposite direction to conserve the total angular momentum. Thus the angular momentum associated with the atomic magnetic moments and the gyromagnetic ratio can be determined by the measurement of the angular momentum of the rod.

3.3 Precession of Magnetic Moment

Let us now consider a magnetic moment $\vec{\mu}$ in an applied magnetic field $\vec{B}$ as shown in Fig. 3.4(a). The energy of this magnetic moment is given by Eqn. 3.17, and that is minimum when the direction of the magnetic moment is along the applied field. The magnetic moment will now be subjected to a torque T_r represented by:

$$T_r = \vec{\mu} \times \vec{B} \tag{3.23}$$

This torque would have turned the magnetic moment towards the applied field if it was not associated with an angular momentum $\vec{L}$ (Eqn. 3.22). Now the torque is equal to the rate of change of the angular momentum, hence with the help of Eqn. 3.22 we can rewrite Eqn. 3.23 as:

$$\frac{d\vec{\mu}}{dt} = \gamma(\vec{\mu} \times \vec{B}) \tag{3.24}$$

This implies that change in $\vec{\mu}$ as a result of the torque will be in a direction perpendicular to both $\vec{\mu}$ and $\vec{B}$. Hence, instead of the magnetic moment $\vec{\mu}$ just

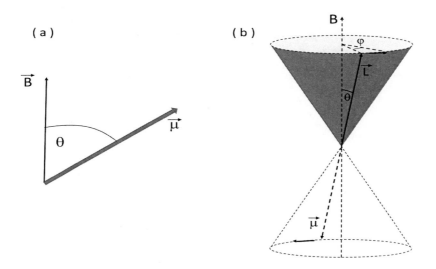

FIGURE 3.4 (a) A magnetic moment $\vec{\mu}$ in the presence of a magnetic field; (b) an electron with magnetic moment $\vec{\mu}$ undergoing Larmour precession around a magnetic filed $\vec{B}$.

getting turned towards the magnetic field $\vec{B}$, the magnetic field makes the direction $\vec{\mu}$ to precess around it. Eqn. 3.24 also indicates that $|\vec{\mu}|$ remains independent of time. We will now discuss a particular case when the magnetic field $\vec{B}$ is applied along the z direction and an electron with magnetic moment $\vec{\mu}$ initially lies in the x–y plane at an angle θ with the magnetic field (Fig. 3.4(b)). The precession angle is labelled as ϕ. Then from Eqn. 3.24 we get:

$$\frac{d\mu_x}{dt} = \gamma \mu_y B \tag{3.25}$$

$$\frac{d\mu_y}{dt} = -\gamma \mu_x B \tag{3.26}$$

$$\frac{d\mu_z}{dt} = 0 \tag{3.27}$$

From this set of equations it is clear that both μ_x and μ_y oscillate, while μ_z is constant with time. Solving these differential equations we can get:

$$\mu_x(t) = |\vec{\mu}| sin\theta cos(w_L t) \tag{3.28}$$

$$\mu_y(t) = |\vec{\mu}| sin\theta sin(w_L t) \tag{3.29}$$

$$\mu_z(t) = |\vec{\mu}| cos\theta \tag{3.30}$$

Here $\phi = \omega_L t$, and $\omega_L = \gamma B$ is called the Larmor precession frequency. The gyromagnetic ratio thus connects the Larmor precession frequency with the magnetic field.

3.4 Magnetization, Magnetic Field, and Magnetic Susceptibility

The magnetization $\vec{M}$ in a magnetic solid consisting of a large number of constituent atoms with magnetic moments is defined as the magnetic moment per unit volume. This is a smooth vector field on a length scale large enough to ignore the discreteness of the individual atomic magnetic moments. This vector field is continuous everywhere except at the edges of the magnetic solid under consideration.

In literature, the magnetic field is described by the vector fields $\vec{B}$ and $\vec{H}$. Historically $\vec{B}$ is called the magnetic induction or magnetic flux density, and $\vec{H}$ is called magnetic field strength (see Appendix C on the units in magnetism). But in common usage, both are termed as magnetic fields, and in free space or vacuum they are linearly related by the expression:

$$\vec{B} = \mu_0 \vec{H} \tag{3.31}$$

Here, $\mu_0 = 4\pi \times 10^{-7}$ Hm^{-1} is the premeability of free space, and $\vec{B}$ and $\vec{H}$ are measured in tesla (T) and in ampere per meter (Am^{-1}), respectively.

The two vector fields $\vec{B}$ and $\vec{H}$ can be very different both in magnitude and direction inside a magnetic solid, where the general relationship between these two fields is expressed as:

$$\vec{B} = \mu_0(\vec{H} + \vec{M}) \tag{3.32}$$

The magnetization $\vec{M}$ is linearly related to the magnetic field $\vec{H}$ in the solids with predominantly linear response, and there one can write:

$$\vec{M} = \chi \vec{H} \tag{3.33}$$

Here χ is a dimensionless quantity termed as magnetic susceptibility. In this special case of linear solid the relation between $\vec{B}$ and $\vec{H}$ can be expressed as:

$$\vec{B} = \mu_o(1+\chi)\vec{H} = \mu_0 \mu_r \vec{H} \tag{3.34}$$

where $\mu_r = 1 + \chi$ is called the relative permeability of the material.

3.5 Orbital and Spin Angular Momentum of Electron and Magnetic Moment

The angular momentum $\tilde{L}$ discussed in Section 3.2 is related to the orbital motion of an electron around the nucleus, and it is called the *orbital angular momentum*. In the frame of quantum mechanics the orbital angular momentum in an atom the orbital angular momentum will depend on the electronic state occupied by the electron defined by the angular quantum number or orbital quantum number l and magnetic quantum number m_L. The component of orbital angular momentum along a fixed axis, say z-axis, is $m_l \hbar$ and the corresponding component of the magnetic moment along the z-axis is $-m_l \mu_B$. On the other hand, the magnitude of the total orbital angular momentum is $\sqrt{l(l+1)}\hbar$ and the magnitude of the total orbital magnetic moment is $\sqrt{l(l+1)}\mu_B$.

In addition to the orbital angular momentum, an electron also has an intrinsic angular momentum called *spin* and an associated intrinsic magnetic moment. The spin, however, is not due to the precession of an electron about its axis, because an electron, in reality, is a point particle. The *spin quantum number* s characterizes the spin of an electron, which takes the value of $\frac{1}{2}$. The component of *spin angular momentum*, which can take only one of $2s+1$ possible values of $s\hbar$, $s-1\hbar$, ... $-s\hbar$, is represented as $m_s \hbar$. Since $s = \frac{1}{2}$ for electrons, there are only two possible values $\pm \frac{1}{2}$ for m_s. This, in turn, leads to the possible values of $\hbar/2$ or $-\hbar/2$ angular momentum along a particular axis. The magnitude of the spin angular momentum for an electron takes the value of $\sqrt{s(s+1)}\hbar = \sqrt{3}\hbar/2$. The magnetic moment associated with the spin angular momentum will have a component of $-g\mu_B m_s$ along a particular axis and a magnitude equal to $\sqrt{s(s+1)}g\mu_B\hbar = \sqrt{3}\hbar/2$. The constant g in these expressions is known as the g-factor, which takes approximately a value of 2. Hence, the component of the intrinsic or spin-magnetic moment of the electron along a particular axis (say z-axis) is $\approx \mp\mu_B$. The $\mp$ sign indicates that the magnetic moment is antiparallel to the angular momentum. The electrons in atoms, in general, may have both orbital and spin angular momenta, which then combine. Hence, the g-factor can take different values in real atoms depending on the relative contributions of spin and angular momentum.

3.6 Magnetism of Isolated Atoms and Ions

In this section, we shall study the effect of an applied magnetic field $\vec{B}$ on the magnetic moments of isolated atoms in the absence of any inter-atomic interactions or any interaction of the isolated moments with their immediate environments.

The energy of an electron spin in a magnetic field parallel to the z-axis can be expressed as:

$$E = g\mu_B B m_s \tag{3.35}$$

where $g \approx 2$ and $m_S = \pm\frac{1}{2}$. This leads to:

$$E \approx \pm\mu_B B \tag{3.36}$$

Electrons in an atom also have orbital angular momentum in addition to the spin angular momentum. The total angular momentum $\hbar\vec{L}_i$ associated with the ith electron in an atom with position $\vec{r}_i$ and linear momentum $\vec{p}_i$ is expressed as:

$$\hbar\vec{L}_i = \vec{r}_i \times \vec{p}_i \tag{3.37}$$

Then the total angular momentum L is obtained by taking sum over all the electrons in the atom as:

$$\hbar\vec{L} = \sum_i \vec{r}_i \times \vec{p}_i \tag{3.38}$$

Let us consider the case of an isolated atom placed in a magnetic field $\vec{B}$ given by:

$$\vec{B} = \nabla \times \vec{A} \tag{3.39}$$

Here $\vec{A}$ is the vector potential and we choose a gauge so that:

$$\vec{A}(r) = \frac{\vec{B} \times \vec{r}}{2} \tag{3.40}$$

We now write down first the Hamiltonian of the atom in the absence of any magnetic field as the sum of the kinetic energy and potential energy of Z number of electrons:

$$H_0 = \sum_{i=1}^{Z}\left(\frac{p_i^2}{2m} + V_i\right) \qquad (3.41)$$

In the presence of the magnetic field $\vec{B}$ the kinetic energy of an electron is modified and, accordingly, the perturbed Hamiltonian is expressed as [4]:

$$H_P = \sum_{i=1}^{Z}\left(\frac{[p_i + e\vec{A}(\vec{r_i})]^2}{2m} + V_i\right) + g\mu_B \vec{B}.\vec{S} \qquad (3.42)$$

With the help of Eqns. 3.38 and 3.40, this can be rewritten as [4]:

$$H_P = \sum_{i=1}^{Z}\left(\frac{p_i^2}{2m} + V_i\right) + \mu_B(\vec{L} + g\vec{S}).\vec{B} + \frac{e^2}{8m_e}\sum_i(\vec{B}\times\vec{r_i})^2$$

$$= H_0 + \mu_B(\vec{L} + g\vec{S}).\vec{B} + \frac{e^2}{8m_e}\sum_i(\vec{B}\times\vec{r_i})^2 \qquad (3.43)$$

or

$$H_P = H_0 + \mu_B(\vec{L} + g\vec{S}).\vec{B} + \frac{e^2}{8m_e}\sum_i(\vec{B}\times\vec{r_i})^2 \qquad (3.44)$$

The second term $\mu_B(\vec{L}+g\vec{S}).\vec{B}$ in Eqn. 3.44 represents the effect of the magnetic moment of the atom and is known as the paramagnetic term, and the third term in Eqn. 3.44 $\frac{e^2}{8m_e}\sum_i(\vec{B}\times\vec{r_i})^2$ arises due to diamagnetic moment.

3.7 Diamagnetism

In a diamagnetic substance, the applied magnetic field $\vec{B}$ induces a magnetic moment that opposes this applied field. This effect can be understood within the framework of classical electromagnetism, where an applied magnetic field on the orbital motion of the electrons causes an induced electromotive force (emf), and according to Lenz's law, this induced emf would oppose the magnetic field causing it. However, as discussed in Section 3.2, in a classical system according to the Bohr–van Leeuwen theorem there is no thermal equilibrium magnetization. Hence, one needs to be cautious with such a classical approach to understanding diamagnetism associated with an atom. Atomic diamagnetism, however, can be easily understood within a quantum mechanical framework. Let us now consider an atom with closed atomic shell. In this case the paramagnetic second term in Eqn. 3.44 can be ignored. If the magnetic field is now applied along the z-axis, then $(\vec{B} \times \vec{r}_i) = B(\text{-}y_i, x_i, 0)$ and

$$(\vec{B} \times \vec{r}_i)^2 = B^2(x_i^2 + y_i^2) \tag{3.45}$$

Within the framework of first-order perturbation theory, the shift in the ground state energy due to the diamagnetic term can be expressed as [4]:

$$\Delta E_0 = \frac{e^2 B^2}{8m_e} \sum_{i=1}^{Z} \langle 0| (x_i^2 + y_i^2) |0\rangle \tag{3.46}$$

where $|0\rangle$ is the ground state wave function. If we consider a spherically symmetric atom, then $\langle x_i^2 \rangle = \langle y_i^2 \rangle = \frac{1}{3} \langle r_i^2 \rangle$ and we can write:

$$\Delta E_0 = \frac{e^2 B^2}{12m_e} \sum_{i=1}^{Z} \langle 0| r_i^2 |0\rangle \tag{3.47}$$

We shall now consider a solid of volume V composed of N atoms (ions) with all atomic shells filled, and derive the magnetization of this solid at $T = 0$. The magnetization of this ensemble of atoms (ions) can be estimated by differentiating the Helmoltz free energy F of this ensemble with respect to the applied field B, i.e. $M = -\frac{\partial F}{\partial B}$. The Helmoltz free energy F can be expressed as:

$$F = -Nk_B T \ln Z \tag{3.48}$$

where Z is the partition function and written as:

$$Z = \sum_i e^{-E_i/k_B T} \tag{3.49}$$

With the help of Eqns. 3.47–3.49 we can write the expression of magnetization of M as:

$$M = -\frac{\partial F}{\partial B} = -\frac{N}{V}\frac{\partial \Delta E_0}{\partial B} = -\frac{Ne^2 B}{6m_e} \sum_{i=1}^{Z} \langle r_i^2 \rangle \tag{3.50}$$

From Eqn. 3.50 we can derive the diamagnetic susceptibility χ as [4]:

$$\chi = -\frac{N}{V}\frac{e^2 \mu_0}{6m_e} \sum_{i=1}^{Z} \langle r_i^2 \rangle \tag{3.51}$$

With an increase in temperature, there will also be a contribution to diamagnetic susceptibility from the states above the ground state. However, that contribution is relatively small and the diamagnetic susceptibilities of solids are largely considered to be temperature-independent.

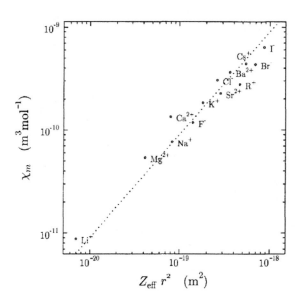

FIGURE 3.5 The experimentally determined diamagnetic molar susceptibilities for a number of ions against $Z_{eff} r^2$, where Z_{eff} is the number of electrons in the outer shell of an ion and r is the measured ionic radius. (From reference [4] with permission from Oxford University Press)

The relation expressed in Eqn. 3.51 can roughly be tested by plotting (Fig. 3.5) the experimentally determined diamagnetic molar susceptibilities for several ions against $Z_{eff} r^2$, where Z_{eff} is the number of electrons in the outer shell of an ion and r is the measured ionic radius. The contributions from the inner shells are ignored and Z_{eff} is an effective atomic number and not the actual atomic number Z of the ions. Further, with the assumption that all the electrons in the outer shells of the ion have approximately the same value of $<r_i>^2$, one can write [4]:

$$\sum_{i=1}^{Z_{eff}} \langle r_i^2 \rangle \approx Z_{eff} r^2 \tag{3.52}$$

Ions are chosen instead of atoms, because Na and Cl atoms, for example, have unpaired electrons, whereas the Na$^+$ and Cl$^-$ ions have closed-shell structures similar to those of Ne and Ar. The experimental values of diamagnetic susceptibilities of various ions shown in Fig. 3.5 are estimated by comparing the measured diamagnetic susceptibility of a range of ionic salts: KCl, KBr, NaCl, NaBr, NaF, etc. The agreement with the calculation is pretty good despite the rather inaccurate approach involving the assumption that all the electrons in an ion have the same mean squared radius. Materials like naphthalene and graphite have delocalized π electrons and they show relatively large and anisotropic diamagnetic susceptibilities.

In fact, diamagnetism is present in all materials. Assuming $r \approx 10^{-10}$ m and number of ions per unit volume $\approx 5 \times 10^{-2}$ m^{-3}, one can obtain $\chi \approx 10^{-6}Z$. This is indeed a rather small negative contribution and can be either ignored or considered as a small correction to the larger positive effects of paramagnetism or long-range magnetic order.

3.8 Paramagnetism

In our discussion on diamagnetism in materials, we considered atoms (ions) with filled atomic shells and without any unpaired electrons. Thus, the atoms (ions) had no intrinsic magnetic moment in the absence of an external magnetic field. We shall now consider solids having atoms with unpaired electrons, hence intrinsic magnetic moment. The application of an external magnetic field will tend to align such moments parallel to the direction of the field. The induced magnetism is termed paramagnetism and it corresponds to a positive magnetic susceptibility. The magnetic moments of the atoms will point to random directions in the absence of a magnetic field. The net magnetization thus is zero. It is assumed that the interaction between the atomic magnetic moments $\vec{\mu}$ is zero or negligible. The applied magnetic field tends to align the magnetic moments and the degree of alignment increases with the increase in field strength. On the other hand, an increase in temperature will randomize the alignments between the magnetic moments. Therefore, it is expected that the alignment of magnetic moments in a paramagnetic solid will depend on the ratio B/T.

In an atom the magnetic moment is associated with its total angular momentum $\vec{J}$, which, in turn, is a sum of orbital angular momentum $\vec{L}$ and the spin angular momentum $\vec{S}$, and measured in units of $\hbar$ is expressed as:

$$\vec{J} = \vec{L} + \vec{S} \qquad (3.53)$$

We will discuss this in more detail while studying the ground state of ions in the next section. Here we will continue mainly with the assumption that each atom has a magnetic moment of magnitude $|\vec{\mu}|$.

3.8.1 Classical theory of paramagnetism

In this classical approach, magnetic moments are allowed to point along any direction with respect to the applied magnetic field instead of pointing only to certain directions because of quantization. Let us consider the magnetic moments lying at an angle between θ and $\theta+d\theta$ to the applied field $\vec{B}$. The energy associated with such moments is $U = -|\vec{\mu}|B\cos\theta$, and the minimum U corresponds to $\theta = 0$. Therefore, all the moments would tend to align themselves in the direction of the magnetic field, which, however, is prevented by the thermal motion. The total magnetic moment of a material is the sum of the projections of magnetic moments of the individual atoms in the direction of the applied field $\vec{B}$. The magnitude of these projections is $|\vec{\mu}|\cos\theta$, and the problem of the quantitative calculation of the magnetization is thus reduced to the calculation of the average value of $|\vec{\mu}|\cos\theta$ corresponding to the equilibrium state, which arises out of the competition between the orientation of the magnetic field and the disrupting effect of the temperature, in other words thermal motion. This problem was solved by Paul Langevin with the help of classical statistical mechanics.

Let us assume that the orientation of the magnetic moment with respect to the magnetic field $\vec{B}$ is arbitrary, and hence θ can take all values. Then the fraction of the moments, which align themselves at an angle in the interval (θ and $\theta+d\theta$) to $\vec{B}$, is proportional to the area $2\pi \sin\theta d\theta$ of the annulus of the sphere of unit radius shown in Fig. 3.6. This fraction is $1/2\sin\theta d\theta$ since the area of the unit sphere is 4π. Then the probability for a moment to align itself at an angle in the interval (θ and $\theta+d\theta$) to the magnetic field $\vec{B}$ at a temperature T is proportional to the product of this statistical factor $1/2\sin\theta d\theta$ and Boltzmann distribution factor $f = Ce^{-U/k_BT} = Ce^{\frac{|\vec{\mu}|B\cos\theta}{k_BT}}$

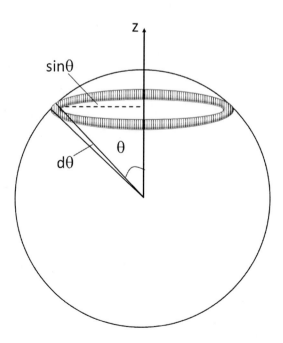

FIGURE 3.6 The probability that the magnetic moment in a paramagnetic material lies between the angles θ and $d\theta$ with respect to z-axis is proportional to the area of the annulus of the unit sphere shown in the figure. This area is $2\pi\sin\theta d\theta$.

where C is a normalization constant and k_B is the Boltzmann constant. The average magnetic moment along $\vec{B}$ can now be expressed as [4]:

$$<|\vec{\mu}|\cos\theta> = <\mu_Z> = \frac{\int_0^\pi |\vec{\mu}|\cos\theta \exp(\mu B \cos\theta/k_B T)\sin\theta d\theta}{\int_0^\pi \exp(\mu B \cos\theta/k_B T)\sin\theta d\theta} \quad (3.54)$$

If we put $y = |\vec{\mu}|B/k_B T$ and $x = \cos\theta$, then the Eqn. 3.54 is expressed as:

$$<|\vec{\mu}|\cos\theta> = <\mu_Z> = \mu \frac{\int_{-1}^{1} x e^{yx} dx}{\int_{-1}^{1} e^{yx} dx} \quad (3.55)$$

or

$$\frac{<\mu_z>}{|\vec{\mu}|} = \coth y - \frac{1}{y} \equiv L(y) \quad (3.56)$$

The function $L(y) = \coth y - 1/y$ is known as the Langevin function.

If there are n number of magnetic moments per unit volume, then the maximum magnetization that can be obtained by alignments of all the magnetic moments, i.e. saturation magnetic moment M_S, is represented as $M_S = n|\tilde{\mu}|$. The magnetization that can actually be obtained is $M = n<\mu_Z>$, and the ratio of the M and M_S is expressed as:

$$\frac{M}{M_S} = \frac{<\mu_Z>}{|\vec{\mu}|} \quad (3.57)$$

Now for a small value of y:

$$\coth(y) = \frac{1}{y} + \frac{y}{3} + O(y^3) \quad (3.58)$$

Hence,

$$L(y) = \coth(y) - \frac{1}{y} = \frac{y}{3} + O(y^3) \quad (3.59)$$

With the help of Eqns. 3.56 and 3.59, the Eqn. 3.57 can be rewritten as:

$$\frac{M}{M_S} = \frac{<\mu_Z>}{|\vec{\mu}|} \approx \frac{y}{3} = \frac{\mu B}{3 k_B T} \quad (3.60)$$

For small magnetic fields $B \approx \mu_0 H$ and accordingly susceptibility χ is expressed as:

$$\chi \approx \mu_0 M/B = \frac{n \mu_0 \mu^2}{3 k_B T} \quad (3.61)$$

The Eqn. 3.61 represents the well-known Curie's law, which states that magnetic susceptibility is inversely proportional to the temperature.

3.8.2 Quantum theory of paramagnetism

In the above discussion of the classical theory of paramagnetism, all orientations of magnetic moments with respect to the applied magnetic field $\tilde{B}$ were allowed. In the quantum theory of paramagnetism there are $2J+1$ ways in which the atomic magnetic moment may align with the applied magnetic field. The probability of each such orientations can be expressed in terms of the Boltzmann distribution function $f = Ce^{\mu_0 g_J m_{JB} B/k_B T}$, where m_{JB} is the projection of m_J on $\tilde{B}$. The average value of m_{JB} is given by:

$$<m_{JB}> = \frac{\sum_{-J}^{+J} m_{JB} exp\left(\frac{\mu_0 g_J m_{JB} B}{k_B T}\right)}{\sum_{-J}^{+J} exp\left(\frac{\mu_0 m_{JB} B}{k_B T}\right)} \quad (3.62)$$

A comparison of the Eqn. 3.62 with the classical expression in Eqn. 3.54 reveals that the integration in this later equation is replaced with a summation over the allowed discrete directions of alignment of the vector m_J. The summations in Eqn. 3.62 can be evaluated to give the following result:

$$<m_{JB}> = JB_J(y) \quad (3.63)$$

where

$$y = \frac{g_J \mu_B J B}{k_B T} \quad (3.64)$$

$$B_J(y) = \frac{2J+1}{2J} coth \frac{2J+1}{2J} y - \frac{1}{2J} coth \frac{y}{2J} \quad (3.65)$$

Here the function $B_J(y)$ is known as the Brillouin function. The magnetization M can now be expressed as:

$$M = M_S B_J(y) \quad (3.66)$$

where saturation magnetization M_S is:

$$M_S = n g_J \mu_B J \quad (3.67)$$

Similarly susceptibility is expressed as:

$$\chi = \frac{n g_J \mu_B J}{B} B_J(y) \quad (3.68)$$

At room temperature and with a magnetic field say $B = 1$ T for $J = 1/2$ and $g_J = 2$ leads to a value of $y \approx 2 \times 10^{-3}$. In such a situation $B_J(y) \approx y(J+1)/3J$ and accordingly susceptibility can be expressed as:

$$\chi = \frac{n\mu_0 \mu_{eff}^2}{3k_B T} \quad (3.69)$$

where

$$\mu_{eff} = g_J \mu_B \sqrt{J(J+1)} \quad (3.70)$$

and

$$g_J = \frac{3}{2} + \frac{S(S+1) - L(L+1)}{2J(J+1)} \quad (3.71)$$

This constant g_J is known as the Lande g-value.

We can see from Eqn. 3.69 that at a low enough magnetic field, the temperature dependence of susceptibility χ behaves in the same way as in Curie's law obtained within a classical framework; susceptibility is inversely proportional to temperature. Thus, the experimental measurement of magnetic susceptibility in a low magnetic field enables one to estimate the value of the effective moment μ_{eff}.

At very low temperatures and strong magnetic field $y \to \infty$. This, in turn, leads to: $coth\frac{2J+1}{J}y \to 1$ and $coth\frac{y}{2J} \to 1$. In such a situation, we can see from Eqns. 3.65 and 3.66 that the magnetization attains the saturation value M_S.

3.8.3 van Vleck paramagnetism

If $J = 0$ in the ground state $|0\rangle$, within the first-order perturbation theory there is no paramagnetic effect [4]:

$$\langle 0| \hat{\mu} |0\rangle = g_J \mu_B \langle 0| \hat{J} |0\rangle = 0 \quad (3.72)$$

This, in turn, indicates that the ground state energy does not change with the application of a magnetic field, and hence there is no paramagnetic susceptibility. A change in the ground state energy E_0 is possible, however, within the framework of second-order perturbation theory by taking into account a mixing with excited states $J \neq 0$. In this situation the change in the ground state energy for an ion with $J = 0$ is expressed as:

$$\Delta E_0 = \sum_n \frac{|\langle 0| (\vec{L} + g\vec{S}) \cdot \vec{B} |n\rangle|^2}{E_n - E_0} + \frac{e^2}{8m_e} \sum_i (\vec{B} \times \vec{r}_i)^2 \quad (3.73)$$

In the first term in Eqn. 3.73 the sum is taken over all the excited states of the system, and the second term arises due to the diamagnetism. The magnetic susceptibility is then expressed as:

$$\chi = n \left(2\mu_B^2 \sum_n \frac{|\langle 0|(L_Z + gS_Z)|n\rangle|^2}{E_n - E_0} - \frac{e^2 \mu_0}{6m_e} \sum_i^Z (<r_i>^2) \right) \quad (3.74)$$

The second term in Eqn. 3.74 is negative and represents the conventional diamagnetic susceptibility, which we have already discussed in Section 3.7 (Eqn. 3.51). The first term is positive since $E_n > E_0$ and represents *Van Vleck paramagnetism*, which is relatively small as well as temperature-independent.

3.9 Ground State of Ions and Hund's Rules

An atom of elements except that of hydrogen contains more than one electron, many of which are placed in filled electron shells with no net angular momentum. In some elements, there may, however, be unfilled electron shells. The electrons in such unfilled shells can combine to give rise to a non-zero total spin angular momentum $\vec{P}_S = \hbar\vec{S}$ and total orbital angular $\vec{P}_L = \hbar\vec{L}$. The orbital and spin angular momentum can be combined in $(2L+1)(2S+1)$ ways, which is the total number of choices for the z-component of $\vec{L}$ multiplied with the total number of choices for the z-component of $\vec{S}$. Now the spin angular momentum influences the spatial part of the electronic wave function and the orbital angular momentum affects the ways the electrons travel around the nucleus. Hence, both the angular momenta influence how well the electrons would avoid each other and influence the Coulomb repulsion energy. This, in turn, implies that different combinations of spin and angular momentum (leading to different configurations) would cost different amounts of energy. We now need to find out the configuration that minimizes the energy.

3.9.1 Fine structure

The spin and orbital angular momentum discussed so far are considered as independent quantities. This, however, is not the case. The spin and orbital angular momentum are weakly coupled through a relativistic effect known as spin–orbit interaction. This effect can be understood by considering an inertial frame co-moving with the electron. In this reference frame, the nucleus appears to be orbiting around the electron, thus constituting a current, which through relativistic effect gives rise to a magnetic field at the origin of the orbit. This magnetic field interacts with the spin of the electron to give rise to an additional term in the Hamiltonian. Because of this, the total angular momentum $\vec{J}$ is conserved instead of $\vec{L}$ and $\vec{S}$ being conserved separately. However, this relativistic spin–orbit coupling effect is often taken as a small perturbation, and in this situation $\vec{L}^2 = L(L+1)$ and $\vec{S}^2 = S(S+1)$ can be considered as being conserved. The states with L and S are split into a number of levels with different values of J. This is known as *fine structure*, and J takes the values from $|L-S|$ to $|L+S|$. The spin–orbit interaction thus causes the splitting of multiplets into different fine structure states labelled by J. There will also be a degeneracy of $2J+1$ for each of these states, and they can be separated into their different m_J values with the application of an external magnetic field. We can see that there can be a lot of possible combinations of angular momentum quantum numbers, and the question is which one is the ground state for a particular ion.

3.9.2 Hund's rules

Hund's rules are a set of empirical rules that can be used to determine the combination of spin and angular momentum quantum numbers that minimize the energy to attain the ground state. These empirical rules are narrated below in the order of decreasing importance. The first rule needs to be satisfied to start with, and subject to that the second and the third rules then become applicable.

1. The electron wave functions are arranged in a manner to maximize S. In this situation, the Pauli exclusion principle ensures electrons with parallel spin remain apart, and thus Coulomb repulsion between electrons is reduced. The Coulomb energy of the system is minimized in this process.
2. Once the wave function is determined by the first rule, in the next step L is maximized. The electrons in the orbits rotating in the same direction avoid each other more effectively, and this again reduces the Coulomb repulsion between them. So this step also minimizes the energy.
3. The third rule is used to find the value of J. It takes the value $J = |L-S|$ if shell is less than half full and $J = |L+S|$ if it is more than half full. The third rule originates from the requirement to minimize the spin–orbit energy.

It may be noted that in the discussions above the magnetic moment on each atom is attributed to the electrons that are localized on that atom. However, the conduction electrons in metals are not localized on a particular atom; those are delocalized and can move over the entire material. Conduction electrons in metals can show Landau diamagnetism and Pauli paramagnetism.

3.9.3 Russel–Saunders coupling versus $\vec{j}$–$\vec{j}$ coupling

The $\vec{L}$ - $\vec{S}$ or more formally Russel–Saunders coupling discussed above is considered as a weak relativistic perturbation. The main energy terms are determined by the electrostatic interactions controlling the values of L and S. The spin and orbital angular momenta of the electrons are separately combined first. The spin–orbit interaction is introduced as a weak perturbation only after the total spin and orbital angular momenta of the atom as a whole are determined. That splits the energy states into fine structure levels, which are labelled by J.

In the case of atoms with the high atomic number Z, the spin–orbit interaction energy is proportional to Z^4 and it can no more be treated as a small perturbation. Here the spin–orbit interaction is the dominant term, and the spin and orbital angular momentum of each electron needs to be first coupled separately. The relatively weak electrostatic effect then may couple the total angular momentum contributions from each electron. This is known as $\vec{j}$–$\vec{j}$ coupling.

3.10 Effect of Crystal Environment on Magnetic Ions

Magnetic ions in solids are not isolated. They interact with their immediate environment and also with other magnetic ions in the solids. In this section, we will see the effects of the crystal environment on magnetic ions.

3.10.1 Crystal fields and their origin

The shapes of the atomic orbitals determine the effect of the crystal environment on the energy levels of the atom. It is thus important first to recapitulate the shapes of the atomic orbitals. The wave functions of each orbital have an angular as well as a radial part. For our present discussion the angular dependences of the electron density of the s, p, and d orbitals are mainly important, and those are shown in Fig. 3.7. We can see that only the s orbitals have spherical symmetry, while angular

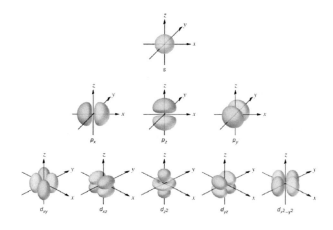

FIGURE 3.7 The shape and angular distribution of s, p, and d orbitals

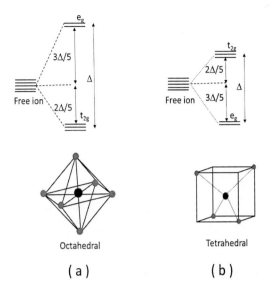

FIGURE 3.8 A metal atom (dark circle) in (a) octahedral (b) tetrahedral environment.

dependence is pronounced in the others. Since the local environments in the crystals usually are not spherically symmetric, it is naturally expected that different orbitals are likely to interact in different ways.

The crystal field acting on an atom is an electric field, which arises due to the neighbouring atoms in the crystal. The magnitude and nature of this crystal field will then depend on the symmetry of the local crystalline environment. In crystal field theory the orbitals of the neighbouring atoms are modelled as negative point charges [4].

In many transition metal compounds, a transition metal ion sits at the centre of an octahedron with an ion such as oxygen on each corner. A picture of such an octahedral environment along with a tetrahedral environment is shown in Fig. 3.8. The crystal field mainly originates from the electrostatic repulsion from the electrons in the oxygen orbitals.

If we see the case of d orbitals, they fall into two classes:

1. The t_{2g} orbitals, which point between x, y, and z axes. They are designated as d_{xy}, d_{xz}, and d_{yz} orbitals (Fig. 3.7).
2. The e_g orbitals, which point along these axes. They are designated as d_{z^2} and $d_{x^2-y^2}$ orbitals (Fig. 3.7). The d_{z^2} orbital has a lobe along the z-axis, whereas $d_{x^2-y^2}$ orbital has lobes pointing along both the x-axis and y-axis.

Now let us consider the case of a cation placed at the centre of a negatively charged sphere of radius r. The presence of the charge will increase the energy

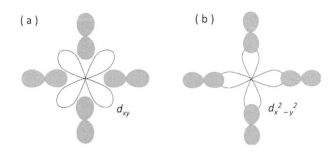

FIGURE 3.9 The effect of octahedral environment on (a) d_{xy}-orbital; (b) $d_{x^2-y^2}$-orbitals

of the total system, but the d-orbitals will remain degenerate in this spherically symmetric environment. We consider a situation where the whole amount of charge is redistributed into six discrete point charges still on the surface of the sphere but at the same time stationed at the vertex of an octahedron. The d-orbitals will no longer be degenerate while the total electronic energy of the d-orbitals will continue remaining the same. This is the effect of an octahedral environment.

To demonstrate further the effect of the octahedral environment on the d-orbitals in Fig. 3.9 we show two different d-orbitals in an octahedral environment, where the crystal field is mostly due to the p-orbitals on neighbouring atoms. One can see that there is a lower overlap between d_{xy}-orbitals and the neighbouring p-orbitals in comparison to the $d_{x^2-y^2}$-orbitals, hence such d_{xy}-orbitals will have lower electrostatic energy. Thus, in the octahedral environment, the three orbitals d_{xy}, d_{yz}, and d_{zx} – orbitals pointing between the x, y, and z axes – will have their energy lowered, whereas $d_{x^2-y^2}$ and d_{z^2} orbitals, which point along the x, y, and z axes, will be raised in energy. As a result, five energy levels are with the two-fold e_g levels raised in energy and the three-fold t_{2g} levels lowered in energy (Fig. 3.8(a)).

The effect of the crystal field is opposite in a tetrahedral environment (see Fig. 3.9). The charge density associated with the atoms is situated here on four corners of the cube describing a tetrahedron, and the orbitals which point along the axes will now mostly avoid the charge density. Thus, in contrast to the octahedral symmetry, the two-fold e_g levels are lowered in energy (Fig. 3.8(b)).

In the case of a transition metal where not all the $3d$-electrons are present, the energy levels with lower energy will get filled first. The precise order in which the orbitals get filled will, however, depend on the crystal field energy and the effect of Coulomb repulsion to put two electrons in the same orbital. The latter (repulsive) energy is called pairing energy and is usually positive [4]. If pairing energy is higher than the crystal field energy, then the added electrons will first singly occupy the available orbitals before any double occupation. This is known as the weak-field case. On the other hand, if the pairing energy is lower than the crystal field energy, the electrons will prefer to doubly occupy the lower energy orbitals before filling the higher energy orbitals. This is known as the strong-field case.

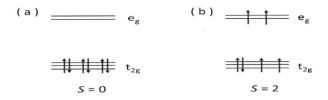

FIGURE 3.10 Electronic configuration of: (a) the low spin state, (b) the high spin state of a $3d^6$-ion Fe^{2+}.

The above principle of filling the orbitals works straightforwardly for 1, 2, 3, 8, 9, and 10 electrons. Interesting situations, however, arise while dealing with 4, 5, 6, and 7 electrons. This situation is illustrated in Fig. 3.10 with the example of a Fe^{2+} ion, which has a $3d^6$ shell. In the strong-field case where the pairing energy is lower than the crystal field energy, six electrons are accommodated into the three t_{2g} orbitals while the e_g orbitals remain unfilled. With no unpaired electrons, the system thus acquires $S = 0$ state, which is known as a low-spin configuration (Fig. 3.10(a)). In the opposite case of weak-field, each orbital is first occupied by a single electron. This leaves behind one electron, which is then paired up with one of the t_{2g} electrons. Thus, there are four unpaired electrons in this configuration, which leads to an $S = 2$ state. This is called a high-spin configuration (Fig. 3.10(b)). It is possible to induce a spin-transition between the low-spin and high-spin configuration in some materials with Fe^{2+} ion by temperature, pressure, or electromagnetic radiation.

3.10.2 Quenching of orbitals

It is a well-known fact that the predicted values of the moment of $3d$-ions given by $g_J\sqrt{J(J+1)}$ do not always agree with the experimentally obtained values, except for the cases of a half-full or full shell of electrons. This is the case of $3d^5$ and $3d^{10}$ with a half-full shell and a full shell of electrons, respectively, where $L = 0$. This discrepancy with the experimental results arises because the crystal field interaction is much stronger in $3d$-ions than the spin–orbit interaction. As a result, Hund's third rule (Section 3.9.2), where the spin–orbit interaction is considered to be the most important energy term after Coulomb interaction, is not strictly valid in such cases. The experimental results seem to suggest that the $3d$-ions choose a ground state such that $L = 0$, which, in turn, implies $J = S$ and $g_J = 2$. This leads to a value of $\mu_{eff} = 2\mu_B\sqrt{S(S+1)}$, and that gives a much better agreement with the experimental results [4]. This phenomenon is known as orbital quenching, as the orbital moment is thought to be quenched.

The $4f$-orbitals in the case of $4f$-ions lie below the $5s$ and $5p$ shells and are much less extended away from the nucleus in comparison to the $3d$-orbitals. The crystal field terms are much less important here and Hund's third rule is followed. The

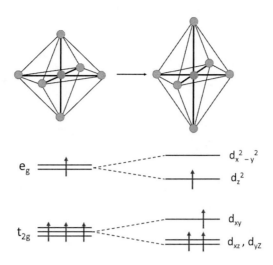

FIGURE 3.11 Jahn–Teller effect for Mn^{3+} ion.

heavier ions in the $4d$ and $5d$ series have a comparable crystal field and spin–orbit interaction, and the situation in those cases is less clear.

3.10.3 Jahn–Teller effect

In the discussions above so far, it is assumed that the knowledge of the symmetry of the local environment is the starting point for deducing electronic structures. Then the magnetic properties of ions are figured out based on how many electrons there are to fill up the energy levels. However, it happens sometimes that the magnetic properties themselves influence the symmetry of the local environment. It is energetically favourable for a crystal structure to get spontaneously distorted when the resultant electronic energy saving due to the lattice distortion balances the energy cost of increased elastic energy. This is known as the Jahn–Teller effect. An example of the Jahn–Teller effect is shown in Fig. 3.11, where an octahedral system having Mn^{3+} ions with $3d^4$ configuration gets spontaneously distorted. The distorted structure shows elongation along the z-axis, and such Jahn–Teller distortion happens when the degeneracy is broken by the lowering in the energy of the d-orbitals with a z component, whereas the orbitals without a z component are higher in energy. In this case distortion splits the t_{2g} and e_g energy levels (see Fig. 3.11). This causes a lowering in the energy because the singly occupied e_g is lowered in energy. The Mn^{4+} ions, however, do not show this effect because there is no net saving of electronic energy by a structural distortion.

4

Exchange Interactions and Magnetism in Solids

The first interaction between magnetic moments, which is expected to play a role in magnetism, is of course the interaction between two magnetic dipoles μ_1 and μ_2 separated by a distance $\vec{r}$. The energy of this system can be expressed as:

$$E = \frac{\mu_0}{4\pi r^3} \left[\mu_1.\mu_2 - \frac{3}{r^2}(\mu_1.\vec{r})(\mu_2.\vec{r}) \right] \quad (4.1)$$

The order of magnitude of the effect of dipolar interaction for two moments each of $\mu \approx 1\mu_B$ separated by a distance of $r \approx 1 \text{Å}$ can be estimated to be approximate $\mu^2/4\pi r^3 \sim 10^{-23}$ J, which is equivalent to about 1 K in temperature. This dipolar interaction is too weak to explain the magnetic ordering observed in many materials at much higher temperatures, even around 1000 K.

4.1 Coupling between Spins

Before we look for suitable interactions between two magnetic moments to explain the magnetic ordering observed in various materials, we shall first discuss the coupling of two spins. We now consider two interacting spin-$\frac{1}{2}$ particles represented by a Hamiltonian:

$$\hat{H} = J\,\vec{\hat{S}}^a.\vec{\hat{S}}^b \quad (4.2)$$

Here $\vec{\hat{S}}^a$ and $\vec{\hat{S}}^b$ represent the spin operators of the two particles. Combining the two particles as a single entity, the total spin operator can be expressed as:

$$\vec{\hat{S}}^{Total} = \vec{\hat{S}}^a + \vec{\hat{S}}^b \tag{4.3}$$

This leads to:

$$(\vec{\hat{S}}^{Total})^2 = (\vec{\hat{S}}^a)^2 + (\vec{\hat{S}}^b)^2 + 2\,\vec{\hat{S}}^a \cdot \vec{\hat{S}}^b \tag{4.4}$$

A combination of two spin-$\frac{1}{2}$ particles gives rise to a single entity with quantum number $s = 0$ or 1. This leads to the eigenvalue of $(\vec{\hat{S}}^{Total})^2$ as $s(s+1)$, which is 0 for $s = 0$ and 2 for $s = 1$. Now the eigenvalues of both $(\vec{\hat{S}}^a)^2$ and $(\vec{\hat{S}}^b)^2$ are $\frac{3}{4}$ [4]. Hence, from Eqn. 4.4 we can write:

$$\vec{\hat{S}}^a \cdot \vec{\hat{S}}^b = \frac{1}{4} \text{ if } s = 1$$
$$= -\frac{3}{4} \text{ if } s = 0 \tag{4.5}$$

The system has two energy levels for $s = 1$ and 0 with energies as follows:

$$E = \frac{J}{4} \text{ if } s = 1$$
$$= -\frac{3J}{4} \text{ if } s = 0 \tag{4.6}$$

Each state will have a degeneracy given by $(2s + 1)$. The $s = 0$ state is a singlet and the z-component of the spin m_S of this state takes the value 0. On the other hand, $s = 1$ state is a triplet and m_S takes one of the three values -1, 0, and 1.

The eigenstates of this two interacting spin-$\frac{1}{2}$ particles can be represented as linear combinations of the following basis states: $|\uparrow\uparrow\rangle$, $|\uparrow\downarrow\rangle$, $|\downarrow\uparrow\rangle$ and $|\downarrow\downarrow\rangle$, where the first (second) arrow corresponds to the z-component of the spin labelled by a (b). The possible eigenstates are presented in Table 4.1. The m_S presented in Table 4.1 is equal to the sum of the z-components of the individual spins. The wave function of these two interacting spin-$\frac{1}{2}$ particles has to be antisymmetric with respect to the exchange of two electrons. This wave function is a product of spatial wave function $\psi_{space}(\vec{r}_1, \vec{r}_2)$ and spin-wave function χ. The spatial wave function can be symmetric or antisymmetric with respect to the exchange of electrons, and can be represented in terms of single-particle wave functions $\psi_a(\vec{r}_i)$ and $\psi_b(\vec{r}_i)$ of the ith electron as:

$$\psi_{space}(\vec{r}_1, \vec{r}_2) = \frac{\psi_a(\vec{r}_1)\psi_b(\vec{r}_2) \pm \psi_a(\vec{r}_2)\psi_b(\vec{r}_1)}{\sqrt{2}} \tag{4.7}$$

The spatial wave function $\psi_{space}(\vec{r}_1, \vec{r}_2)$ expressed in Eqn. 4.7 can be symmetric (+) or antisymmetric (-) with respect to exchange of electrons depending on the ±

TABLE 4.1 The eigen-states and eigen values of $\hat{\vec{S}}^a \cdot \hat{\vec{S}}^b$ and the respective values of m_s and s.

Eigen state	m_S	s	$\hat{\vec{S}}^a \cdot \hat{\vec{S}}^b$
$\lvert\uparrow\uparrow\rangle$	1	1	$\frac{1}{4}$
$\frac{\lvert\uparrow\downarrow\rangle + \lvert\downarrow\uparrow\rangle}{\sqrt{2}}$	0	1	$\frac{1}{4}$
$\lvert\downarrow\downarrow\rangle$	-1	1	$\frac{1}{4}$
$\frac{\lvert\uparrow\downarrow\rangle - \lvert\downarrow\uparrow\rangle}{\sqrt{2}}$	0	0	$-\frac{3}{4}$

sign. The spin eigen-states χ are presented in Table 4.1. To make the the overall wave function antisymmetric with respect to the exchange of electrons, the spatial and spin wave functions must have opposite exchange symmetry, i.e., $\psi_{space}(\vec{r}_1, \vec{r}_2)$ must be antisymmetric when χ is symmetric and vice versa. From Table 4.1 we can see that the spin eigen-state $\frac{\lvert\uparrow\downarrow\rangle + \lvert\downarrow\uparrow\rangle}{\sqrt{2}}$ is symmetric under the exchange of electrons along with the other two $s = 1$ states $\lvert\uparrow\uparrow\rangle$ and $\lvert\downarrow\downarrow\rangle$, whereas the $\frac{\lvert\uparrow\downarrow\rangle - \lvert\downarrow\uparrow\rangle}{\sqrt{2}}$ eigen-state is antisymmetric.

A consequence of the asymmetry with respect to the exchange of electrons is the Pauli exclusion principle, which states that two electrons cannot reside in the same quantum state. If the electrons were in the same spatial quantum state (say $\psi_a(\vec{r})$) and same spin quantum state (say, spin-up state), then the spin wave function χ must be symmetric under exchange of electrons. This, in turn, dictates that the spatial wave function must be antisymmetric under exchange and will be represented by:

$$\psi_{space}(\vec{r}_1, \vec{r}_2) = \frac{\psi_a(\vec{r}_1)\psi_a(\vec{r}_2) - \psi_a(\vec{r}_2)\psi_a(\vec{r}_1)}{\sqrt{2}} = 0 \qquad (4.8)$$

The state thus vanishes, implying that the two electrons cannot reside in the same quantum state.

Eqn. 4.6 indicates that the ground state of the two-spin system will be a singlet with energy $-\frac{3J}{4}$ if $J > 0$, and there will be a triplet of excited states with energy $\frac{J}{4}$ at energy J above the singlet state. This triplet state can be split with the applications of an external magnetic field into three states with different values of m_s. The triplet state will be the lowest level if $J < 0$ in Eqn. 4.6.

4.2 Origin of Exchange Interactions

We can see from above that to make the overall wave function of an interacting two-electron system antisymmetric under exchange, the spin part of the wave function must be an antisymmetric singlet ($s = 0$) state χ_S in the case of symmetric spatial state or a symmetric triplet ($s = 1$) state χ_T in the case of the antisymmetric spatial state. We can therefore write the overall wave function for the singlet (triplet) case ψ_S (ψ_T) as:

$$\psi_S = \frac{1}{\sqrt{2}}[\psi_a(\vec{r}_1)\psi_b(\vec{r}_2) + \psi_A(\vec{r}_2)\psi_B(\vec{r}_1)]\chi_S \tag{4.9}$$

and

$$\psi_T = \frac{1}{\sqrt{2}}[\psi_a(\vec{r}_1)\psi_b(\vec{r}_2) - \psi_A(\vec{r}_2)\psi_B(\vec{r}_1)]\chi_T \tag{4.10}$$

The energies of these singlet and triplet states are expressed as:

$$E_S = \int \psi_S^* \hat{H} \psi_S dr_1 dr_2 \tag{4.11}$$

and

$$E_T = \int \psi_T^* \hat{H} \psi_T dr_1 dr_2 \tag{4.12}$$

It is assumed here that the spin parts of the wave function χ_S and χ_T are normalized. The difference between the two energies is [4]:

$$E_S - E_T = 2 \int \psi_a^*(\vec{r}_1)\psi_b^*(\vec{r}_2)\hat{H}\psi_a(\vec{r}_2)\psi_b(\vec{r}_1) dr_1 dr_2 \tag{4.13}$$

We have seen earlier from Eqn. 4.5 that $\vec{S}_a \cdot \vec{S}_b$ = -3/4 and 1/4 for a singlet and triplet state, respectively. The Hamiltonian accordingly can be written as:

$$\hat{H} = \frac{1}{4}(E_S + 3E_T) - (E_S - E_T)(\vec{S}_a \cdot \vec{S}_b) \tag{4.14}$$

This effective Hamiltonian now consists of a constant term and an interesting spin-dependent term. We now define the *exchange constant* or *exchange integral* as:

$$J = \frac{E_S - E_T}{2} = \int \psi_a^*(\vec{r}_1)\psi_b^*(\vec{r}_2)\hat{H}\psi_a(\vec{r}_2)\psi_b(\vec{r}_1) dr_1 dr_2 \tag{4.15}$$

and accordingly the spin-dependent term in the Hamiltonian is written as:

$$\hat{H}^{Spin} = -2J\vec{S}_a \cdot \vec{S}_b \tag{4.16}$$

If the exchange constant $J > 1$, then $E_S > E_T$ and the triplet state $S = 1$ is favoured. On the other hand, if $J < 1$, then the singlet state $S = 0$ is favoured. In the early days of quantum mechanics the Eqn. 4.16 for two-spin particles system was generalized to a many-body system involving interactions between all the neighbouring atoms leading to the famous Heisenberg Hamiltonian:

$$\hat{H} = -\sum_{ij} J_{ij} \vec{S}_i \cdot \vec{S}_j \qquad (4.17)$$

where J_{ij} represents the exchange constant between the ith and j^{th} spins. The summation in Eqn. 4.17 involves each pair of spins twice, hence factor 2 is omitted.

The exchange integral is usually positive if the two electrons reside on the same atom, which stabilizes the triplet-spin state. This, in turn, leads to the antisymmetric spatial wave function and that minimizes the Coulomb repulsion between the electrons by keeping them apart. This arrangement is consistent with Hund's first rule.

The situation is quite different when the two electrons reside on the neighbouring atoms. The state of this electron system will be the combination of a state centred on one atom and another centred on the other atom. We shall now qualitatively discuss the nature of this energy state by recalling the problem of a particle in a one-dimensional box of length L. The energy of such a particle is proportional to L^{-2}, implying that a particle confined within a small box will have large kinetic energy associated with it. Thus, in this two-electron system if each electron is allowed to access both the atoms by forming a bond, then the box they are confined in will be relatively large and that will lead to a saving of kinetic energy. In this situation, the appropriate state to consider would be molecular orbitals instead of atomic orbitals. This molecular orbital state can be a spatially symmetric bonding orbital or spatially antisymmetric antibonding orbital. These states are shown schematically in Fig. 4.1. However, we can see from Fig. 4.1 that the antibonding orbital goes through zero and changes sign midway between the two atoms. This indicates that the probability of finding the electron in this region is relatively small, and hence offers less accessible space for electrons in comparison with those regions with spatially symmetric bonding orbital. This means that the antibonding orbital

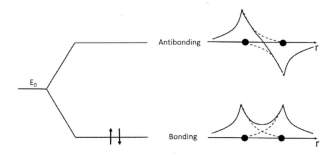

FIGURE 4.1 Molecular orbital state with spatially symmetric bonding orbital or spatially antisymmetric antibonding orbital.

will have higher kinetic energy. So an antisymmetric spin-singlet state with spatial symmetric bonding orbital will be energetically favourable, and thus the exchange integral is usually negative.

4.3 Physical Meaning of Exchange Energy

After the above rather mathematical description of the quantum-mechanical exchange interaction, it may be worth summarizing the physical origin of exchange energy. We may begin by explaining why two hydrogen atoms come together to form a stable molecule [2]. Each of these hydrogen atoms possesses a single electron moving around the nucleus consisting of a single proton. Considering a pair of hydrogen atoms separated by a certain distance, there are (i) electrostatic attractive forces between the electrons and protons and (ii) repulsive forces between two electrons and between two protons. Coulomb's law can estimate these electrostatic forces. In addition, there is a quantum-mechanical force, which depends on the relative orientation of the spins of two electrons. This is the exchange force, and it is a consequence of the *Pauli exclusion principle* applied to the system consisting of the pair of hydrogen atoms.

Pauli exclusion principle states that two electrons can be in the same energy state only if they have opposite spins. In the event the two electrons in the pair of hydrogen atoms have opposite spins, the hydrogen atoms can come so close together that their two electrons can have the same velocity and occupy very nearly the same small region of space, i.e., have the same energy [2]. On the other hand, the two electrons will tend to stay far apart if their spins are parallel. So, it can be said that if the spins are antiparallel, the sum of the electrostatic and exchange forces is attractive and a stable hydrogen molecule forms. The two hydrogen atoms repel one another if the electron spins are parallel. The ordinary Coulomb electrostatic energy is therefore modified by the electron spin orientations, which means that the exchange force is fundamentally electrostatic in origin [2].

The term *exchange* can be perceived in the following manner. When the two hydrogen atoms are nearby, the electron in each hydrogen atom is moving around the respective proton in all likelihood. However, electrons are indistinguishable, and hence one must consider the possibility that the two electrons exchange places, so that the electron in atom 1 moves around the proton in atom 2 and the electron in atom 2 moves around the proton in atom 1. This introduces an additional term, that of the exchange energy, into the expression for the total energy of the two atoms. This exchange of electrons takes place at a very high frequency, about 10^{18} times per second in the hydrogen molecule [2]. It is important to note here that the exchange energy forms an important part of the total energy of many molecules and in the formation of covalent bond. This fact was recognized well before Heisenberg showed that exchange interaction also plays a crucial role in the magnetic ordering in solids.

4.4 Direct Exchange

When the electrons on neighbouring magnetic atoms interact through an exchange interaction, it is known as a direct exchange. It is expected to be the most obvious route for exchange interaction to take place, but in real materials this is seldom the case. In most magnetic materials the overlap between the neighbouring magnetic orbitals is not sufficient enough to make the direct exchange important for the mechanism to explain the observed magnetic properties. In rare earth metals, for example, the $4f$ electrons are strongly localized and reside quite near to the electrons with little chance of the overlap of $4f$ orbitals of the neighbouring rare earth atoms. Thus, direct exchange interaction is unlikely to be effective in rare earth metals. Even in the case of magnetic $3d$-transition metals like Fe, Co, and Ni, where the $3d$ orbitals are relatively extended further from the nucleus, it is quite difficult to correlate direct exchange with their magnetic properties. These materials are metallic, and the conduction electrons are expected to play an important role. A correct description of magnetism in such materials needs to account for both the localized and the itinerant character of the electrons. Thus, in many magnetic materials, some kind of indirect exchange interaction needs to be considered for a proper understanding of their magnetic properties.

4.5 Indirect Exchange in Insulating Solids

4.5.1 Superexchange

Many insulating solids like transition metal oxides and fluorides display long-range magnetic order. Antiferromagnetism has been observed in insulating solids like MnO and MnF_2, which at first sight is a bit surprising since there is no direct overlap of unpaired $3d$ electron orbitals on Mn^{2+} ions in these systems. This implies that long-range magnetic interactions must be taking place here via some intermediary. Taking the example of MnO we shall see below that the physical basis for this long-range antiferromagnetic interaction is that the electrons with antiparallel spins can gain energy by spreading into non-orthogonal overlapping orbitals, whereas electrons with parallel spins cannot.

This kind of interaction is known as superexchange because such interaction involves the relative interaction of magnetic ions over large distances in the medium occupied by diamagnetic ions and molecules. The idea of superexchange was first introduced by Kramers [6] in 1934 in the context of paramagnetic salts. Kramers suggested that the magnetic ions could induce spin-dependent perturbations in the wave functions of the ions in-between and transmit the exchange effect over large distances. Subsequently, P. W. Anderson used this earlier idea of Kramers to explain

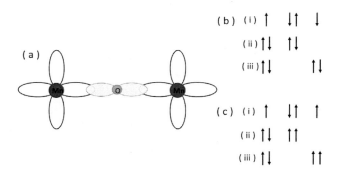

FIGURE 4.2 Superexchange process in magnetic insulator with the example of manganese monoxide (MnO). (a) Collinear arrangement of two Mn^{2+} and one O^2 ions. (b) Spin configurations leading to antiferromagnetic coupling between two manganese ions: (i) ground state; (ii) and (iii) excited states. (c) Spin configurations leading to ferromagnetic coupling between two manganese ions: (i) ground state; (ii) and (iii) excited states. (Adapted from reference [4] with permission from Oxford University Press.)

the antiferromagnetic order in MnO [6]. For illustration, consider two Mn^{2+} and one O^{2-} ions arranged collinearly, and with a simple model involving four electrons [4]. The ground state will consist of a single unpaired electron in the d-orbital on each Mn^{2+} ion and two p electrons on the O^{2-} ion in its outermost occupied states. The dumbbell-shaped oxygen p-orbitals lie in the same axis joining the two Mn^{2+} ions (Fig. 4.2(a)). If the spins on Mn ions are coupled antiferromagnetically, then the ground state is represented by the spin (shown by arrows) configuration (i) in Fig. 4.2(b). It is now possible for one of the two p electrons from the O^{2-} ion to go across to the other Mn^{2+} ions, because of the finite overlap of their wave functions. The two allowed excited states are shown in the configurations (ii) and (iii) of Fig. 4.2(b). The kinetic energy advantage of antiferromagnetism in this spin arrangement is quite apparent since the antiferromagnetic ground state configuration (i) in Fig. 4.2(b) can mix with the two allowed excited states shown in the configurations (ii) and (iii) of Fig. 4.2(b). This makes it clear that the antiferromagnetic coupling between spins (or moments) in Mn ion allows the delocalization of electrons over the whole Mn-O-Mn units, and, in that process, there is an overall reduction in the kinetic energy of the system. We also present in Fig. 4.2(c) the ground state and the excited states if the Mn ions are coupled ferromagnetically for the sake of comparison. It can be seen immediately that the Pauli exclusion principle would have prevented mixing between these states, thus making the ferromagnetic configuration energetically more costly.

There are, however, certain problems in this theory of Anderson to explain the antiferromagnetism in magnetic insulators. The exchange effect appears as third order in this perturbation theory with large early order terms, which do not describe magnetic effects. This results in a poor convergence of the theory. This problem was resolved in a new theory of superexchange interaction proposed by Anderson in 1959 [7]. In this theory the wave functions of the d (or f) shell electrons in the magnetic insulators were assumed to be exact solutions of the problem of a single d-electron in the presence of the full diamagnetic lattice. Inclusion of interactions amongst the electrons gave rise to three spin-dependent effects, which were termed by Anderson as (i) superexchange, always antiferromagnetic, (ii) direct exchange, always ferromagnetic, and (iii) indirect polarization effect. Anderson considered molecular orbitals formed by the mixing of the localized $3d$ orbitals and the p orbitals of the surrounding negative ions. This resulted in two orbitals: (i) the bonding orbital, mainly occupied by p electron of a negative ion, and (ii) the antibonding orbital, which is partially occupied by $3d$ electrons and leading to the magnetism of the system. The wave function of the localized d orbitals thus extends over to the neighbouring negative ion, and there is a probability of transferring one $3d$ electron of the magnetic ion to the neighbouring $3d$ orbitals. Such a transition, however, will be opposed by the repulsive Coulomb interaction between two electrons. This situation is described with the help of the model Hamiltonian in the second quantized form in terms of creation (annihilation) and number operators [8]:

$$H = \sum_{(i,j)\sigma} t_{ij} c_{i,\sigma}^{+} c_{j,\sigma} + \sum_{i} U n_{i\uparrow} n_{i\downarrow} \quad (4.18)$$

where the summation is taken over the pair (i,j), $c_{j\sigma}^{+}$ and $c_{j\sigma}$ are the creation and annihilation operators of an electron with spin σ on the ion at j site, t_{ij} is the amplitude for the electron to hop from the lattice site i to the lattice site j, and U represents the Coulomb interaction energy between two electrons with different spin directions in the same atom. One recognizes this as the seminal Hubbard model if the hopping matrix elements are restricted to the nearest neighbour, and all of them have the same value t [9]. There is an energy increase by U when one d electron of the magnetic ions hops into the unoccupied site of the neighbouring magnetic ions.

In this perturbation theory of Anderson, the first order of the perturbation gives rise to a usual ferromagnetic exchange interaction. The superexchange is a second-order process since it involves both O ions and Mn ions. This is derived from second-order perturbation theory, where the energy involved is obtained from the square of the matrix element of the transition divided by the energy cost of getting the excited state. The transition matrix element here is controlled by the amplitude t_{ij} for the electron to hop from one lattice site to the other, and the energy cost of making an excited state is given by the Coulomb interaction energy U between two electrons with different spin directions. The resultant is an antiferromagnetic exchange interaction, which is expressed as [4, 8]:

$$J_{ij} = -\frac{2t_{ij}^2}{U} \quad (4.19)$$

When $U \ll t$, electrons can propagate in the crystal; this is the situation in a metal. On the other hand, in the limit of $U \gg t$, electrons are localized at the lattice points in the solid. This is the case of an insulator. Starting from the insulator as a limiting case, the superexchange interaction arises from perturbation energy. The exchange interaction can be expressed in the more general form $J_{ij}\tilde{S}_i.\tilde{S}_j$ for both direct and superexchange [10]. The exchange integral consists of two parts: potential exchange and kinetic exchange. The potential exchange represents electron repulsion and favours the ferromagnetic ground state. This term, however, is small when the ions are well separated. The kinetic exchange favours the antiferromagnetic ground state as is elaborated above with the case of Mn-O-Mn. This kinetic exchange term depends on the extent of orbital overlap. These terminologies of kinetic exchange and potential exchange are chosen to emphasize that antiferromagnetism is the result of a gain in kinetic energy, and ferromagnetism is the result of a gain in potential energy. In transition metal oxide systems like Mn-O-Mn, the hopping integral t is of order 0.1 eV and the Coulomb interaction U is in the range 3–5 eV [1]. The exchange interaction J depends sensitively on the ionic separation, and also on the Mn-O-Mn bond angle. In summary, the exchange interaction in various magnetic insulators is predominantly caused by the superexchange arising from the overlap of the localized orbitals of the magnetic electrons with those of intermediate diamagnetic atoms/molecules.

In the discussion on superexchange, so far it has been explicitly assumed that the oxygen ion lies between the two d-orbitals, i.e., the 180^0 geometry. However, the situation changes entirely when the oxygen bridge between the two d-orbitals is 90^0 instead of 180^0. The hopping between the d- and the p-orbital is possibly by symmetry only when they point towards each other. The energy for the system, however, will depend on the relative orientation of the electron spins in the two d-orbitals, and the superexchange can also be ferromagnetic under certain circumstances. If the bond is between a filled orbital and a half-filled orbital or between a half-filled orbital and an empty orbital, the hopping of electron will be influenced by Hund's rule. There will be an energy advantage if the e_g electron arrives in the unoccupied orbital with its spin aligned with the spin of t_{2g} electrons. In this case, superexchange will be ferromagnetic. But this interaction is relatively weak and less common than the antiferromagnetic superexchange.

Taking account of the occupation of the various d levels as dictated by the crystal field theory, some empirical rules were also developed [11, 12] for magnetic insulators, which are related to the formalism of Anderson [6] about the sign of superexchange. These empirical rules (widely known as Goodenough–Kanamori-Anderson rules) are quite useful in practice. These rules are narrated below [1]:

1. The exchange is strong and antiferromagnetic ($J < 0$) when two magnetic (M) ions possess lobes of singly occupied 3d-orbitals pointing towards each

other and have a large overlap and hopping integrals. This is the case of 120–180^0 M-O-M bonds as discussed above with the example of Mn-O-Mn.
2. The exchange is ferromagnetic ($J > 0$) but relatively weak when two M ions have an overlap integral between singly occupied $3d$-orbitals, which is zero by symmetry. This is the case for $\sim 90^0$ Mn-O-Mn bonds.
3. The exchange is also ferromagnetic and relatively weak when two M ions have an overlap between singly occupied $3d$ orbitals and empty or doubly occupied orbitals of the same type.

4.5.2 Double exchange

Double exchange is usually found in mixed-valence compounds. In the insulating systems discussed above the lowest energy state had essentially the same number of electrons on every site, and in the presence of strong electron correlation the hopping of electrons was strongly suppressed by the Coulomb repulsion energy U. In contrast, in a mixed-valence system, the number of electrons per site is a non-integer, so despite the presence of large Coulomb repulsion, some sites will have more electrons than others. Electron hopping between such sites will be allowed without involving expenditure of energy U. This can lead to a new exchange mechanism called a double exchange. We shall now elaborate on this double exchange interaction between magnetic ions in such mixed-valence systems taking the example of a Mn-oxide mixed-valence compound containing a Mn ion which can exist in oxidation state 3+ or 4+, i.e., as Mn^{3+} or Mn^{4+}. The resulting exchange interaction is a combination of potential and kinetic exchange.

The Mn-oxide compound with which we choose to discuss the double exchange mechanism is Sr-doped $LaMnO_3$, which forms in a perovskite structure. $LaMnO_3$ is a Mott-insulator, and it undergoes antiferromagnetic ordering via a superexchange mechanism [13]. A Mn^{3+} ion in $LaMnO_3$ has the electronic configuration of $t_{2g}^3 e_g^1$. The e_g^1-electrons with spin 1/2 strongly hybridizes with the oxygen $2p$ state, and they are subject to the electron correlation effect. On the other hand, the t_{2g} electrons are less hybridized with the oxygen $2p$ electrons, and they are stabilized by crystal field splitting. They form a local spin 3/2, which is strongly coupled to spin 1/2 of e_g electrons via strong on-site ferromagnetic Hund's coupling. This gives a total spin S = 2 for $LaMnO_3$. When doped with divalent Sr^{2+} a fraction x of manganese atoms in $La_{1-x}Sr_xMnO_3$ compound goes to the Mn^{4+} state, while the remaining (1-x) fraction remain in Mn^{3+} state. With Sr-concentration approximately above x = 0.175 the system becomes ferromagnetic around room temperature and shows metallic conductivity in the ferromagnetic state.

The ferromagnetism in the Sr-doped $LaMnO_3$ can be explained in terms of the double exchange mechanism giving rise to the ferromagnetic coupling between Mn^{3+} and Mn^{4+} ions. This is elaborated with the help of Fig. 4.3. The e_g electron can

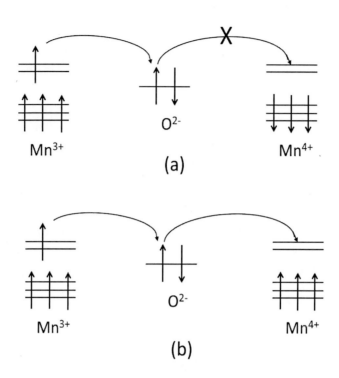

FIGURE 4.3 Double exchange mechanism leading to electron transfer and ferromagnetic coupling between Mn^{3+} and Mn^{4+} ions in Sr-doped $LaMnO_3$: (a) forbidden antiferromagnetic configuration of Mn-spins; (b) energetically favourable ferromagnetic configuration of Mn-spins.

hop to another Mn-ion site only when that site is not occupied by an electron with the same spin direction. There no spin-flip process is allowed during hopping. A Mn^{4+} site with no e_g electron thus allows an electron to hop there from a Mn^{3+} site. However, the strong intra-site Hund coupling demands that the e_g electron on arrival must be ferromagnetically aligned with the three existing t_{2g} electrons. Thus, the hopping of an e_g electron to a neighbouring Mn-ion site where the t_{2g} electrons would have an antiparallel spin arrangement with respect to arriving e_g electron (Fig. 4.3(a)) would be energetically unfavourable. The high-spin arrangement arising out of strong intra-site Hund's coupling, therefore, requires that both the donor and acceptor Mn ions in this mixed valence $La_{1-x}Sr_xMnO_3$ compound must be ferromagnetically aligned. The ability of the electrons to hop causes the saving of kinetic energy. Thus, the ferromagnetic configuration is shown in Fig. 4.3(b) reduces the overall energy of the system by allowing the hopping process to take place [4]. The double exchange mechanism between the Mn ions via oxygen comprises both the kinetic and potential components. Moreover, this hopping of electrons in this ferromagnetic state of $La_{1-x}Sr_xMnO_3$ compound gives rise to a metallic behaviour.

4.5.3 Anisotropic exchange interaction: Dzyaloshinski–Moriya interaction

In some materials, spin–orbit interaction can play a similar role as intervening oxygen atoms in materials showing superexchange interaction. Instead of being linked with oxygen, the excited state arises due to the spin–orbit interaction in one of the magnetic ions. This generates the possibility of exchange interaction between the excited state of one magnetic ion and the ground state of the other. This is variously known as the Dzyaloshinski–Moriya interaction or anisotropic exchange interaction. This interaction between two spins leads to a term in the Hamiltonian $\hat{H}_{DM}$:

$$\hat{H}_{DM} = \vec{D}.\vec{S}_1 \times \vec{S}_2 \tag{4.20}$$

This vector $\vec{D}$ vanishes only in the case of the crystal field having an inversion symmetry with respect to the centre between the two magnetic ions. The $\vec{D}$, in general, may not vanish and depending upon the symmetry will lie parallel or perpendicular to the line connecting the two spins. Fig. 4.4 shows schematically the mechanism of the Dzyaloshinski–Moriya interaction. This interaction tries to force the spins $\vec{S}_1$ and $\vec{S}_2$ to be at right angles in a plane perpendicular to the vector $\vec{D}$ in such an orientation that it ensures negative energy. The net effect results in a slight rotation of the spins by a small angle. In scientific terms, it is often mentioned as canting. Such canting is usually observed in antiferromagnets, which results in a small ferromagnetic component of the moments perpendicular to the spin-axis of the antiferromagnet. This effect is also known as weak ferromagnetism.

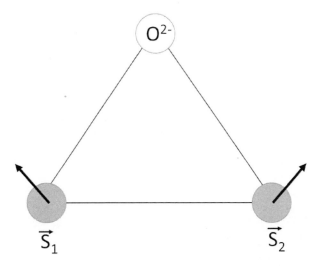

FIGURE 4.4 Schematic representation of the Dzyaloshinski–Moriya interaction.

4.6 Indirect Exchange in Metals

The conduction electrons in metals can mediate an exchange interaction between magnetic ions. A localized magnetic moment in a magnetic ion can cause a polarization of spins in the surrounding conduction electrons. This induced spin-polarization of the conduction electrons, in turn, can couple with the localized moment on another magnetic ion at a distance r away from the first one. This exchange interaction is an indirect one since it does not directly couple the magnetic ions, and is known as a Ruderman–Kittel–Kasuya–Yosida (RKKY) interaction. Assuming the metal to have a spherical Fermi surface of radius k_F, this r-dependent interaction can be expressed for a large distance as:

$$J_{RKKY} \propto \frac{cos(2k_F r)}{r^3} \qquad (4.21)$$

Fig. 4.5 shows a schematic representation of an $RKKY$ interaction. It is evident that RKKY interactions are long-range in nature and have an oscillatory dependence on the distance between the interacting magnetic ions. Thus, depending on the distance that separates them, the interaction between the magnetic ions can be both ferromagnetic and antiferromagnetic.

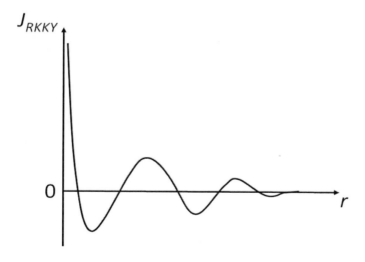

FIGURE 4.5 Schematic representation of damped oscillatory nature of RKKY interaction.

5

Magnetically Ordered States in Solids

In this chapter, we shall study different types of ordered magnetic states that can arise as a result of various kinds of magnetic interactions as discussed in the previous section. In Fig. 5.1 we present some of these possible ground states: ferromagnet, antiferromagnet, spiral and helical structures, and spin-glass. There are other more complicated ground states possible, the discussion of which is beyond the scope of the present book. For detailed information on the various magnetically ordered states in solids, the reader should refer to the excellent textbooks by J. M. D. Coey [1] and S. J. Blundell [4].

5.1 Ferromagnetism

In a ferromagnet, there exists a spontaneous magnetization even in the absence of an external or applied magnetic field, and all the magnetic moments tend to point towards a single direction. The latter phenomenon, however, is not necessarily valid strictly in all ferromagnets throughout the sample. This is because of the formation of domains in the ferromagnetic samples. Within the individual domains, the magnetic moments are aligned in the same direction, but the magnetization of each domain may point towards a different direction than its neighbour. We will discuss more on the magnetic domains later on.

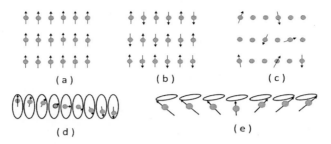

FIGURE 5.1 Arrangements of magnetic moments in various kinds of magnetic systems: (a) ferromagnet, (b) antiferromagnet, (c) spin-glass, (d) helical magnets and (e) spiral spin system.

The Hamiltoninan for a ferromagnet in an applied magnetic field can be expressed as:

$$\hat{H} = -\sum_{ij} J_{ij} \vec{S}_i . \vec{S}_j + g\mu_B \sum_j \vec{S}_j . \vec{B} \tag{5.1}$$

The exchange interaction J_{ij} involving the nearest neighbours is positive, which ensures ferromagnetic alignment. The first term on the right-hand side of Eqn. 5.1 is the Heisenberg exchange energy, and the second term is the Zeeman energy. In the discussion below it is assumed that one is dealing with a system with no orbital angular momentum, so that $L = 0$ and $J = S$.

In order to solve the equation it is necessary to make an assumption by defining an effective molecular field at the ith site by:

$$\vec{B}_{mf} = -\frac{2}{g\mu_B} \sum_{ij} J_{ij} S_j \tag{5.2}$$

Now the total energy associated with i^{th} spin consists of a Zeeman part $g\mu_B \vec{S}_i . \vec{B}$ and an exchange part. The total exchange interaction between the ith spin and its neighbours can be expressed as:

$$-2\vec{S}_i . \sum_{ij} J_{ij} \vec{S}_j = -g\mu_B \vec{S}_i . \vec{B}_{mf} \tag{5.3}$$

The factor 2 in Eqn. 5.3 arises due to double counting. The exchange interaction is essentially replaced by an effective molecular field $\vec{B}_{mf}$ produced by the neighbouring spins. The effective Hamiltonian can now be expressed as:

$$\hat{H} = g\mu_B \sum_i \vec{S}_i . (\vec{B} + \vec{B}_{mf}) \tag{5.4}$$

This now looks like a Hamiltonian of a paramagnet in a magnetic field $\vec{B} + \vec{B}_{mf}$. The moments can be aligned by the internal molecular field $\vec{B}_{mf}$ at low temperatures, even in the absence of any externally applied field. The thermal fluctuations, however, will tend to progressively disturb the alignments of the magnetic moments as the temperature is increased, and at a critical temperature, the ferromagnetic order will be destroyed completely. This model is known as the Weiss model of ferromagnetism.

The molecular field measures the effect of ordering of the system, hence it can be assumed that:

$$\vec{B}_{eff} = \lambda \vec{M} \tag{5.5}$$

Here λ is a constant, which parameterizes the strength of $\vec{B}_{mf}$ as a function of magnetization, and for a ferromagnet $\lambda > 0$. The molecular field can be extremely large in ferromagnets since exchange interaction involves large Coulomb energy.

To find a solution to the Weiss model the following equations are to be solved simultaneously [4]:

$$\frac{\vec{M}}{\vec{M}_S} = B_J(y) \tag{5.6}$$

and

$$y = \frac{g_J \mu_B J(\vec{B} + \lambda \vec{M})}{k_B T} \tag{5.7}$$

Eqns. 5.6 and 5.7 are solved graphically as illustrated in Fig. 5.2. In the restricted case of $\vec{B} = 0$, we get $\vec{M} = k_B T y / g_J \mu_B J \lambda$. The plot $\vec{M}$ against y is a straight line with a gradient proportional to temperature T as shown in Fig. 5.2. There is no simultaneous solution of the above equations except at the origin, where $y = 0$ and $\vec{M}_S = 0$. This situation, however, changes below a critical temperature T_C. In this temperature regime, the gradient of the $\vec{M}$ versus y line is less than the Brillouin function $B_J(y)$ at the origin, and there are three solutions, one at $\vec{M}_S = 0$ and another two for $\vec{M}_S$ at $\pm$ some non-zero value. It is found that the non-zero solutions are stable, while the zero solution is unstable. Thus, non-zero spontaneous magnetization appears in ferromagnetic materials below a critical temperature T_C even in the absence of an external magnetic field. The spontaneous magnetization increases with the decrease in temperature. Above this critical temperature T_C the ferromagnetic materials behave like a paramagnet.

This critical temperature in the ferromagnetic materials is known as the Curie temperature. This can be estimated by determining when the gradients of the line $\vec{M} = k_B T y / g_J \mu_B J \lambda \vec{M}_S$ and the curve $\vec{M} = \vec{M}_S B_J(y)$ are equal at the origin [4]. For small values of y, $B_J(y) = (J+1)y/3J + O(y^3)$ and the Curie temperature T_C

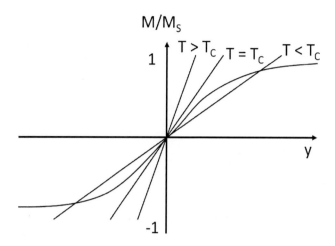

FIGURE 5.2 The solutions of Eqns. 5.6 and 5.7 obtained graphically for $\vec{B} = 0$.

is then expressed as

$$T_C = \frac{g_J \mu_B (J+1) \lambda M_S}{3 k_B} = \frac{n \lambda \mu_{eff}^2}{3 k_B} \quad (5.8)$$

The molecular field is given by

$$B_{mf} = \lambda M_S = \frac{3 k_B T_C}{g_J \mu_B (J+1)} \quad (5.9)$$

For a ferromagnet with $J = 1/2$ and $T_C \approx 1000$ K, B_{mf} is estimated to be about 1500 T [4]. One can see that this effective field is an enormous magnetic field. This effective field reflects the strength of the exchange interaction.

The mean-field magnetization $\vec{M}$ as a function of temperature is shown in Fig. 5.3 for different values of J. The form of the magnetization curve is slightly different for different J, but the general features remain the same. Magnetization is non-zero in the temperature regime $T < T_C$ and becomes zero for $T \geq T_C$.

The effect of an applied magnetic field in ferromagnetic materials is to shift the straight line in the graphical solution of the equations (Fig. 5.2) to the right. This results in a non-zero solution for magnetization $\vec{M}$ for all temperatures. Energetically, in an applied magnetic field there is always an advantage in ferromagnetic materials to have a non-zero magnetization with moments lining up along the magnetic field.

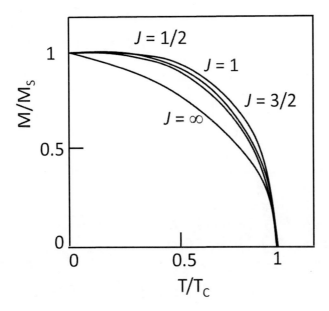

FIGURE 5.3 Mean-field magnetization plotted against reduced temperature (T/T_C) for different values of exchange interaction.

5.1.1 Magnetic susceptibility in a ferromagnet

The application of a small field $\vec{B}$ at temperatures much higher than the Curie temperature, i.e., $T \gg T_C$, will induce a small magnetization in a ferromagnetic sample. In this situation the approximation $y \ll 1$ can be used in the Brillouin function. This leads to:

$$\frac{M}{M_S} \sim \frac{g_J \mu_B (J+1)}{3k_B} \left(\frac{\vec{B} + \lambda \vec{M}}{T} \right) \tag{5.10}$$

So that:

$$\frac{M}{M_S} \approx \frac{T_C}{\lambda \vec{M}_S} \left(\frac{\vec{B} + \lambda \vec{M}}{T} \right) \tag{5.11}$$

Eqn. 5.11 can be rearranged and written as:

$$\frac{M}{M_S} \left(1 - \frac{T_C}{T} \right) \approx \frac{T_C B}{\lambda M_S} \tag{5.12}$$

One can then write the expression for susceptibility as:

$$\chi = Lt_{B \to 0} \frac{\mu_0 M}{B} \propto \frac{1}{T - T_C} \tag{5.13}$$

This is known as the Curie-Weiss law.

5.2 Antiferromagnetism

We will now consider the case of a material with negative exchange interaction. The molecular field here will be oriented in such a way that it will be favourable for the nearest neighbour moments to be aligned antiparallely to each other. This leads to a magnetic order known as antiferromagnetism. Such antiferromagnetic order is commonly found in insulating materials, and may be represented in terms of two interpenetrating sublattices (Fig. 5.4). In some of these interpenetrating sublattices the magnetic moments point up, while pointing down in others. As we see in Fig. 5.4, the nearest neighbours of each magnetic moment reside entirely on the other sublattice. We shall now assume that the molecular field on one sublattice is proportional to the magnetization of the other sublattice, and also that there is no applied magnetic field. Labelling the sublattice with spin-up (spin-down) magnetization as + (-) one can express the molecular field on each sublattice as:

$$B_+ = -|\lambda|M_-$$
$$B_- = -|\lambda|M_+ \tag{5.14}$$

Here λ is the molecular field constant, and its value is now negative. The magnetization on each sublattice is expressed as:

$$M_\pm = M_S B_J \left(\frac{-g_J \mu_B J |\lambda| M_\mp}{k_B T} \right) \tag{5.15}$$

The two sublattices are equivalent in every respect except for the fact that the magnetic moments are in opposite directions. Hence, one can write:

$$|M_+| = |M_-| \equiv M \tag{5.16}$$

and

$$M = M_S B_J \left(\frac{g_J \mu_B J |\lambda| M}{k_B T} \right) \tag{5.17}$$

This is quite similar to the equation for magnetization for ferromagnetism (Eqns. 5.6 and 5.7), and hence the molecular field in each sublattice will have a

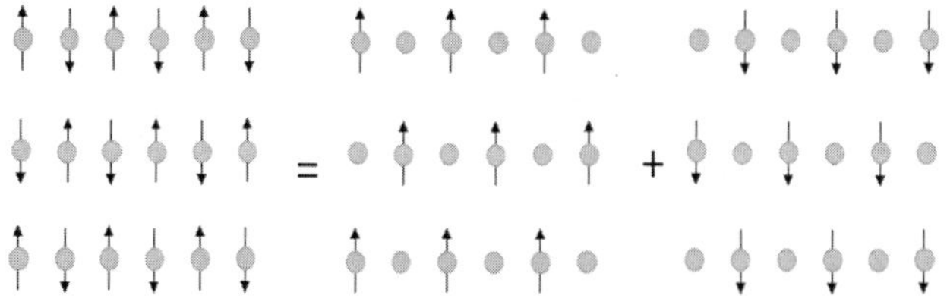

FIGURE 5.4 Schematic representation of an antiferromagnet comprising of two interpenetrating sublattices.

similar temperature dependence as shown in Fig. 5.3 and will approach zero in the temperature regions above the antiferromagnetic transition temperature or Néel temperature T_N. This temperature T_N is expressed as:

$$T_N = \frac{g_J \mu_B (J+1)|\lambda| M_S}{3k_B} = \frac{n|\lambda| \mu_{eff}^2}{3k_B} \quad (5.18)$$

Since the magnetization in the two sublattices will be in opposite directions, the total magnetization $\vec{M}_+ + \vec{M}_-$ of the antiferromagnet will be zero. A concept called "staggered magnetization" is defined now, as the difference of the level of magnetization in each sublattice $\vec{M}_+ - \vec{M}_-$. This staggered magnetization is then non-zero in the temperature region below T_N and is akin to the spontaneous magnetization of ferromagnets. More on the antiferromagnetism will be discussed in section 9.2.6 in connection with neutron scattering experiments.

5.2.1 Magnetic susceptibility in an antiferromagnet

In a very similar way as we have done earlier in Section 5.1.1 for a ferromagnet, we can calculate the magnetic susceptibility for an antiferromagnet by expanding the Brillouin function with the approximation $y \ll 1$. This gives $B_J(y) = (J+1)/3J + O(y^3)$, and leads to an expression for the susceptibility:

$$\chi = Lt_{B \to 0} \frac{\mu_o M}{B} \propto \frac{1}{T + T_N} \quad (5.19)$$

This is the same as the Curie–Weiss law with the term $+T_N$ replacing $-T_C$.

The magnetic susceptibility expressed in Eqns. 5.13 and 5.19 provides a way of interpreting the experimental results in the temperature region above the magnetic transition temperature (i.e., in the paramagnetic state). The experimental results on the temperature dependence of magnetic susceptibility can be fitted with a generalized Curie–Weiss law:

$$\chi \propto \frac{1}{T - \theta} \quad (5.20)$$

where θ is known as the Weiss temperature. The material under investigation is identified as a paramagnet if $\theta = 0$. On the other hand, if the material is a ferromagnet (antiferromagnet), its Weiss temperature becomes $\theta = T_C$ ($\theta = -T_N$). Fig. 5.5(a) shows the expected temperature dependence of the susceptibility for paramagnet, ferromagnet, and antiferromagnet. Usually, it is the reciprocal susceptibility $(1/\chi)$ that is plotted as a function of temperature, and the intercept of the straight line on the temperature axis yields the value of the Weiss temperature (Fig. 5.5(b)).

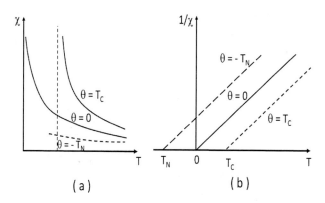

FIGURE 5.5 (a) The Curie–Weiss law of magnetic suceptibility shown for three cases: paramagnet ($\theta = 0$), ferromagnet ($\theta = T_C$) and antiferromagnet ($\theta = -T_N$). (b) Reciprocal susceptibility ($1/\chi$) versus temperature straight line plots with the intercept on the temperature axis yielding the Wiess temperature θ.

The Weiss temperature obtained experimentally in many antiferromagnets is found to be quite different from $-T_N$. This is attributed mainly to the assumption that the molecular field in one sublattice is solely dependent on the magnetization of the other sublattice.

Unlike in the case of a ferromagnet, the application of an external magnetic field in an antiferromagnet does not lead to an energetic advantage for the moments to line up along the applied magnetic field. This is because of the reason that any energy saving in one sublattice would be cancelled by the energy expense in the other sublattice since the magnetization of the two sublattices is equal and opposite.

Let us first discuss the case of the external magnetic field applied parallel to one of the sublattices in an antiferromagnet at $T = 0$, where thermal fluctuations can be ignored. Sublattice magnetizations $|M_+| = |M_-| = M_S$ (saturation magnetization). In the presence of a small applied magnetic field along the magnetization direction of one sublattice, the field will be antiparallel to the magnetization direction of the other sublattice. This will cause a small term to be added to or subtracted from the local field of each sublattice. Overall this, however, has no effect since both the sublattices are already saturated. The net induced magnetization in the antiferromagnetic material will be zero so that parallel susceptibility $\chi_{||} = 0$. In the event of increasing the temperature (and $T < T_N$), thermal fluctuations will decrease the molecular field in each sublattice. The applied magnetic field will now enhance the magnetization of the sublattice to which it is parallel but will reduce the magnetization on the other. The parallel susceptibility $\chi_{||}$ will now attain a non-zero value and will rise further with the increase in temperature and reach a maximum value at $T = T_N$ (Fig. 5.6).

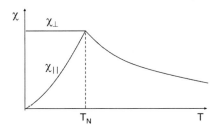

FIGURE 5.6 Temperature dependence of parrallel and perpendicular susceptibility χ in an antiferromagnet.

In the case of a small magnetic field applied perpendicularly to the magnetization direction of one of the sublattices, the field will cause the magnetization in each of the sublattices to tilt slightly. This, in turn, will result in the development of a small component of magnetization along the applied magnetic field, and the perpendicular susceptibility $\chi_\perp$ will acquire a non-zero value. Increase in the temperature, however, will have little effect on $\chi_\perp$ since sublattice magnetizations will be reduced equally and also will be affected symmetrically by the small applied magnetic field. Hence, $\chi_\perp$ will remain independent of temperature in the temperature region $T < T_N$ (Fig. 5.6).

5.3 Ferrimagnetism

In the discussion on antiferromagnetism, it was assumed that the two sublattices were equivalent. However, if for some crystallographic reasons the two sublattices are not equivalent, the sublattice magnetizations then will not be equal. The material will show a net magnetization because the sublattice magnetizations are not cancelled out. This behaviour is called ferrimagnetism. The spontaneous magnetization of each sublattice will have different temperature dependences since the molecular field on each sublattice is different. This will lead to a net magnetization in ferrimagnets with complicated temperature dependences. In some ferrimagnets, one sublattice can dominate in low-temperature regions and the other at a higher temperature. In such cases, magnetization can go to zero and change sign at a temperature known as the compensation temperature. The magnetic susceptibilities in general do not follow the Curie–Weiss law.

5.4 Helical Magnetic Order

Many rare earth metals form in such a crystal structure where the constituent atoms are arranged in layers. In dysprosium, for example, there are ferromagnetic alignments within the layers and the interaction between the layers can be described by a nearest-neighbour exchange constant J_1 and a next-nearest-neighbour exchange constant J_2. Let us now consider that the angle between the magnetic moments in the successive planes is θ. In this case, the energy of the system can be expressed as:

$$E = -2NS^2(J_1 cos\theta + J_2 cos2\theta) \qquad (5.21)$$

where N is the number of atoms in each plane. The energy of the system will be minimum when $\partial E/\partial \theta = 0$, and that will lead to:

$$(J_1 + 4J_2 cos\theta)sin\theta = 0 \qquad (5.22)$$

One of the solutions of the above equation is $sin\theta = 0$. This implies either $\theta = 0$ indicating ferromagnetic order or $\theta = \pi$, which indicates antiferromagnetic order. The other possible solution is:

$$cos\theta = -\frac{J_1}{4J_2} \qquad (5.23)$$

The above solution gives rise to helical magnetic order or helimagnetism. This kind of order is favoured over either ferromagnetism or antiferromagnetism when $J_2 < 0$ and $|J_1| < 4|J_2|$. In general, the pitch of the spiral may not be commensurate with the lattice parameter, and hence no two layers in a crystal of helimagnetic system will be exactly in the same spin directions.

Among the many magnetic systems showing helical magnetic structure, the most prominent ones are the rare earth metals, many of which form in hexagonal close-packed structure. The axis of the helix is along the c-axis, which is perpendicular to the hexagonally close-packed planes. In rare earth metals Tb, Dy, and Ho the plane in which the spin rotates is the hexagonally close-packed planes. On the other hand, in Er and Tm the easy axis for the spins is the c-axis, and the c-component of the spins is modulated sinusoidally over a certain temperature range. More on the helical magnetic structures will be discussed in section 9.2.6 in connection with neutron scattering experiments.

5.5 Spin-Glass Order

We have so far discussed the different possible magnetic orderings in systems with source of one or more magnetic moments in each unit cell. It is also possible to have metallic alloys where magnetic elements like Fe, Mn, Co, and Ni are randomly dispersed in a non-magnetic metal matrix-like Au, Ag, Cu, etc. Of course, the first important question here is whether this magnetic atom carries a magnetic moment at all. In fact, in a very small concentrated set of less than 1% of magnetic elements, the magnetic atoms may not carry a magnetic moment. This is the famous Kondo effect, which is a subject by itself and beyond the scope of this book. Assuming that the concentration of magnetic elements is large enough so that they carry a magnetic moment in the non-magnetic metallic hosts, the question is what kind of magnetic ordering can be expected in such systems. Beyond a certain critical percolation concentration, the magnetic elements can form an infinite percolative chain/cluster. The magnetic moments would interact via direct exchange interaction, and the system is expected to form a long-range magnetic order, ferromagnetic or antiferromagnetic depending on the nature of the solute magnetic element.

A more interesting situation arises in the concentration regime below the percolation concentration. In this concentration regime, the localized inner d-electron spins in $3d$-solute atoms undergo Ruderman–Kittel–Kasuya–Yosida (RKKY) interactions through conduction electrons of the dilute alloy system. RKKY interactions oscillate in sign as a function of distance. Since the solute magnetic atoms are distributed randomly in the lattice, the magnetic moment at a particular lattice site can be subjected to competing ferromagnetic and antiferromagnetic interactions arising from the oscillating signs of RKKY interaction depending on their separation distance from the other magnetic moments. This competition gives rise to a great deal of built-in chaos with no well-defined ground state. Instead of a single ground state, a large number of degenerate states become possible. The magnetic moments undergo thermal fluctuations at high temperatures and turn into a paramagnetic state. The dynamics of magnetic moments slow down with a decrease in temperature, and then below a certain characteristic temperature T_F the moments get entangled in one of the degenerate states and form a frozen magnetic state called *spin-glass*. In a crude sense, while the random spin configuration of a paramagnet changes rapidly in time, the configuration of the magnetic moments in spin-glass is also random but is frozen in time (Fig. 5.1).

The spin-glass order in a binary magnetic alloy with one magnetic component is depicted in Fig. 5.7 with the help of the magnetic phase diagram of Au-Fe alloys, which has been created with the help of results obtained with various experimental techniques that are both macroscopic as well as microscopic [14]. Above the concentration of approximately 16% Fe in Au matrix, long-range ferromagnetic orders form in the low-temperature region. Below that concentration and down to about 8% Fe, clusters of iron moments with short-range correlation freeze randomly below a characteristic temperature and give rise to a magnetic state commonly known as cluster spin-glass or cluster-glass. Further down in the dilute concentration range it is the individual Fe moments that are frozen randomly giving rise to a low-temperature canonical spin-glass state.

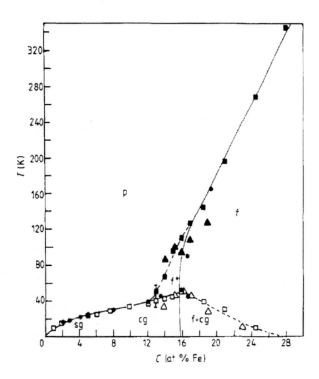

FIGURE 5.7 Experimentally obtained magnetic phase diagram of the Au-Fe alloys. In this phase digram, p stands for paramagnet, f for ferromagnet, sg for spin-glass, cg for cluster glass and f+cg for a mixed phase with coexistence of cluster glass and ferromagnetic orders. (From reference [14] with permission from IOP Publishing)

5.6 Spin Waves in Ferromagnets

A ferromagnet is perfectly ordered at $T = 0$ with all the magnetic moments or spins aligned in the same direction. However, there will always be some misalignments of the spins at any non-zero temperature because of thermal activation. This will result in a decrease in the spontaneous magnetization of a ferromagnet with increase in temperature. The simplest excited state of the ferromagnet will be that in which the spontaneous magnetization is reduced by the misalignment of just a single spin. This misalignment will not, however, be static, since an exchange interaction of some kind would ensure that it travels through the lattice of the ferromagnetic crystal. If the misalignment of spin is confined to a particular lattice site, it would be a single particle excitation and this would require a considerable amount of energy. It will be energetically favourable if the spin misalignment is associated with all the atoms/ions in the ferromagnetic crystal. The spin-misalignment will then be in the form of a collective excitation [15]. Such collective excitations are known as spin waves. By analogy with phonons and photons, the quantized spin wave is often

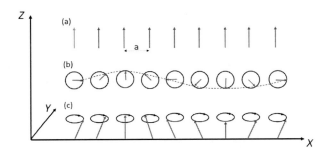

FIGURE 5.8 (a) Linear chain of classical spins in the ground state. (b) Top view of the spins of a spin wave in a linear chain of classical spins propagating in the $+x$ direction. (c) Side view of the spins in the same spin wave.

referred to as a magnon. It requires just a vanishingly small amount of energy to create a magnon, provided the magnon has a sufficiently long wavelength.

There exists a rigorous quantum mechanical theory of spin waves [4, 15, 16]. But to visually conceptualize spin waves it is easier to follow a semi-classical approach [4, 15, 16]. This semi-classical theory was first put forward by Heller and Kramers [17] and later by Herring and Kittel [18], where the spin $\vec{S}$ is considered as a classical vector and the spin vectors precess about the direction of a small external magnetic field. We now consider a one-dimensional chain having N uniformly spaced spins with magnitude S separated by a distance a (Fig. 5.8(a)), and which are coupled through nearest-neighbour exchange interaction. Assuming only the Heisenberg interaction, the energy of the spin chain can be expressed as [16]:

$$U_{exc} = -2J \sum_i \vec{S}_i \cdot \vec{S}_{i+1} \qquad (5.24)$$

Here J is the nearest-neighbour exchange parameter and $\vec{S}_i$ denotes the classical spin at the coordinate $x_i = ia$. All the spins are parallel in the ground state (Fig. 5.8(a)) so that with $\vec{S}_i \cdot \vec{S}_{i+1} = S^2$. The exchange energy of the linear chain is then $U_0 = -2JNS^2$.

We assume that the motion of spin waves in a linear chain of spins is governed by the classical mechanical equation for the torque. We further consider that the spin $\vec{S}_i$ has an associated magnetic moment $\vec{\mu}_i = -g\mu_B \vec{S}_i$ and that the torque acting on the moment has the form $\tau = \vec{\mu}_i \times \vec{B}_T$, where $\vec{B}_T$ is an effective magnetic induction representing all interactions on the spin $\vec{S}_i$. The effective magnetic field can be evaluated by considering that the energy of the magnetic moment has the form $U_i = -\vec{\mu}_i \cdot \vec{B}_T$. Comparing with Eqn. 5.24 we can write down this effective magnetic field $\vec{H}_{exc}^{Eff}$ arising from the exchange interaction in the linear spin chain as:

$$\vec{H}_{exc}^{Eff} = -\frac{2J}{g\mu_B\mu_0}(\vec{S}_{i-1} + \vec{S}_{i+1}) \quad (5.25)$$

Apart from the exchange interactions with the neighbouring spins, the spins are also subject to an applied static magnetic field H, so that the total magnetic induction on the spins is $\vec{B}_T = \mu_0(\vec{H} + \vec{H}_{exc}^{Eff})$.

With the help of torque equation $\hbar d\vec{S}/dt = \tau$, we can write down the equation of motion for the spin $\vec{S}_i$ as [16]:

$$\frac{d\vec{S}_i}{dt} = -\gamma\mu_0 \vec{S}_i \times (\vec{H} + \vec{H}_{exc}^{Eff}) \quad (5.26)$$

Here $\gamma = g\mu_B/\hbar$ is the gyromagnetic ratio. The magnetic moment has a direction opposite to the spin. Hence, to be consistent with the ground state shown in Fig. 5.8(a), it is considered that the external magnetic field is applied in the $-\hat{z}$ direction, so that $\vec{H} = -\hat{z}H$. Accordingly, the equation for the spin component $\vec{S}_i^x$ can be expressed as:

$$\frac{d\vec{S}_i^x}{dt} = \gamma\mu_0 S_i^y \left[H + \frac{2J}{g\mu_B\mu_0}(S_{i-1}^z + S_{i+1}^z)\right] - \gamma\mu_0 \frac{2J}{g\mu_B\mu_0}(S_{i-1}^y + S_{i+1}^y)S_i^z \quad (5.27)$$

With the assumption that the amplitude of the spin wave excitation is small, it is possible to linearize the above equation using $S_i^x, S_i^y << S_i^z \approx S$, and the equations for the two transverse spin components can be expressed as:

$$\frac{d\vec{S}_i^x}{dt} = \gamma\mu_0 H S_i^y + \frac{2JS}{\hbar}(2S_i^y - S_{i-1}^y - S_{i+1}^y) \quad (5.28)$$

$$\frac{d\vec{S}_i^y}{dt} = -\gamma\mu_0 H S_i^x - \frac{2JS}{\hbar}(2S_i^x - S_{i-1}^x - S_{i+1}^x) \quad (5.29)$$

Eqs. 5.28 and 5.29 show that the motion of the spin in any site in the linear chain is coupled to the motions of the neighbouring spins. This indicates that their solutions must be in the form of collective excitations. Let us consider for possible solutions excitations in the form of harmonic traveling waves:

$$S_i^x = A_x e^{(ikx_i - wt)}, S_i^y = A_y e^{(ikx_i - wt)} \quad (5.30)$$

Here ω is the angular frequency and k is the wavenumber. Substitution of Eqn. 5.30 in Eqn. 5.28 leads to [16]:

$$-i\omega A_x = A_y \left[\gamma\mu_0 H + \frac{2JS}{\hbar}(2 - e^{-ika} - e^{ika})\right] \quad (5.31)$$

Eqn. 5.31 can be rewritten as:

$$-i\omega A_x = A_y \left[\gamma\mu_0 H + \frac{4JS}{\hbar}(1 - coska)\right] \quad (5.32)$$

Similarly, starting from Eqn. 5.29 one can obtain:

$$-i\omega A_y = -A_x \left[\gamma\mu_0 H + \frac{4JS}{\hbar}(1 - coska)\right] \quad (5.33)$$

Eqns. 5.32 and 5.33 can be expressed in matrix form [16]:

$$\begin{bmatrix} iw & \left[\gamma\mu_0 H + \frac{4JS}{\hbar}(1 - coska)\right] \\ -\left[\gamma\mu_0 H + \frac{4JS}{\hbar}(1 - coska)\right] & iw \end{bmatrix} \begin{pmatrix} A_x \\ A_y \end{pmatrix} = 0 \quad (5.34)$$

The solution of Eqn. 5.34 is obtained by equating the main determinant to zero, which can be expressed for the frequency:

$$\omega_k = \gamma\mu_0 H + \frac{4JS}{\hbar}(1 - coska) \quad (5.35)$$

This Eqn. 5.35 is known as the dispersion relation, which represents the dependence of the spin-wave frequency on the wavenumber. The relation between the amplitudes of the spin components can be obtained by substituting Eqn. 5.35 in Eqns. 5.32 and 5.33. This gives: $A_y = -iA_x = -iA_0$. The longitudinal spin component is $S_i^z \approx S$ and the real parts of the transverse spin components become [16]:

$$S_x^i = A_0(coskx_i - \omega_k t), S_i^y = A_0(sinkx_i - \omega_k t) \quad (5.36)$$

These equations indicate that the classical picture of a spin wave in one dimension consists of spins precessing circularly about the equilibrium direction at a certain instant of time. This is illustrated schematically in Figs. 5.8(b) and 5.8(c). The spin precession has the same amplitude along the chain in a traveling wave propagating in the $+x$ direction, and the phase varies with the position as $\phi_i = k_x i$. The wavelength of the spin wave is the shortest distance between two spins that precess with the same phase and it is related to the wavenumber $\lambda = 2\pi/k$.

The generalization to a three-dimensional cubic lattice with nearest-neighbour interactions is [1]:

$$\omega_k = \frac{4JS}{\hbar}\left[Z - \sum_{\delta} cos\vec{k}.\delta\right] \quad (5.37)$$

Here the sum is over the Z vectors δ connecting the central atom to its nearest neighbours.

A quantum mechanical treatment shows that the spin excitations are quantized and the quanta of spin waves are termed magnons [4, 15, 16]. In the one-dimensional chain we have studied earlier the energy of one magnon is expressed as:

$$\hbar\omega_k = \hbar\gamma\mu_0 H + 4JS(1 - coska) \tag{5.38}$$

Eqn. 5.38 indicates that at $k = 0$, the magnon energy is determined only by the magnetic field intensity $\hbar\omega_0 = \hbar\gamma\mu_0 H$. This happens because there is no contribution from the exchange energy if all spins precess in phase [16]. The phase difference of precession for neighbouring spins increases with the increase in wavenumber, and so does the exchange energy. The magnon dispersion relation over the positive side of the first Brillouin zone is shown Fig. 5.9(a). The Zeeman contribution to the energy is many orders of magnitude smaller than the exchange energy for magnetic field intensities typically available in the laboratories. Thus, the energy gap at $k = 0$ is not visible in Fig. 5.9(a). The dispersion relation for the spin waves with long wavelength, i.e., low wave number magnons, can be obtained with the binomial

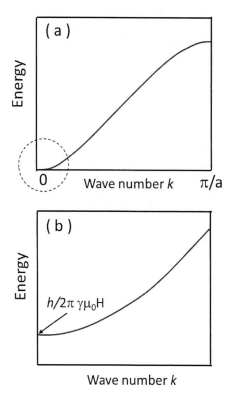

FIGURE 5.9 Dispersion relation for spin wave in a linear spin chain: (a) view of the full Brillouin zone, (b) expanded view near the origin ($ka \ll 1$).

expansion of the exchange term in Eqn. 5.38 for $ka \ll 1$:

$$\hbar\omega_k = \hbar\gamma\mu_0 H + 2JSa^2 k^2 \quad (5.39)$$

This quadratic dispersion relation with an energy gap is shown in Fig. 5.9(b). It may be noted that the energy scale in Fig. 5.9(b) is several orders of magnitude smaller than that in Fig. 5.9(a).

The derivation of the spin-wave dispersion discussed above was based on the existence of a ferromagnetic state in an isotropic spin chain, or a three-dimensional lattice. These assumptions need some scrutiny. The number of magnons excited at a temperature T can be expressed as [1]:

$$n_m = \int_0^\infty \frac{N(\omega_k) d\omega_k}{e^{\hbar\omega_k/k_B T} - 1} \quad (5.40)$$

Here $N(\omega_k)$ is the density of states for magnons, which in one, two, and three dimensions varies as $\omega_k^{-1/2}$, $\omega_k^0 = $ constant and $\omega_k^{1/2}$, respectively [1]. If we put $x = \hbar\omega_k/k_B T$, the integral in three dimensions in Eqn. 5.40 varies as $(\kappa_B T/\hbar)^{3/2} \int_0^\infty \frac{x^{1/2}}{e^x - 1} d^3 x$. This leads to the Bloch $T^{3/2}$ law, which describes the reduction in magnetization in a ferromagnet at low temperatures due to the excitation of magnons. However, in one and two dimensions the integrals diverge at finite temperature. This means that the ferromagnetic state is unstable in dimensions lower than three. This is the Mermin–Wagner theorem, which indicates that magnetic order is possible in the Heisenberg model in three dimensions, but not in one or two [1]. Thus, the linear chain we discussed above cannot order except at $T = 0$ K. The consequences of the Mermin–Wagner theorem, however, are not as drastic as it appears at first sight. The divergence of the integral is avoided if there is some anisotropy in the system, which creates a gap in the spin-wave spectrum at $k = 0$. This makes the lower limit of integration greater than zero, and the divergence is thus avoided. We shall see in Section 5.9 that some kind of anisotropy always exists in real ferromagnetic materials due to the crystal field and/or finite shape of the sample, i.e., dipolar interactions.

5.7 Domains and Domain Wall

The Curie temperature of iron is above 1000 K. However, a sample of iron at room temperature and in zero magnetic field will appear demagnetized. On the application of a small magnetic field, say, with a bar magnet, this Fe sample would be attracted more strongly than a paramagnetic sample. This phenomenon can be understood by considering the dipolar interaction between spins. We know from our earlier discussions that dipolar interaction is a thousand times smaller than exchange

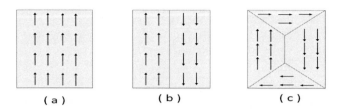

FIGURE 5.10 Schematic representation of domain structures: (a) single-domain structure, (b) two-domain structure, and (c) four-domain structure.

interaction. The dipolar interaction is a long-range interaction and decreases as the inverse cube of the distance between two spins. On the other hand, exchange coupling is a short-range interaction and falls off exponentially with spin separation.

A uniformly magnetized state of a ferromagnet is uneconomical in dipolar energy. This can be understood from Fig. 5.10. Reorganization of a ferromagnet structure with uniform magnetization into a complex structure of uniformly magnetized domains of macroscopic size can result in a significant reduction of dipolar energy. In such magnetic domain structures, magnetization vectors point in random directions. The two-domain structures in Fig. 5.10(b) have lower dipolar energy in comparison to the uniformly magnetized single-domain structure of the ferromagnet shown in Fig. 5.10(a). This can be understood by considering the configuration presented in Fig. 5.10(b) as two bar magnets. The orientations of these bar magnets have to be reversed to get the single-domain structure in which the like poles of the bar magnets are near one another. Thus, the single-domain configuration of Fig. 5.10(a) will have higher dipolar energy than the two-domain configuration shown in Fig. 5.10(b). The dipolar energy of the two-domain configuration can be reduced further by introducing more domains as shown in Fig. 5.10(c). Such a reconfiguration into a multidomain structure, however, will have an energy cost because of unfavorable exchange between the spins in the neighbouring misaligned domains. However, the exchange interaction is short-ranged, and, as a result, only the spins near the domain boundaries will have their energy raised. In contrast, because of the long-range nature of dipolar interaction, the dipolar energy of every spin is reduced in the formation of the domains. Therefore, the formation of domains is energetically favoured despite the greater strength of the exchange interaction unless the sizes of domains are not too small. Only a few spins near the domain boundaries have their large exchange energy raised, whereas every spin lowers its small dipolar energy, resulting in an overall gain in energy.

A ferromagnet thus consists of several small regions known as domains. The local magnetic moment within each domain reaches the saturation value, but the directions of magnetization within different domains are not necessarily parallel. There exist boundaries called domain walls separating the domains.

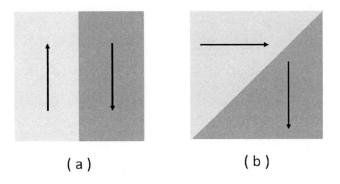

FIGURE 5.11 Schematic representation of: (a) 180^0 domain structure, (b) 90^0 domain structure.

In ferromagnetic samples, the magnetization can be zero in the absence of an applied magnetic field. This is because while the magnetization within each domain is saturated, the directions of the magnetization of each domain can be random. This results in a net-zero magnetization for the whole ferromagnetic sample. On the other hand, in some ferromagnetic samples it is possible to reach saturation magnetization (μ_0M of the order of 1 T) of the whole sample at room temperature with the application of a very weak field of, say, 10^{-6} T. This kind of low field is expected to have a negligible effect on a paramagnet. The surprisingly large effect in a ferromagnet arises because the applied magnetic field does not order magnetic moment in a macroscopic scale; instead, they just cause the domain (inside which the magnetic moments are already ordered) to align. This alignment is achieved by the energy-efficient process of domain wall motion.

A domain wall between the adjacent domains can be classified by the angle between the magnetization in the two domains involved. The domains of opposite magnetization are separated by a 180^0, whereas a 90^0 domain wall separates domains of perpendicular magnetization (Fig. 5.11). A Bloch wall is the most common type of domain wall, where the magnetization rotates in a plane parallel to the plane of the wall. The Néel wall is the other possible configuration, in which the magnetization rotates in a plane perpendicular to the plane of the domain wall. Fig. 5.12 schematically presents a Bloch wall and a Néel wall.

It expends energy to rotate the neighbouring spins in a ferromagnet. So, if the boundary between two domains is very sharp, it will use up significant amounts of exchange energy. The surface energy of the domain wall, however, can be reduced by having the reversal of spin direction spread out over many spins. We shall now estimate the width of a domain wall by taking the case of the Bloch wall as an example. The energy involving two spins $\vec{S}_1$ and $\vec{S}_2$ with an angle θ between them is given by $-2J\vec{S}_1.\vec{S}_2 = -2JS^2\cos\theta$. The energy of the configuration is $-2JS^2$ if the angle $\theta = 0$. If θ is small (i.e., $\theta \ll 1$), then using $\cos\theta \approx 1 - \theta^2/2$, we can estimate

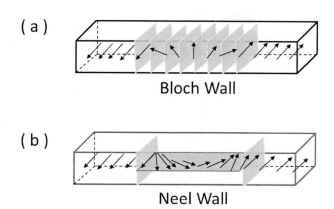

FIGURE 5.12 Schematic representation of: (a) Bloch wall, (b) Néel wall.

that the energy expended for having a configuration of spins at a small angle is approximately $JS^2\theta^2$.

If the spin reversal is spread over N spins, then each spin will be seen to differ in orientation from its neighbour by a small angle π/N as one passes through the domain wall. Here the energy cost of a line of spins consists of N contributions of $JS^2\theta^2$. The Bloch wall consists of planes of spins, and the important parameter here is the energy σ_{BW} per unit area (a^2) of the Bloch wall. Hence,

$$\sigma_{BW} = JS^2 \frac{\pi^2}{Na^2} \tag{5.41}$$

This will tend to zero as $N \to \infty$. This result implies that a domain wall would just unwind itself and grow in size throughout the entire system. This is because it takes up more energy to have a twisted spin configuration, and therefore all the spins will tend to untwist unless some other interaction stops them from doing that.

5.8 Magnetic Skyrmion

A magnetic skyrmion is a small swirling topological defect of the spin configuration in a magnetic material [19, 20]. Skyrmions can be defined by the topological number or skyrmion number S. It is a measure of the winding of the normalized local magnetization $\vec{M}$. Fig. 5.13 shows the magnetic texture of a skyrmion, where the spins rotate progressively with a fixed chirality from the up direction at one edge to the down direction at the centre, and then to the up direction again at the other edge [19]. There can be four typical types of magnetic skyrmions: Bloch-type skyrmion (Fig. 5.13(a)), Néel-type skyrmion (Fig. 5.13(b)), Bloch-type antiferromagnetic skyrmion (Fig. 5.13(c)), and Néel-type antiferromagnetic skyrmion (Fig. 5.13(d))

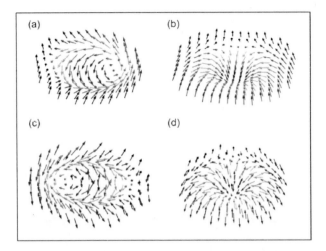

FIGURE 5.13 (a) Bloch-type magnetic skyrmion; (b) Néel-type magnetic skyrmion; (c) Bloch-type antiferromagnetic skyrmion; (d) Néel-type antiferromagnetic skyrmion. (From reference [21] Creative Commons Attribution License)

[21]. These are attributed to different symmetries of the interaction between spins, which can originate from the underlying crystal lattice or an interface.

The magnetic skyrmion lattice is a periodic array of tubular skyrmions protected against decay by their topological winding number. The magnetic skyrmions were first observed in single crystals of magnetic compounds with a non-centrosymmetric lattice [22, 23]. Those observations were explained by the existence of the Dzyaloshinski–Moriya interaction induced by spin–orbit coupling in the absence of inversion symmetry in the crystal lattice. Later on, skyrmions were found in ultrathin magnetic films epitaxially grown on heavy metals [24]. Such systems are subjected to large Dzyaloshinski–Moriya interaction induced by the breaking of inversion symmetry at the interface and to the strong spin–orbit coupling of the neighbouring heavy metal [19].

5.9 Magnetic Anisotropy

The interaction that can stop the domain wall from unwinding is magneto-crystalline anisotropy. All magnetic crystals have a magnetic easy axis, along which crystallographic direction it is easy to magnetize the crystal. On the other hand, there is also a hard axis in the crystal, along which it is difficult to magnetize the material. This anisotropy gives rise to additional energy, which in Co, for example, can be expressed in the form:

$$E = K_1 sin^2\theta + K_2 sin^4\theta \tag{5.42}$$

Here K_1 and K_2 are anisotropy constants and θ is the angle between the magnetization and the stacking direction of the hexagonally close-packed planes of Co.

Eqn. 5.42 represents the case of uniaxial anisotropy, where the energy depends on the angle to a single axis; in this case, it is the stacking axis of the hexagonally closed packed planes of cobalt. In this equation E, K_1, and K_2 are energy densities and they are measured in the unit of Jm^{-3}. The constants are positive, and hence the energy is minimized when the magnetization lies along the stacking direction. These anisotropy constants are usually found to be depending on temperature.

In a cubic system the appropriate expression is [4]:

$$E = K_1(m_x^2 m_y^2 + m_y^2 m_z^2 + m_z^2 m_x^2) + K_2 m_x^2 m_y^2 m_z^2 + \ldots \quad (5.43)$$

where $\vec{m} = (m_x, m_y, m_z) = \vec{M}/|\vec{M}|$

Eqn. 5.43 can be rewritten in spherical coordinates as:

$$E = K_1 \left(\frac{1}{4} sin^2\theta sin^2 2\phi + cos^2\theta\right) sin^2\theta + \frac{K_2}{16} sin^2 2\phi sin^2 2\theta sin^2\theta + \ldots \quad (5.44)$$

The anisotropy energy in magnetic materials arises from the spin–orbit interaction and the partial quenching of the angular momentum. Anisotropy energy densities are usually in the range of 10^2 to 10^7 Jm^{-3}, which corresponds to an energy per atom in the range 10^{-8} to 10^{-3} eV. The anisotropy energy is smaller in the lattices of high symmetry and larger in the lattices of low symmetry. Cubic Fe and Ni, for example, have K_1 equal to $4.8 \times 10^4 Jm^{-3}$ and $-5.7 \times 10^3 Jm^{-3}$, respectively, whereas hexagonal Co has $K_1 = 5 \times 10^5 Jm^{-3}$. On the other hand, the anisotropy constant K_1 in the permanent magnetic materials $Nd_2Fe_{14}B$ and $SmCo_5$ is equal to $5 \times 10^6 Jm^{-3}$ and $1.7 \times 10^7 Jm^{-3}$, respectively.

There is another kind of anisotropy possible in the magnetic systems, and that is known as "shape anisotropy". This additional energy term arises from the demagnetizing energy associated with the sample shape. In thin-film samples, this shape anisotropy is expressed as $\frac{1}{2}\mu_0^2 cos^2\theta$, where θ is the angle between the magnetization $\vec{M}$ and the normal to the film. The presence of this term causes an energy saving in keeping the magnetization in the plane of the film.

In practice, the thickness of a domain wall is determined by a balance between exchange and magnetic anisotropy energies.

5.10 Magnetization Process in Ferromagnets

In a ferromagnetic sample in the temperature region below Curie temperature (T_{Curie}), magnetization with the application of an external magnetic field involves a process where magnetic domains are rearranged and reoriented. In the small applied field, the domains oriented in the direction of the field grow at the expense of the domains of the other orientation by smooth movement of the domain walls. This growth of the favourably oriented domain is shown in the second picture panel from the bottom on the right-hand side of Fig. 5.14. The magnetization process in the region of the small applied field is reversible. The domains go back to their previous configuration on the withdrawal of the magnetic field with net bulk magnetization zero for the whole sample (bottom-most picture panel on the right-hand side of Fig. 5.14). If the applied field is not very weak, the growth of favourably oriented domains may take place through irreversible processes. Crystalline defects and imperfections may impede domain wall motion in the low field, but the domain walls can pass through such defects in strong enough fields if the gain in energy in the external field is sufficiently large. On further increase in the applied magnetic field, the magnetization direction within the domains tends to rotate to the direction of the applied field.

In a very large applied magnetic field, despite the expenditure of magnetic anisotropy energy, the rotation of the entire domain towards the applied field direction may become energetically favourable. The sample in such applied fields tends to reach the state of saturation magnetization. Once magnetized this way, it may be quite difficult for a ferromagnetic material to get back to the multi-domain configuration on the removal of the external magnetic field. However, in real cases, some residual domain structures are often left within the material during the magnetization cycle, and such structures act as nucleation centres for the growth of reverse domains through the motion of the walls. However, when the external field is removed, crystalline defects will impede the motion of domain to return to their original unmagnetized configuration in zero fields. This gives rise to magnetic history dependence or hysteresis in magnetization (Fig. 5.15). A residual magnetization or

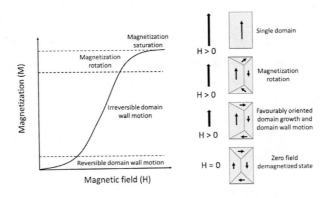

FIGURE 5.14 Magnetization process of a ferromagnetic sample with the variation of magnetization with applied magnetic field, and the associated changes in the domain configuration.

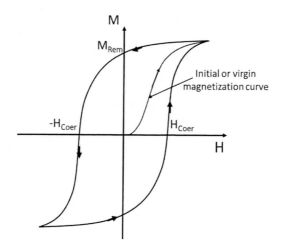

FIGURE 5.15 Schematic presentation of isothermal field variation of magnetization in real ferromagnetic materials with crystalline defects. Magnetization shows distinct field hysteresis characterized by remanent magnetization M_{Rem} and coercive field H_{Coer}.

remanent magnetization M_{Rem} is left with on reducing the applied magnetic field back to zero. To restore the zero magnetization state, a strong negative magnetic field called "coercive field (H_{Coer})" (Fig. 5.15) is needed. The actual value of the coercive field depends on the metallurgical state of the ferromagnetic sample. It may also be noted that in a certain class of ferromagnets known as soft ferromagnets hysteresis can be very small with a low value of "coercive field", and the isothermal field dependence may appear almost reversible. It may be noted here that cluster spin-glass materials can often show isothermal hysteretic magnetization. Hence, the presence of magnetic hysteresis alone by no means can be taken as a sign of the long-range ferromagnetic order in a magnetic sample.

It is worth mentioning here that the magnetic dipolar interactions can also give rise to a strong internal magnetic field at the site of each spin. Thus, the local magnetic field experienced by a spin can be significantly different from the applied magnetic field. The value of this magnetic field depends in a complicated way on the shape and geometry of the sample. This is an important matter in the experimental measurement of magnetization of a ferromagnetic sample, and a "demagnetization factor" is introduced to convert the applied magnetic field to the true local field. The "demagnetization effect" is discussed in more detail in Appendix C.

Bibliography

[1] Coey, J. M. D. (2010). *Magnetism and Magnetic Materials.* Cambridge: Cambridge University Press.

[2] Cullity, B. D., and Graham, C. D. (2009). *Introduction to Magnetic Materials*. Hoboken: Willey.

[3] Jiles, D. (1991). *Introduction to Magnetism and Magnetic Materials*. New York: Springer.

[4] Blundell, S. J. (2001). *Magnetism in Condensed Matter*. Oxford: Oxford University Press.

[5] Ashcroft, N. W., and Mermin, N. D. (1976). *Solid State Physics*. San Diego: Harcourt College Publishers.

[6] Anderson, P. W. (1950). Antiferromagnetism: Theory of Superexchange Interaction. *Physical Review* 79: 350.

[7] Anderson, P. W. (1959). New Approach to the Theory of Superexchange Interaction. *Physical Review* 115: 2.

[8] Anderson, P. W. (1978). Local Moments and Localized States. *Science,* 201: 307.

[9] Roy, S. B. (2019). *Mott Insulators: Physics and Applications*. Bristol: IOP Publishing.

[10] Anderson, P. W. (1963). Chapter 2. In G. T. Tado and H. Suhl, eds. *Magnetism*. New York: Academic Press.

[11] Goodenough, J. B. (1955). Magnetism and the Chemical Bond. *Physical Review*, 100 (2): 564.

[12] Kanamori, J. (1959). Superexchange interaction and symmetry properties of electron orbitals. *Journal of Physics Chemistry of Solids*, 10 (2–3): 87–98.

[13] Tokura, Y., and Nagaosa, N. (2000). Orbital Physics in Transition-Metal Oxides. *Science,* 288 (5465): 462–468.

[14] Sarkissian, B. V. B. (1981). The Appearance of Critical Behaviour at the Onset of Ferromagnetism in AuFe Alloys. *Journal of Physics F: Metal Physics*, 11(10): 2191.

[15] Phillips, T. G., and Rosenberg, H. M. (1966). Spin Waves in Ferromagnets. *Reports on Progress in Physics*, 29(1): 285.

[16] Rezende, S. M. (2020). *Fundamentals of Magnonics*. Springer.

[17] Hellers, G., and Kramers, H. A. (1934). Spin Waves in Ferromagnets. *Proc. K. Akad. Wet.*, 37: 378.

[18] Herring, C. and Kittel, C. (1951). On the theory of spin waves in ferromagnetic media. *Physical Review*, 81 (5): p. 869.

[19] Fert, A., Reyren, N. and Cros, V. (2017). Magnetic Skyrmions: Advances in Physics and Potential Applications. *Nature Reviews Materials*, 2 (7): 1–15.

[20] Lancaster, T. (2019). Skyrmions in magnetic materials. *Contemporary Physics*, 60 (3): 246–61.

[21] Giustino, Feliciano, et al. (2020). The 2021 Quantum Materials Roadmap. *Journal of Physics Materials*, 3 (4): 042006.

[22] Mühlbauer, S., Binz, B., Jonietz, F., et al. (2009). Skyrmion Lattice in a Chiral Magnet. *Science,* 323 (5916): 915–19.

[23] Yu, X. Z., Onose, Y., Kanazawa, N., et al. (2010). Real-space Observation of a Two-dimensional Skyrmion Crystal. *Nature*, 465 (7300): 901–904.

[24] Heinze, S., Von Bergmann, K., Menzel, M., et al. (2011). Spontaneous Atomic-scale Magnetic Skyrmion Lattice in Two Dimensions. *Nature Physics*, 7 (9): 713–718.

Part III
Experimental Techniques in Magnetism

In this part of the book we will study various experimental techniques currently employed in the study of magnetism and magnetic materials. We shall start with a very popular experimental technique called magnetometry in Chapter 6. This technique is used to study macroscopic or bulk magntic properties of materials in research laboratories all over the world. This is a versatile technique to investigate magnetic properties over a wide range of sample environments that differ from each other in terms of temperature, pressure and applied magnetic field. Magnetometry is usually the first experiment to perform for the general magnetic characterization of materials, which apart from providing a plethora of useful information enables one to identify and focus on the particular temperaure, pressure and magnetic field regime of interest for deeper investigation using more specialized experimental techniques.

Electromagnetic (EM) radiation is an interesting probe to study magnetic properties of materials. A material containing magnetic dipoles is expected to interact with the magnetic component of EM radiation. If the material of interest is subjected to a static magnetic field, it is possible to observe a resonant absorption of energy from an EM wave tuned to an appropriate frequency. This phenomenon is known as *magnetic resonance*. Depending upon the type of magnetic moment involved in the resonance, a number of different experimental techniques can be employed to study *magnetic resonance*. In Chapter 7 we shall discuss such experimental techniques as nuclear magnetic resonance (NMR), electron paramagnetic resonance (EPR), ferromagnetic resonance (FMR), and Mössbauer spectroscopy, along with muon spin rotation (μSR).

In Chapter 8 we will study experimental techniques employing interaction of EM waves with magnetic material without any resonant absorption. Magneto-optical effects can arise due to an interaction of electromagnetic radiation with magnetic material having either spontaneous magnetization or magnetization induced by the presence of an external magnetic field. Experiments based on the magneto-optical Kerr effect (MOKE) and scanning near-field optical microscopy (SNOM) will be discussed. In addition we will discuss Brillouin light scattering (BLS) or Brillouin spectroscopy, which is based on the phenomenon of inelastic scattering of light. Brillouin spectroscopy provides insights to spin waves in ferromagnetic material without exciting them explicitly.

In Chapter 9 we will provide an introduction to neutron scattering techniques as a powerful tool to study the microscopic magnetic structure of magnetic materials. The neutrons for the study of physical properties of materials are usually obtained by slowing down energetic neutrons after their production in neutron sources by nuclear reaction. Such thermal neutrons are very suitable for investigations of the magnetic structures of solids with neutron diffraction experiments. The kinetic energy of thermal neutrons is of the order of 25 meV, and that is a typical energy for collective spin excitations in magnetic solids. Both wavelength and energy of thermal neutrons are thus very suitable for the studies of spin dynamics in magnetic materials through inelastic neutron scattering experiments.

Traditionally X-ray diffraction is mainly used for determining the crystal structure of solids through the interaction of X-rays with the electronic charge distribution. However, the absorption cross section and the scattering amplitudes of electromagnetic waves are related through the optical theorem, hence magnetic effects in principle are expected to be observed as well in X-ray scattering. In early 1970s, theoretical calculation of the amplitude of X-rays elastically scattered by a magnetically ordered solid was carried out within the framework of the relativistic quantum theory, which indeed predicted that the effect could be observed. However, the intensity of the magnetic Bragg peaks was expected to be $\approx 10^8$ smaller than the intensity of the Bragg peaks originating from electronic charge. This makes the magnetic Bragg peaks difficult to be obseved with the conventioal laboratory based X-ray sources. Subsequently with the advent of more powerful X-ray sources with high photon flux such as synchrotron-radiation facilities, the magnetic X-Ray scattering technique has now established itself as a powerful technique complementary to neutron scattering for the study of magnetic properties of materials. In addition to the normal magnetic X-ray scattering, there is also a resonant energy one in which the X-ray energy is tuned close to the characteristic energies of electrons bound to the atom in the solid, so that both the Bragg scattering experiments and absorption measurements becocme sensitive to electron spin densities. This provides the basis of resonant X-ray scattering experiments. Further, X-ray magnetic circular dichroism (XMCD) spectroscopy is a powerful technique that measures differences in the absorption of left- and right-circularly polarized X-rays by a magnetic sample. XMCD techniques can provide quantitative information about the distribution of spin and orbital angular momenta with the help of simple sum rules. All such magnetic X-ray scattering techniques will be discussed in Chapter 10

Magnetic imaging techniques are used to get a direct view on magnetic properties on a microscopic scale. In Chapter 11 we will discuss three different classes of magnetic imaging techniques namely (i) electron-optical methods, (ii) imaging with scanning probes and (iii) imaging with X-rays from synchrotron radiation sources.

In Chapter 12 we will introduce a relatively new technique of magnetometry based on the electron spin associated with the nitrogen-vacancy (NV) defect in diamond. A very impressive combination of capabilities has been demonstrated with NV magnetometry, which sets it apart from the other magnetic sensing techniques.

6
Conventional Magnetometry

Magnetization and magnetic susceptibility are the physical quantities that describe the response of a magnetic material to the application of an external magnetic field. Magnetic susceptibility χ is defined as the ratio between magnetization $\vec{M}$ and the intensity of the externally applied magnetic field $\vec{B}$,

$$\vec{M} = \chi \vec{B} \qquad (6.1)$$

These magnetic properties can be studied by determining the force on a magnetized sample of the material under study, or the magnetic induction or a perturbation of the field in the neighbourhood of the sample.

There is a generalized approach for the magnetic measurements techniques [1]. In an externally applied magnetic field, the material under study produces a force (F), or magnetic flux (ϕ), or an indirect signal (I). These phenomena are usually sensed by a detector, which results in output usually in the form of an electrical signal, either DC or AC. These techniques are in general termed magnetometry. The heart of a magnetometer is the detector, and it defines the principle involved in a particular type of magnetometer. A magnetic sample placed in a uniform magnetic field affects the magnetic flux distribution. This change can be sensed by a flux detector, which is usually in the form of a coil but can also be detected through a variety of sensors. On the other hand, a sample placed in a non-uniform magnetic field experiences a force, which can be detected by a force transducer. Such experimental techniques involving force or flux detections are classified as direct techniques, and they measure macroscopic or bulk magnetic properties of the material. These experimental techniques assume only the validity of Maxwell's electromagnetic equations and the thermodynamic equilibrium of the sample. The magnetic properties can also be measured by various indirect techniques, namely the

Hall effect, magneto-optical Kerr/Faraday effects, NMR, FMR, Mössbauer, Neutron Scattering, μSR, and others, which take advantage of known relationships between the phenomenon detected and the microscopic magnetic properties of the specimen. In this chapter, we will focus on the direct techniques, and the indirect techniques will be the subject of later chapters in the book. It may be mentioned here that each magnetic measurement technique has distinct merits as well as limitations, and no single technique is universally applicable to study all types of magnetic phenomenon [1]. It needs to be considered whether the planned measurement technique has natural constraints built into it. Specifically, are the measurements limited to low or high temperatures, low magnetic fields or high magnetic fields, fixed fields or swept fields, very stable field environments, large or small magnetic moments, ambient pressure or high pressure, very short or long time observations, a broad range of investigations or a very narrow area of research with narrowly defined requirements? Also for all these techniques, it should be recognized that these peripheral aspects may play an important role in deciding the best technique for the given experimental requirements.

6.1 Force Method

In this method, the principle of measurement involves the determination of force or torque experienced by a sample of magnetic material placed in the environment of a magnetic field. A balance is usually employed to measure this force. That is the reason why the equipment involving the force method is often called a balance. The magnetic component of the force is estimated from the change in the apparent sample weight with the externally applied magnetic field switched on and off.

6.1.1 Gouy and Faraday balance

The Gouy method is the most common and simple force technique for magnetization studies. In this method, the sample used for investigation is in the shape of either a long cylinder or a sample packed in a cylindrical sample holder. The sample is suspended with its axis vertical and positioned in such a way that its one end is in a strong uniform field $\vec{B}$ near the centre of a pole gap of an electromagnet, while the other end stays outside the pole gap in a much weaker field. The axial force F acting on the sample is given by:

$$F = \frac{1}{2}(\chi - \chi_0) A (B_1^2 - B_0^2) \qquad (6.2)$$

Here χ and χ_0 are susceptibilities of the sample and surrounding medium, respectively. B_1 and B_0 are the magnitudes of the magnetic fields in the centre

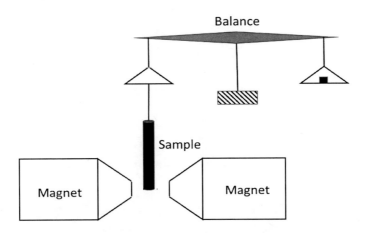

FIGURE 6.1 Schematic diagram depicting the working principle of a Gouy balance.

and outside, respectively, of the pole gap of the electromagnet, and A is the cross-sectional area of the sample. The implicit assumption in a Gouy balance is that the sample is of uniform density and its presence does not distort the magnetic field appreciably. This force can be measured using various kinds of microbalances as a weight differential between the conditions when the externally applied magnetic field is on and off. The working principle of a Gouy balance is shown in Fig. 6.1. The minimum length of the cylindrical sample (or sample container) is determined by the dimensions of the magnet. The sample positioning is not quite critical if the field at the centre of the magnet gap is reasonably homogeneous. The method is useful for weakly magnetic materials, but for the powder samples or small chips, the assumption of uniform packing of the sample in its container introduces a major intrinsic source of error. The main drawbacks of the Gouy balance, however, are the requirement of large size samples and the difficulties in obtaining uniform temperature over the sample during the temperature-dependent study of magnetic properties. In addition, inaccuracy can also arise if the magnetic response of the material under study is anisotropic.

The Faraday method is usually suitable for susceptibility measurement of a small magnetic sample in the form of small chips or powder pellets. The force F acting on such a sample in a magnetic field with the gradient in z direction is

$$F = -\frac{dE}{dz} = \frac{1}{2} V \frac{d(MB - \chi_0 B^2)}{dz} \qquad (6.3)$$

where M is magnetization, B is the magnitude of the applied magnetic field $\vec{B}$, χ_0 is the susceptibility of the surrounding medium and V is sample volume. The working principle of a Faraday balance is presented in Fig. 6.2. For a linear dependence of this force F on χ, $B\frac{dB}{dz}$ needs to be constant. Electromagnets with suitably shaped

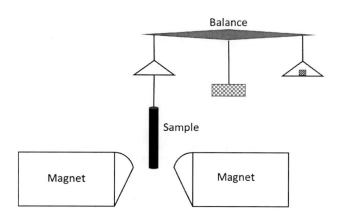

FIGURE 6.2 Schematic diagram showing the working principle of a Faraday balance.

(tapered) poles are used to keep the $B\frac{dB}{dz}$ product constant. It is also assumed that $\vec{M}$ is isotropic, volume is small enough so that the presence of the sample does not change $\vec{B}$ significantly, and there is negligible displacement of the sample. Both B and $\frac{dB}{dZ}$ need to be measured at the same position of the sample, and $\frac{dB}{dz}$ is a function of magnetic field for gradient pole pieces [1]. This problem is circumvented with the help of an independent gradient coil. However, in this case, the magnitude of $\frac{dB}{dz}$ will be limited by the power dissipation in the coils and also the coil geometry. In addition, the magnetization in principle cannot be measured unless the magnetic field is non-uniform, so that $\vec{B}$ is the field averaged over sample volume V. A sufficiently sensitive balance, which can maintain the sample in a standard position, is required for the experimental determination of F. The Sucksmith balance was originally developed for this purpose, and more modern and sophisticated balances evolved over the period that also take into account the need for variable temperature measurement down to the low-temperature regime. A relatively high sensitivity is expected in a Faraday balance because very small forces can be measured with the help of a sensitive microbalance. The difficulty of accurate calibration of the field gradient, however, limits the wide usage of the Faraday balance for absolute measurement of magnetic parameters. Fig. 6.3 presents susceptibility of an $NiTiO_3$ sample as a function of temperature measured with a Faraday balance [2]. The result shows a susceptibility peak around 23 K, indicating the presence of antiferromagnetic transition in the material.

6.1.2 Alternating gradient magnetometer

In an alternating gradient magnetometer, an alternating field gradient is used to produce a periodic force on a magnetic sample.

Conventional Magnetometry   97

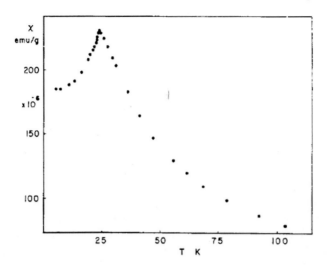

FIGURE 6.3 Temperature-dependent magnetic susceptibility of NiTiO$_3$ measured with a Faraday balance. (From reference [2] with permission from AIP Publishing.)

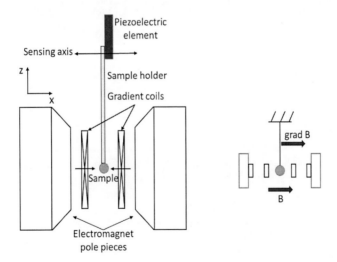

FIGURE 6.4 Schematic diagram showing the working principle of an alternating gradient magnetometer.

Fig. 6.4 presents a schematic diagram with the working principle of an alternating gradient magnetometer. A set of series-opposition coils generates a vertical AC force perpendicular to the magnetization of the sample. The sample is mounted on the tip of a vertical rod, which lies along the z-axis, and the gradient magnetic field $\vec{B}$ is along the x-axis. The alternating gradient magnetometer operates by subjecting the magnetic sample and sample rod to an alternating field gradient. The force on the sample is proportional to [3]:

$$F = M_z \int_v \frac{\partial N_z}{\partial x} dv \qquad (6.4)$$

Here N is a geometric field function of the gradient coils. Thus the measurements of magnetization can be performed if $\frac{\partial N_z}{\partial x}$ can be integrated over the volumes relevant to both the sample and to a calibrating coil [3]. The integration is trivial if the value of the gradient is constant. In practice, the following options are available to achieve this condition: (i) using the instrument only for comparing magnetic samples of the same volume, (ii) using small enough samples and calibrating coil, so that the field over their volumes is uniform within the required accuracy, and (iii) compute the integrals using the known dimensions of coils and samples for the large dimension samples. Fig. 6.5 shows a typical example of a gradient coil system for clamping between the pole pieces of the electromagnet [3]. The four coils with horizontal axes form a set whose proportions have been optimized to produce the most uniform field of $\frac{\partial N_z}{\partial x}$ in the region near the origin, the field being constant within 8% in a 1 cm diameter spherical region. This type of coil set, with circular coils configured like a cartwheel, is suitable for accurate and rigid construction.

FIGURE 6.5 Gradient coils for producing magnetic field gradient. (From reference [3] with permission from IOP Publishing).

Such coils also allow easy access to the sample under investigation. The sample rod is designed in such a way as to have a mechanical resonance typically at a frequency of a few hundred Hz. The other end of the sample rod is attached to a piezoelectric element. The force due to the magnetic field gradient on the sample produces a bending moment on the piezoelectric element, which in turn generates an electrical signal proportional to the force on the sample. The output from this piezoelectric sensing element is synchronously detected at the frequency of the AC gradient field [4]. The force can be enhanced by a factor equal to the mechanical quality factor Q by operating at the resonant frequency of the mechanical system. The amplitude of the resultant electrical voltage is proportional to the magnetic moment of the sample. For more detailed information on the alternating gradient magnetometer, the readers are referred to the articles by Flander [4, 5].

As an example in Fig. 6.6, we present the result of a magnetization study on two nanostructured iron oxide (γFe_2O_3) particles (using zeolites as the matrix of synthesis) ZIO-1 and ZIO-2 obtained with an alternating gradient magnetometer [6]. The sample ZIO-1 (ZIO-2) is supposedly paramagnetic (superparamagnetic) in nature. In the experiment, a small amount of the sample powder was placed in a region between two electromagnets and subjected to a gradient field strength alternating from 10 kOe to -10 kOe [6].

6.1.3 Cantilever beam magnetometer

The working principle of a cantilever beam magnetometer (CBM) is based on the simple cantilever beam theory. A conducting or semiconducting substrate in the form of a cantilever plate is fixed at one end. The deflection of the free end of the cantilever

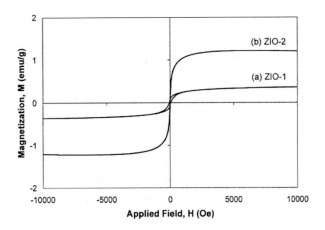

FIGURE 6.6 Magnetization curves for (a) ZIO-1 and (b) ZIO-2 measured with an alternating gradient magnetometer. (From reference [6] with permission from Springer).

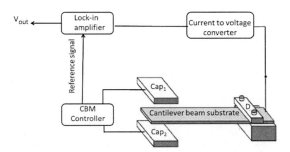

FIGURE 6.7 Schematic representation of a cantilever beam magnetometer for the measurement of cantilever deflection. Cap_1 and Cap_2 represent the upper and lower capacitor plates and D is the substrate holder from where the signal is being derived. V_{Out} is the output lock-in amplifier voltage.

is proportional to stress on the cantilever plate. This deflection can be measured with high sensitivity and long time stability by employing a differential capacitance method [7] or through an optical method involving phase-sensitive detection of a laser beam [8]. If a magnetic film is deposited on the cantilever substrate or a small sample is attached to the cantilever substrate, then under the influence of an external magnetic field $\vec{B}$ the free end of the substrate will be subjected to a torque $\tau = \mu_0 \vec{M} \times \vec{B}$. This torque may again lead to the bending of the cantilever substrate, and the torque (and, in turn, sample magnetization) can be determined by measuring the amount of deflection of the free end of the cantilever substrate. Fig. 6.7 presents a schematic of a lock-in amplifier-assisted differential capacitance technique to measure the deflection of the cantilever substrate. The free end of the cantilever substrate is inserted between two parallel plates of a capacitor. The capacitor plates are supplied with an AC voltage, and a reference signal with the same frequency of supply voltage is fed to the lock-in amplifier. The cantilever substrate is kept in the floating ground, which acts as a common capacitor plate. The system is considered to be in the initial condition when the substrate is equidistant from the top and the bottom plates. In this condition the net induced charge on the substrate is nil. There would be a net induced charge on the substrate for any deflection Δ of the cantilever substrate from its equidistant position between the plates. This will result in the flow of current through the substrate. This current is then converted into voltage using an I-V converter-cum-preamplifier, and the amplified voltages are detected by the lock-in-amplifier. It is, however, not practically possible to place the substrate exactly equidistant from the capacitor plates. At the starting of the experiment, a control unit is used to initialize the current/voltage signal coming from the substrate to zero by controlling the relative voltage supplied to the top and the bottom capacitor plates.

Summarizing, we can say that the CBM works on the principle of capacitative force or torque detection and is based on a capacitor with a certain capacitor

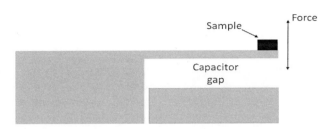

FIGURE 6.8 Schematic diagram of the working principle of a cantilever magnetometer working under force method.

gap or inter-plate distance d. Fig. 6.8 is a schematic diagram of a cantilever magnetometer working under the force method. In this method, when a sample is mounted on the top plate of the capacitor, the force experienced by the sample in the externally applied inhomogeneous magnetic field $\vec{B}$ causes a variation in the inter-plate distance. This, in turn, results in a change in the capacitance, which can be accurately determined by a sensitive capacitative bridge. Assuming that the direction of the $\vec{M}$ of the sample is parallel to the direction of the external magnetic field $\vec{B}$, $\vec{M}$ is related to the change in capacitance by the relation:

$$\vec{M} = \frac{\Delta C}{\mu_0 \beta \nabla \vec{B}} \quad (6.5)$$

where μ_0 is the permeability of the free space, β is the calibration constant, and $\nabla \vec{B}$ is magnetic field gradient. Such a magnetometer is particularly good for measuring the magnetic properties of thin-film and small-size samples.

In the cantilever torque magnetometer, which is particularly suitable for magnetic sample with anisotropic response, the sample experiences a torque $\tau = \mu_0 \vec{M} \times \vec{B}$. Fig. 6.9 is a schematic diagram of a cantilever magnetometer working under the torque method. In this method magnetization $\vec{M}$ is related to the change in capacitance by the relation:

$$\vec{M} = \frac{\Delta C}{\mu_0 \beta' \vec{B} \sin \theta_{MB}} \quad (6.6)$$

where μ_0 is the permeability of the free space, β' is the calibration constant, and θ_{MB} is the angle between the applied magnetic field direction and the magnetic anisotropy axis of the sample (Fig. 6.9). For more detailed information on the cantilever torque magnetometry, the readers are referred to the paper by Martín-Hernández et al. [10] and Rossel et al. [11].

Cantilever magnetometry methods can be further categorized into cantilever torque magnetometry (CTM) and dynamic cantilever magnetometry (DCM) [12]. In CTM, the static equilibrium deflection of the cantilever is measured, whereas in DCM the magnetic field-dependent resonance frequency shifts of the cantilever are

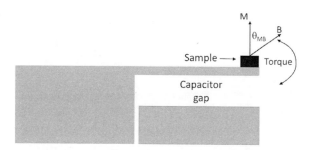

FIGURE 6.9 Schematic diagram of a cantilever magnetometer working under torque method.

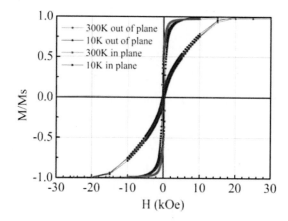

FIGURE 6.10 Isothermal field dependence of magnetization of a Co film for two different orientations of the sample measured with dynamic cantilever magnetometry technique. (From reference [12] with permission from IOP Publishing; Copyright (2018) The Japan Society of Applied Physics.)

measured. Cantilever torque magnetometry has some advantage over the induction method-based magnetometry (to be discussed in the next sections) in studying the magnetic response of nanostructures of various materials and geometries. Fig. 6.10 shows experimentally measured isothermal field variation of magnetization of a Co film (1.8 mm × 2.5 mm × 100 nm) with DCM technique for two different orientations of the sample [12].

6.2 Induction Method

Magnetic flux through a coil with surface area S is given by $\Phi = \int_S \vec{B}.\mathbf{dS}$. If the flux linkage varies with time, by Faraday's law of electromagnetic induction a loop voltage is generated, which can be expressed as:

$$V(t) = -\frac{d\Phi}{dt} = -\frac{d}{dt}\left[\int \vec{B}.d\vec{S}\right] \quad (6.7)$$

The induction method involves the detection of the voltage induced in a detection coil by a change in magnetic flux when the applied magnetic field, the coil position, or sample position is changed. Thus, the output of the detector is an electrical signal, which can be DC or AC.

The detector is the heart of a magnetometer, and it determines the principle employed and the response obtained. In a magnetometer, the sample under study produces a distortion in the field, thus affecting the flux distribution. A flux detector senses the changes in flux distribution. The sensor is usually a coil. For various measurement techniques, the focus of interest is the magnetization $\vec{M}$ of materials, which is a bulk property and is measured in emu/gm or emu/cc (Gaussian or CGS units) or A/m (SI units). When attempting to compare samples of different materials in a variety of sizes, the actual parameter of interest is the magnetic moment μ.

The DC magnetometers measure the magnetization of the sample. If there is no spontaneous or permanent magnetic moment associated with the sample being measured, an applied field is required to magnetize the sample. A detection coil is then used to detect the change in magnetic flux due to the presence of the magnetization of the sample. A typical experimental arrangement is shown in Fig. 6.11 where an applied DC field $\vec{B}$ magnetizes the sample to a magnetization $\vec{M}$. In actual measurements, a system of coils (of various geometry) is used to minimize

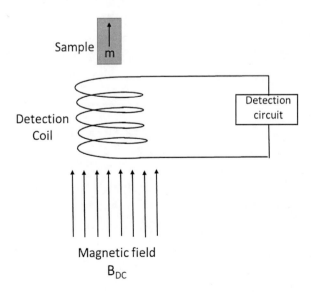

FIGURE 6.11 Typical experimental arrangement in DC induction method.

background effects and noise. However, there will be no output in the detection circuit of Fig. 6.11 unless there is a change in flux in the detection coil. Since the applied field $\vec{B}$ is constant, there will be no signal associated with it. Therefore, some relative movement between the sample and the detection coil system is necessary for the variation of sample flux coupled to the detection coil.

One disadvantage of the induction method is that one has to use an air-core solenoid type magnet; otherwise the usual laboratory electromagnet needs to be suitably modified. In the latter case, the arrangement has to be there to drive either the sample or the detection coil with a rod that passes through one of the magnet pole faces. In addition, an extremely uniform magnetic field is required. However, even in a uniform magnetic field, a large correction needs to be made in the measurement of samples with a small magnetic moment for the magnetic effects of materials in the surrounding environment where the sample is placed. There are mainly three kinds of magnetometers based on the induction method, which is widely used in laboratories all over the world:

1. Vibrating sample magnctometer: The sample is vibrated and this vibration generates an AC signal at a frequency determined by the frequency of vibration. There were also some early works on vibrating coil magnetometer (VCM) where the detection coil was vibrated instead of the sample [13]. However, the major problem faced in producing a practical VCM was the elimination of the signal induced by the curvature of the magnetizing field. In this respect, VSM turned out to be a better option.
2. Superconducting quantum interference device (SQUID) magnetometer: In a SQUID-magnetometer the sample is made to move through the detection coil system. The output from the magnetometer, when properly calibrated, can provide the numerical value of the magnetic moment of a given sample. This measurement process can be repeated at various temperatures (T) and applied fields ($\vec{B}$) to obtain the ($T, \vec{B}$) dependence of the magnetic response of a sample.
3. Extraction magnetometer: In an extraction magnetometer, the sample under study is initially placed in the centre of the detection coil, and then it is moved out of the coil away to a large distance. The resulting flux-variation induces a voltage in the coil circuit which is measured by integration of the signal generated. The magnetic moment is thus measured from the difference in magnetic induction in a space region with and without a sample.

6.2.1 Vibrating sample magnetometer

The idea of a vibrating sample magnetometer (VSM) is mainly attributed to Simon Foner [14], and and there is much to be learnt from those early works, which formed

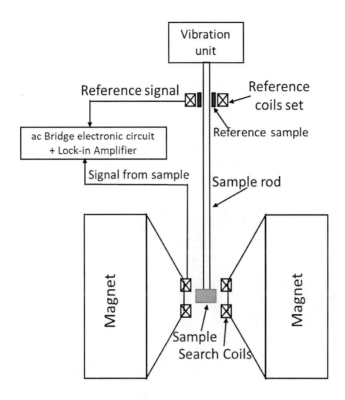

FIGURE 6.12 Schematic representation of the basic principle of a VSM.

the basis of modern-day commercial VSMs. The basic principle of the instrument [14] is presented schematically in Fig. 6.12. The sample is vibrated in the applied magnetic field by a loudspeaker assembly. The oscillating magnetic field resulting from the vibrating sample induces a voltage in the stationary detection coils. This induced voltage E for a small sample being vibrated at a frequency f in a suitable detection coil array can be expressed as [15]:

$$E = \mu G A 2\pi f \cos 2\pi f t \qquad (6.8)$$

where μ is the magnetic moment of the sample, G is the sensitivity function, and A is the amplitude of the sinusoidal vibration. The sensitivity function G represents the spatial distribution of detection coil sensitivity in a VSM, i.e., the dependence of VSM output on sample position. In the original work of Foner [14], a second voltage was induced in a similar stationary set of reference coils by a reference sample. The reference sample can be a small permanent magnet [14]. The unknown sample and the reference sample were driven synchronously by a common method, hence the phase and amplitude of the resulting voltages were directly related. If ratios of reference and sample signal are null-detected at the sensor input, to first order the null balance is independent of changes in A, f, amplifier gain or linearity, waveform, magnetoresistance of the detector coils, and small magnetic field instabilities [15].

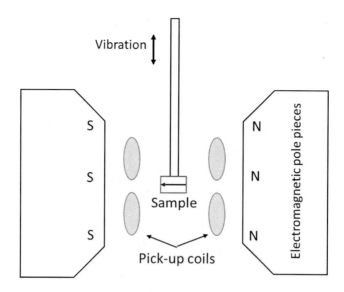

FIGURE 6.13 Mallinson pair detection coils system used as a pickup coil in a vibrating sample magnetometer.

The null detector is an electronic AC bridge often involving a lock-in amplifier. In subsequent developments, a reference vibrating capacitor has been used in the feedback balance arrangement. Practical detection coil arrangements for the VSM are designed to ensure a linear response over the length of vibration and eliminate the signal from the applied DC field [15]. A typical arrangement for the pickup coil is a Mallinson pair shown in Fig. 6.13. The proportionality constant for the magnetometer output of a unit dipole moment is given by the sensitivity functions. These along with the detection coil geometries have been discussed in detail by Zieba and Foner [16] and Pacyna and Ruebenauer [17].

The availability of inexpensive digital computers, automation, and packaged systems has made measurements in VSM a relatively easy exercise. Before starting the experiments to measure magnetic moments of an unknown sample, the instrument needs to be calibrated, usually using a Ni standard sample obtained from a standards body such as the National Institute of Standards and Technology (NIST). VSMs are extremely versatile and flexible instruments, which can be configured with conventional electromagnets for working in moderate magnetic fields (< 2.5 T) or superconducting magnets where larger magnetic fields are required. Pickup coil arrangements can be designed to detect magnetic moments along two orthogonal axes, thus allowing vector measurements to be performed. In addition, both cryogenic (down to 2 K) and high temperature (up to 1,000 K) environments can be readily provided, thus enabling measurement over a very wide range of temperatures. While VSMs are nowadays made by many commercial companies,

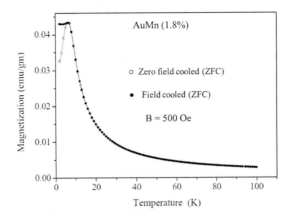

FIGURE 6.14 Magnetization versus temperature plot of AuMn(1.8%) spin-glass measured with a VSM. ([20])

a low-cost precision VSM can also be built rather easily in a standard laboratory [18, 19].

Fig. 6.14 shows as a representative example the results of magnetization measurements on a canonical spin-glass AuMn(1.8%) measured using a VSM in the zero field cooled (ZFC) and field cooled (FC) mode with an applied field of 500 Oe. In the ZFC mode the sample is cooled to a temperature below the spin-glass transition temperature before the magnetic field is switched on, and the measurement is then performed while warming up the sample. In the FC mode the magnetic field is switched on well above the spin-glass transition temperature and measurements are taken while cooling the sample across the transition temperature. The results in Fig. 6.14 clearly shows the characteristic peak in magnetization at the spin-glass transition temperature along with the onset thermomagnetic irreversibility, i.e., $M_{ZFc}(T) \neq M_{FC}(T)$ below the transition point. Fig. 6.15 presents the temperature dependence of inverse susceptibility in the paramagnetic regime of the same AuMn(1.8%) sample, which clearly shows the characteristic Curie–Weiss behaviour.

One of the many subsequent developments of VSM is the very low-frequency VSM or VLFVSM, which is a rugged differential low-frequency flux integration device [15]. At the heart of a VLFVSM is a driving oscillator attached with the sample rod, which varies the frequency and amplitude of the sample displacement in the pickup coils. The driving frequency is between 50 and 100 Hz, and amplitude is typically 1–3 mm. A type of VLFVSM now incorporates a SQUID detector and electronics instead of a conventional integrating circuit. We will discuss such VSMs in more detail later on in this chapter. But first, we will discuss the SQUID detector in connection with the SQUID magnetometer.

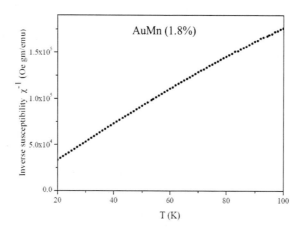

FIGURE 6.15 Inverse susceptibility versus temperature plot of AuMn(1.8%) spin-glass measured with a VSM. ([20].)

The flux distribution produced by a sample may get distorted by any highly permeable medium in its neighbourhood; this affects the detected signal. This effect is called the image effect [15]. The image effects in VSMs using superconducting magnets have been discussed by Zeiba [21] and Zeiba and Foner [22]. If the VSM is calibrated with a known test sample, this image effect automatically can be corrected to first order, because both the flux from calibration and measured samples are affected in the same way.

6.2.2 Superconducting quantum interference device magnetometer

The heart of a Superconducting Quantum Interference Device (SQUID) magnetometer is the SQUID sensor. In 1962 Brian Josephson predicted that if two superconducting materials are separated by a thin insulating barrier layer of insulating materials, a current can transmit from one superconductor to the other even without the application of an external voltage between the two superconductors. Theoretical calculations by Josephson showed that the flowing supercurrent was proportional to the sine of the quantum mechanical phase difference between the wave functions of the two superconductors. Together with the relationship between the phase of the superconducting wave function and the magnetic field, this dependence of the supercurrent on the relative phase difference is the key to the functioning of the SQUID sensor.

There can be two types of SQUID sensors: DC-SQUID and RF-SQUID. A DC-SQUID consists of two matching Josephson junctions formed into a superconducting ring. The current flowing in the SQUID ring is very sensitive to the flux through this ring and this property is utilized to detect any change in flux in the vicinity of the SQUID sensor. The maximum zero-voltage current or the critical current of the DC SQUID changes sinusoidally with the integral of the magnetic field through the area of the SQUID loop owing to interference of the quantum phases of the two junctions. The period of modulation is the superconducting flux quantum $h/2e = 2.07 \times 10^{-15}$ Tm^2. This phase can be measured to a few parts per million with a one-second integration time with the help of flux modulation, phase-sensitive detection, and flux feedback. In practice, one does not measure the current but rather the voltage across the SQUID, which swings back and forth under a steadily changing magnetic field. The SQUID in essence is a flux-to-voltage transducer, converting a tiny change in magnetic flux into a voltage. While the DC-SQUID is generally more sensitive, the RF-SQUID is easier to fabricate, since it requires only a single Josephson junction. For a detailed discussion on SQUID sensors and devices, the readers are referred to the articles by Fagaly [23] and by Kirtley and Wiksow [24] and the references therein.

In practice, the magnetometers are coupled to an input circuit called a flux-transformer (see Fig. 6.16) to take advantage of the extraordinary sensitivity of the SQUID-sensor. The flux-transformer consists of a loop of superconducting material coupled to a SQUID sensor. An external magnetic field causes a persistent supercurrent to circulate in the loop, which induces a flux in the SQUID. The flux transformer boosts the field sensitivity because the loop encloses a much larger area than can a SQUID.

In the widely used commercial SQUID magnetometers (such as from Quantum Design Inc.), the SQUID sensor is the source of the instrument's remarkable sensitivity. But it does not directly detect the magnetic field from the sample. A schematic representation of the experimental configuration in a SQUID magnetometer is shown in Fig. 6.17. The measurement procedure involves a movement of the sample under study through the superconducting detection coils, which are located at the centre of a superconducting magnet. When the sample moves through the superconducting coils, the magnetic moment of the sample induces an electric current in the detection coils. The detection coils, the connecting

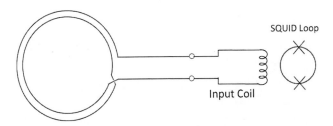

FIGURE 6.16 Flux transformer for coupling external flux.

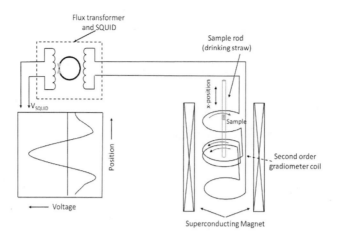

FIGURE 6.17 A schematic representation of the experimental configuration in a SQUID magnetometer.

wires, and the SQUID input coil together form a closed superconducting loop, which includes electronics for a locked-loop operation of an RF-SQUID. Any change of magnetic flux in the detection coils produces a change in the persistent current in the detection circuit, which is proportional to the change in magnetic flux. The SQUID functions as a highly linear current to voltage converter. The variations in the current in the detection coils produce corresponding variations in the SQUID output voltage, which are proportional to the magnetic moment of the sample.

For a SQUID magnetometer, it is useful to introduce a sensitivity function $G(r)$ which represents the proportionality constant between the magnetometer output and the measured magnetic moment m placed at the point r. From the reciprocity principle, $G(r)$ has a dual meaning, which can be understood alternatively as a flux of the real magnetic moment or as a fictitious magnetic field produced by the unit current flowing through the pickup coil array. For a sample with uniform magnetization, the signal detected at the pickup coil assembly is expressed as [25]:

$$S(R) = constant \int d^3 r G(r+R) m(r) \qquad (6.9)$$

where $m(r)$ is the magnetization at a distance r from the centre of the sample and R is the distance of the pickup coil array from the centre of the sample. When the magnetization is not uniform throughout the sample, one needs to expand the sensitivity function G about the origin of the sample in powers of r, and then relate the series to a multipole expansion $m(r)$ within the sample. With the assumptions of the axial symmetry of the pickup coil assembly and the small sample size (so that the z-component of the hypothetical field G_z varies slowly over the sample) G can be written as:

$$G_z = \Sigma \frac{l}{l!} \left(\frac{\partial^l G_z}{\partial z}\right)_{z=z_0} r^l P_l \cos\theta) \qquad (6.10)$$

Equation 6.9 is then rewritten as

$$S(z_0) = \left[\int m_z(r)d^3r\right] G_z(z_0) + \left[\int z m_z(r)d^3r\right] \left(\frac{\partial G_z}{\partial z}\right)_{z=z_0}$$
$$+ \frac{1}{2}\left[\int (3z^2 - r^2)m_z(r)d^3r\right] \left(\frac{\partial^2 G_z}{\partial z^2}\right)_{z=z_0} + \ldots \qquad (6.11)$$

where z_0 is the mean sample position. The different terms in the above equation represent contributions from the dipole, quadrupole, octupole moments, and higher order moments of the sample and they are distinguished from each other by their different dependence on z_0 as dictated by successive higher derivatives of G_z. The voltage signal at the SQUID output (see Fig. 6.17) of a SQUID magnetometer is usually used to fit the response of an ideal magnetic dipole. To eliminate the influence of all kinds of external magnetic fields, the pickup coil in SQUID magnetometers is made as a second derivative gradiometer (see Fig. 6.18). The response function G of an ideal magnetic dipole in a second derivative gradiometer coil assembly is given

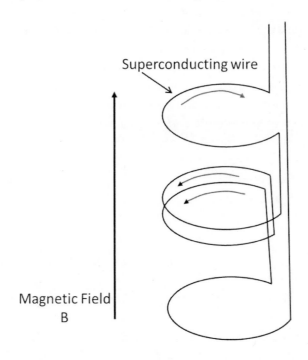

FIGURE 6.18 Schematic diagram of a second derivative gradiometer coil assembly.

by

$$G(x) = -(R^2 + (x - A)^2)^{-3/2} + 2(R^2 + x^2)^{-3/2} - (R^2 + (x + A)^2)^{-3/2} \quad (6.12)$$

where x stands for the position of the dipole and A the distance of outer coils with respect to the centre of the pickup coil assembly, and R is the radius of the coils. During the measurement, a constant background and a drift (linear in x) is superimposed on this response. The total output signal is represented by

$$S(x) = a + bx + G_T(x) \quad (6.13)$$

The software developed and supplied with a commercial magnetometer (Quantum Design Inc.) generates the values for a and b such that $G_T(x)$ approximates as $cG(x)$. The parameter c is directly proportional to the magnetic moment of the sample. The raw data $S(x)$ is not shown as the output of an experiment; rather $G_T(x)$ is displayed as the SQUID profile, and the best fit c is presented as the magnetic moment. The parameter b includes temporal drift in the electronics and is chosen to make $G_T(x)$ symmetric about $x = 0$ and it is for this reason that a scan during an experiment usually collects data on either side of $x = 0$. In Fig. 6.19 we show some typical results obtained with a commercial Quantum Design (MPMS-XL) SQUID magnetometer [26]. Fig. 19(a) presents temperature dependence of magnetization and inverse susceptibility of two weak ferromagnetic samples of Co-doped FeSi, and Figs. 19(b) and (c) present isothermal field dependence of the same samples measured at various temperatures both below and above the magnetic transition temperature.

The ultimate sensitivity of any SQUID system is limited by random noise voltages generated either in the input superconducting circuit or by the associated electronic amplifiers. It is often expressed as an equivalent flux-noise in a bandwidth of 1 Hz at the input to the detection system. For a typical commercial SQUID magnetometer (Quantum Design Inc.) operating at 4K and 20 MHz, the flux resolution is about $10^{-4} \, \Phi_0 B^{-1/2}$. A major contribution to the noise in SQUID magnetometry is the slowly changing background signal that arises from flux creep in the superconducting magnet (of the magnetometer). This is minimized by using a second derivative gradiometer geometry (Fig. 6.18). Apart from this, there exist other problems associated with SQUID magnetometry, which are listed below:

1. Image effect and sample geometry effect refer to the influence of superconducting elements in the vicinity of the sample and the detection coil, and change in the instrument calibration due to the sample size, shape, and location. This can be an avoidable problem in any open-circuit measurement made with an electromagnet or superconducting magnet. The image effect has been discussed in some detail by Zeiba [21] and also in the book by Cullity and Graham [27]. Interested readers may refer to these.

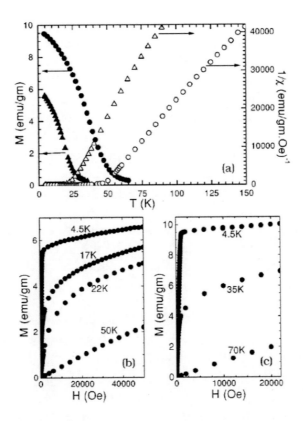

FIGURE 6.19 (a) Temperature dependence of magnetization and inverse susceptibility for weak ferromagnetic $(Fe_{1-x}Co_x)Si$ alloys with $x = 0.15$ and 0.35, (b) isothermal field dependence of magnetization at various temperatures for the same $(Fe_{1-x}Co_x)Si$ alloys with $x = 0.15$ and 0.35, respectively. (From reference [26] with permission from APS.)

2. A small amount of oxygen can leak into the sample chamber during the measurement procedure. Oxygen solidifies around 50 K and orders magnetically. A small amount of solid oxygen can easily contribute to the signal output of a SQUID magnetometer, particularly in the high sensitivity scale ($\leq 10^{-5}$ emu) [28]. A regular flush and purge routine of the sample chamber with helium gas at $T > 200K$ before any experimental cycle can effectively get rid of this problem.

3. In many commercial SQUID magnetometers a clear drinking straw serves as a sample holder where a sample is mounted by clamping it in-between the walls of the straw. It should be noted here that the sample holder itself needs to be homogeneous across the entire length of the gradiometer while moving from the bottom to the top-most position. Any holes, cuts, or small dents present in the straw sample holder act like locally missing diamagnetic material, which is recorded in the pickup system as a spurious

net-paramagnetic signal. Also, any labelling on the sample holder, especially close to the sample, may be avoided. This is because the text markers many a time contain paramagnetic pigments.

4. In the superconducting magnets used in the SQUID magnetometers, there is almost always a remanent or offset field that originates from trapped magnetic flux pinned at defects in the material of the superconducting coil. This remanent field is directed antiparallely to the last applied strong field generated by the magnet. This remanent field cannot be avoided or corrected after a measurement is performed in high magnetic fields. The remanent field gives rise to an apparent residual hysteresis for diamagnetic samples and an inverted hysteresis for paramagnetic samples. This leads to residual signals for low moment samples on diamagnetic substrates like sapphire (Al_2O_3) and limits the ultimate detection sensitivity in a SQUID magnetometer. It may also be noted that the sapphire substrates can exhibit ferromagnetism themselves, which arises from residual contaminations originating from the manufacturing sources.

5. The superconducting magnets are designed to have a constant field strength in the central region. In practice, however, there is always a small inhomogeneity. This inhomogeneity of the field experienced by the sample depends on the distance (or scan length) through which the sample moves during a measurement. This causes a problem for samples with an irreversible magnetic response, e.g., type-II superconductors with flux-line pinning, since such a movement through an in-homogeneous magnetic field causes the sample to traverse a minor hysteresis loop [29], and can lead to a wrong estimation of the experimentally determined irreversibility field. The field inhomogeneity experienced by a sample in a 2 cm scan length in commercial SQUID magnetometers of Quantum Design Inc. is about 0.2 mT in the field range of 1–2 T. It is now known that if the field for full penetration for a superconductor is much higher than this value in the concerned field range, a measurement with this 2 cm scan length will not introduce any appreciable error in the measurements [30].

6. In the measurement of thin-film samples with small magnetic moments like dilute magnetic semiconductors (DMS), magnetic signals of a DMS layer needs to be separated from the large diamagnetic signals from the substrate. To this end, the second-order gradiometer is already susceptible to errors due to the finite sample size because the standard measurement protocol of the commercial SQUID magnetometers assumes the sample as a point dipole, and some correction factors are provided to take care of this finite sample size. One possible solution may be to deposit the DMS layer in the centre point of a suitable diamagnetic substrate long enough to span the standard scan length employed in a SQUID magnetometer. In this way, the

contribution from the diamagnetic substrate can be eliminated with the use of a second-order gradiometer coil.

The readers are referred to these works [31, 32] for some discussion on such artifacts and pitfalls of measurements with a SQUID magnetometer taking examples of some real magnetic thin film and nano materials.

6.2.3 SQUID-VSM

A SQUID-VSM is an instrument that combines the sensitivity of a conventional SQUID magnetometer with the data acquisition speed of a conventional VSM. The experimental configuration of a SQUID-VSM is quite similar to that of a SQUID magnetometer as shown in Fig. 6.17. This involves a superconducting detection coils system configured as a second-order gradiometer, which is very similar to the one shown in Fig. 6.18 and used in a SQUID magnetometer. In contrast with conventional VSMs with copper-detection-coil systems, a changing magnetic flux is not strictly required to generate a signal in the SQUID-VSM. A current is generated in the superconducting detection coils system of a SQUID-VSM in response to local magnetic field disturbances, and this current is inductively coupled to the SQUID sensor of the instrument. We shall see below that the vibration of the sample is necessary only to create a signal at a known modulation frequency, which helps in the separation of the sample signal from instrumental artifacts, and not to generate the signal itself [33]. During the experiment, no current flows in the coils system, since the SQUID feedback circuit anulls the current in the detection coils. This feedback current then generates the actual SQUID voltage for analysis.

In a SQUID-VSM experiment, the sample is vibrated at frequency ω around the very centre of the detection coils. In this position the signal peaks as a function of sample position z. The generated SQUID signal V as a function of time t can be expressed as [33]:

$$V(t) = AB^2 sin^2(\omega t) \quad (6.14)$$

In the equation above A is a scaling factor relating to the magnetic moment of the sample, and B is the amplitude of sample vibration. With the known relation of $sin^2(\omega t) = 1/2(1 - cos(2\omega t))$ a lock-in amplifier can be utilized to detect and quantify the signal occurring at frequency 2ω. If the vibration frequency is selected wisely, this signal can be safely identified with the signal originating from the sample. The measured signal is multiplied with a phase-corrected reference signal at 2ω and then the DC component of the result is extracted. The 2ω component of the measured signal is proportional to this DC component. It is prudent to avoid vibration frequencies that are half of the power line frequency in the laboratory and all its higher harmonics. The size of the signal does not depend on the vibration frequency and higher vibration frequencies do not give a better signal-to-noise ratio.

The default vibration frequency may be a mechanically quiet frequency for the system as a whole [33]. The sample needs to be fixed securely to the sample holder, and the sample holder must be connected securely to the sample rod. In case the sample does not oscillate synchronously with the VSM head encoder, then the instrument rejects much of the sample signal as out-of-phase signal [33].

6.2.4 Extraction magnetometer

In this method, the sample under study is placed initially at the centre of one of the two counter-wound coils of equal turns and then moved to a large distance away from it. The pickup coils are placed symmetrically at the centre of an electromagnet or superconducting solenoid, which is the source of an externally applied DC magnetizing field. Fig. 6.20 presents a schematic diagram of the working principle of an extraction magnetometer. The movement of the sample through the pickup coils causes variation in the magnetic flux (Φ) which is sensed in the pickup coil circuit in the form of an inductive voltage:

$$V = -\frac{d\Phi}{dt} \tag{6.15}$$

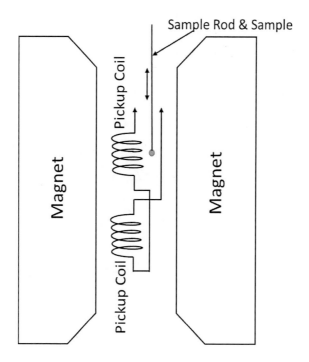

FIGURE 6.20 Schematic representation of the working principle of an extraction magnetometer.

During the experiment the difference in the magnetic induction in the sample region with and without a sample is obtained in terms of the total flux variation Φ by integrating the induced signal in the pickup coil circuit:

$$\Phi = -\int V(t)dt \qquad (6.16)$$

The magnetic moment of the sample can be deduced from the relation $M = \alpha\Phi$, where α is the calibration constant obtained experimentally using a standard sample. The extraction method does not allow continuous recording of $\vec{B}$ and $\vec{M}$ in a simple way. Extraction magnetometers are no longer widely used and have been surpassed by more user-friendly and popular VSMs and SQUID magnetometers.

6.3 AC Susceptibility

In an AC susceptometer, the applied magnetic field is an AC field ($\vec{B}_{ac}$) produced in an induction coil system usually called the primary. There will now be a change in flux related to $\vec{B}_{ac}$ itself. The detection circuitry often includes two identical but counter-wound empty coils known as secondary coils which nullify the flux changes related to time-varying applied magnetic field $\vec{B}_{ac}$. Unlike in a DC magnetometer where the magnetic moment of the sample does not change, in an AC susceptometer, the magnetic moment of the sample changes with time in response to the applied AC magnetic field. Here what one measures is not the magnetic moment of the sample, but the magnetic response of the sample to the applied magnetic field, namely the magnetic susceptibility χ, which is generally defined by Eqn. 6.1 as $M = \chi B$. In many types of magnetic materials including ferromagnets, the magnetic response of the sample to the applied magnetic field is non-linear, and $\chi = \frac{M}{B}$ is magnetic field dependent. Here, one defines the differential susceptibility as:

$$\chi_{diff} = \frac{dM}{dB} \qquad (6.17)$$

and χ_{diff} is field dependent. The amplitude of applied $\vec{B}_{ac}$ is usually small ($< \pm$ 10 Oe), and the field dependent $\chi(\vec{B})$ is measured under a DC-bias field $(B)_{dc}$ (Fig. 6.21).

Another useful quantity, which is regularly measured with an AC susceptometer is the initial susceptibility:

$$\chi_{ini} = Lt_{B\to 0}\frac{dM}{dB} \qquad (6.18)$$

This is the susceptibility that is regularly measured for characterizing magnetic and superconducting materials. This shows a sharp rise at the Curie temperature in a

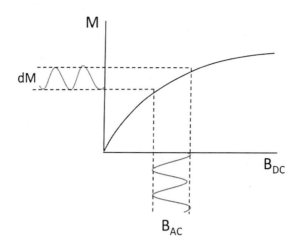

FIGURE 6.21 Differntial susceptibility $\chi_{diff} = \frac{dM}{dB_{AC}}$ as a function of DC field.

ferromagnet, the characteristic peak at the Néel temperature of an antiferromagnet, and the sharp onset of diamagnetism at the superconducting transition temperature of the superconductors.

If a sample is placed in an externally applied periodic magnetic field $\vec{B}(t) = \vec{B}_{dc} + \vec{B}_A \sin(wt)$, the sample will develop a periodic magnetization, which can be expressed as:

$$M(t) = B_A \sum_{n=1}^{\infty} (\chi'_n \sin nwt - \chi''_n \cos nwt) \qquad (6.19)$$

In the above expression, $\chi_n = \chi'_n - \chi''_n$ is the complex magnetic susceptibility of the nth harmonic, where:

$$\chi'_n = \frac{w}{\pi B_A} \int_{-\frac{\pi}{w}}^{\frac{\pi}{w}} M(t) \sin nwt\, dt$$

and

$$\chi''_n = -\frac{w}{\pi B_A} \int_{-\frac{\pi}{w}}^{\frac{\pi}{w}} M(t) \cos nwt\, dt$$

We have mentioned above that the susceptibility χ_n will be in general a function of both applied DC and AC fields. While higher harmonic susceptibilities provide a wealth of information both on the magnetic [34, 35] and superconducting materials [36, 37], the higher harmonic susceptibilities become negligible in the limit $\vec{B}_A \to 0$, and the AC susceptibility becomes essentially equal to the initial susceptibility χ_{ini}. We shall here focus mainly on the fundamental component of

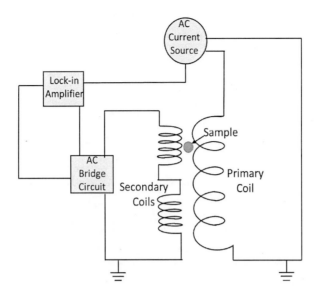

FIGURE 6.22 Schematic representation of the working principle of an AC-susceptometer.

magnetization/susceptibility:

$$M(t) = B_A(\chi' \sin wt + \chi'' \cos wt) \qquad (6.20)$$

The real (imaginary) part of the susceptibility χ' (χ'') is in phase (90^0 out of phase) with the applied field $\vec{B}(t)$ and corresponds to the inductive response (resistive losses or energy dissipations) in the sample.

The working principle of an AC susceptometer is shown schematically in Fig. 6.22. The primary or drive coil is energized by an alternating current source to produce an AC magnetic field $\vec{B}_{ac}$. A DC bias field can also be superimposed on it by passing a direct current through this coil. There is a set of two oppositely wound secondary coils around the primary coil, which are ideally of equal turns. The time variation of the magnetic flux in the coils will induce an electrical signal (voltage) $v(t)$ in the series of secondary coils as follows:

$$v(t) = (L_1 - L_2)\frac{di_p}{dt} \qquad (6.21)$$

where i_p is the primary current and L_1 and L_2 are mutual inductances between each of the secondary coils and the primary coil. The negative sign between the mutual inductances indicates that the secondary coils are wound in the opposite directions. In the ideal case of the secondary coils with an exactly equal number of turns, the induced voltage $v(t) = 0$. However, even after a careful winding of the secondary coils and placing them symmetrically with respect to the primary coil, there is always a small residual voltage in a real situation that needs to be compensated externally for an accurate measurement using the susceptometer. After achieving a reasonable balance in the secondary coils, if a sample is placed in one of the secondary coils, there will be an imbalance in the secondary circuit. This imbalance signal can be represented as:

$$v(t) = \mu_0 N A \frac{dM(t)}{dt} \tag{6.22}$$

where N is the number of turns in each of the secondary coils, A their cross-sectional area, and μ_0 the permeability of the free space. This Eqn. 6.22 in combination with Eqn. 6.20 can be rewritten as:

$$v(t) = \mu_0 w N B_A \sum_{n=1}^{\infty} n(\chi'_n \cos nwt + \chi''_n \sin nwt)$$

$$= v_0 \sum_{n=1}^{\infty} n(\chi'_n \cos nwt + \chi''_n \sin nwt) \tag{6.23}$$

where $v_0 = \mu_0 w N B_A$.

The above expression for the induced voltage is valid for a homogeneously magnetized sample filling the detection coil. In practice, it will depend on the geometry of the sample as well as the relative orientation of the sample and detection coil. This information, however, can be obtained through the proportionality constant α between the experimentally determined induced voltage and $\frac{dM}{dt}$. While α can be calculated mathematically with the detailed information of the sample and coil geometry and coil parameters, the more practical way is to estimate α experimentally through the calibration of the detector coils using a standard sample of known magnetic susceptibility. In the subsequent measurements, on the unknown sample, however, one needs to have the same sample dimensions and also the same position in the secondary coil as that of the standard sample. For high sensitivity measurements, often an electronic bridge involving variable mutual inductance and a lock-in amplifier as a phase-sensitive signal detector are used for an accurate determination of the induced voltage. For a detailed description of the electronic detection circuits of AC susceptometers including the various types of inductance bridges, the readers should refer to the book by Hein, Francavilla, and Liebenberg [38]. Sensitive AC susceptometers with a high degree of accuracy can be built relatively easily and at low cost in any standard laboratory [39, 40].

In Fig. 6.23 we present the AC susceptibility of a canonical spin-glass sample AuMn(1.8%) measured with an AC magnetic field of 1 Oe and a driving frequency of 532 Hz. The figure highlights the well-known AC susceptibility peak that marks the onset of spin-glass transition. Fig. 6.24 shows the AC susceptibility of the same AuMn(1.8%) sample at various driving frequencies of AC magnetic field and also in the presence of a DC magnetic field [41]. The figure reveals the well-known frequency dependence of the AC susceptibility peak, which is also a characteristic feature of spin-glass transition.

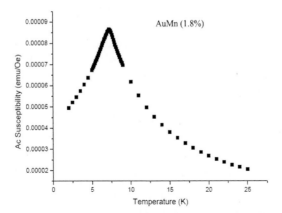

FIGURE 6.23 The AC susceptibility of canonical spin-glas AuMn(1.8%) measured with an AC field of $H_{ac} = 1$ Oe and driving frequency 532 Hz. ([20].)

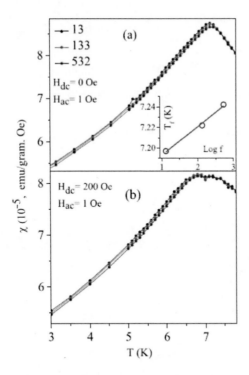

FIGURE 6.24 Frequency dependent of AC susceptibility of AuMn(1.8%) measured with an AC field of $H_{ac} = 1$ Oe in: (a) ZFC state and (b) FC state obtained with a DC bias field of $H_{dc} = 200$ Oe. The inset shows the frequency dependence of spin-glass transition temperature T_f. (From reference [41] with permission from IOP Publishing.)

6.4 Summary

In this chapter we have discussed a variety of experimental techniques involving both *force method* and *induction method* to study bulk magnetic properties of condensed matter. In addition to the primary magnetic characterization of a new material, such experiments generate a wealth of information that forms the basis of the deeper investigation and understanding of the magnetic properties with complementary experiental techniques to be discussed in the subsequent chapters. With the advent of rather ubiquitous commercial vibrating sample magnetometers (VSM) and SQUID magnetometers, the induction method has become more popular in the last few decades than the force method. The pitfalls of using such commercial instruments without a proper understanding of their functional principles have been pointed out especially in the case of thin-film magnetic samples. It is also possible to build some of these instruments, especially AC susceptometers, in the laboratory starting from scratch at relatively low costs and with high enough sensitivity comparable to commercial instruments.

Bibliography

[1] Foner, S. (1981). Review of Magnetometry. *IEEE Transactions on Magnetics*, 17 (6): 3358–3363.

[2] Morris, B. L., and Wold, A. (1968). Faraday Balance for Measuring Magnetic Susceptibility. *Review of Scientific Instruments*, 39 (12): 1937–1941.

[3] Reeves, R. (1972). An Alternating Force Magnetometer. *Journal of Physics E: Scientific Instruments*, 5 (6): 547.

[4] Flanders, P. J. (1988). An Alternating-gradient Magnetometer. *Journal of Applied Physics*, 63 (8): 3940–3945.

[5] Flanders, P. J. (1990). A Vertical Force Alternating-gradient Magnetometer. *Review of Scientific Instruments*, 61 (2): 839–847.

[6] Yee, M. and Yaacob, I. I. (2004). Synthesis and Characterization of Iron Oxide Nanostructured Particles in NaY Zeolite Matrix. *Journal of Materials Research*, 19 (3): 930–936.

[7] Höpfl, Th., Sander, D. Höche, H., and Kirschner, J. (2001). Ultrahigh Vacuum Cantilever Magnetometry with Standard Size Single Crystal Substrates. *Review of Scientific Instruments*, 72: 1495. https://doi.org/10.1063/1.1340560

[8] Adhikari, R., Kaundal, R., Sarkar, A., et al. (2012). The Cantilever Beam Magnetometer: A Simple Teaching Tool for Magnetic Characterization. *American Journal of Physics*, 80: 225.

[9] Sarkar, A., Adhikari, R., and Das, A. K. (2013). Development of a High Vacuum Cantilever Beam Magnetometer for Measurement of Mechanical and Magnetic Properties of Thin Films. *Current Science*, 104 (7): 826–834.

[10] Martín-Hernández, F., Bominaar-Silkens, I. M., Dekkers, M. J. and Maan, J. K. (2006). High-field Cantilever Magnetometry as a Tool for the Determination of the Magnetocrystalline Anisotropy of Single Crystals. *Tectonophysics*, 418 (1–2): 21–30.

[11] Rossel, C., Willemin, M., Gasser, A., et al. (1998). Torsion Cantilever as Magnetic Torque Sensor. *Review of Scientific Instruments*, 69 (9): 3199–3203.

[12] Yu, Y., Xu, F., Wang, N., et al. (2018). Dynamic Cantilever Magnetometry of Individual Co Nanosheets and Applicable Conditions of Uniaxial Magnetic Anisotropy Assumption. *Japanese Journal of Applied Physics*, 57 (9): 090312.

[13] Smith, D. O. (1956). Development of a Vibrating-coil Magnetometer. *Review of Scientific Instruments*, 27 (5): 261–268.

[14] Foner, S. (1959). Versatile and Sensitive Vibrating-sample Magnetometer. *Review of Scientific Instruments*, 30 (7): 548–557.

[15] Foner, S. (1996). The Vibrating Sample Magnetometer: Experiences of a Volunteer. *Journal of Applied Physics*, 79 (8): 4740–4745.

[16] Zieba, A. and Foner, S. (1982). Detection Coil, Sensitivity Function, and Sample Geometry Effects for Vibrating Sample Magnetometers. *Review of Scientific Instruments*, 53 (9): 1344–1354.

[17] Pacyna, A. W., and Ruebenbauer, K. (1984). General Theory of a Vibrating Magnetometer with Extended Coils. *Journal of Physics E: Scientific Instruments*, 17 (2): 141.

[18] Niazi, A., Poddar, P., and Rastogi, A. K. (2000). A Precision, Low-cost Vibrating Sample Magnetometer. *Current Science*, 79 (1): 99–109.

[19] Burgei, W., Pechan, M. J., and Jaeger, H. (2003). A Simple Vibrating Sample Magnetometer for Use in a Materials Physics Course. *American Journal of Physics*, 71 (8): 825–828.

[20] Experimental results courtesy of Sudip Pal.

[21] Zięba, A. (1993). Image and Sample Geometry Effects in SQUID Magneto-meters. *Review of Scientific Instruments*, 64 (12): 3357–3375.

[22] Zięba, A., and Foner, S. (1983). Superconducting Magnet Image Effects Observed with a Vibrating Sample Magnetometer. *Review of Scientific Instruments*, 54 (2): 137–145.

[23] Fagaly, R. L. (2006). Superconducting Quantum Interference Device Instruments and Applications. *Review of Scientific Instruments*, 77 (10): 101101.

[24] Kirtley, J. R., and Wikswo Jr, J. P. (1999). Scanning SQUID Microscopy. *Annual Review of Materials Science*, 29 (1): 117–148.

[25] McElfresh, M. (1994). Fundamentals of Magnetism and Magnetic Measurements. *Quantum Design Technical Note*. https://www.qdusa.com/siteDocs/appNotes/Fund Primer.pdf

[26] Chattopadhyay, M. K., Roy, S. B., and Chaudhary, S. (2002). Magnetic Properties of $Fe_{1-x}Co_xSi$ Alloys. *Physics Reviews. B* 65: 132409.

[27] Cullity B. D., and Graham, C. D. (2009). *Introduction to Magnetic Materials*. Hoboken: Willey.

[28] *Quantum Design Technical Advisory, MPMS No. 8.* (1991). https://www.qdusa.com/siteDocs/appNotes/FundPrimer.pdf

[29] *Quantum Design Technical Advisory MPMS No. 1.* (1989) https://www.qdusa.com/siteDocs/appNotes/FundPrimer.pdf

[30] Roy, S. B., and Chaddah, P. (1996). Unusual Superconducting Phenomenon in $CeRu_2$: Effects of Nd Substitutions. *Physica C: Superconductivity*, 273 (1–2): 120–126.

[31] Ney, A., Kammermeier, T., Ney, V., Ollefs, K., and Ye, S. (2008). Limitations of Measuring Small Magnetic Signals of Samples Deposited on a Diamagnetic Substrate. *Journal of Magnetism and Magnetic Materials*, 320: 3341–3346.

[32] Buchner, M., Hfler, K., Henne, B., Ney, V., and Ney, A. (2018). Tutorial: Basic Principles, Limits of Detection, and Pitfalls of Highly Sensitive SQUID Magnetometry for Nanomagnetism and Spintronics. *Journal of Applied Physics*, 124: 161101.

[33] Simon, Randy W., Zaharchuk, G., Burns, M. J., et al. (2007). Chapter 4. *Quantum Design SQUID VSM Users Manual*. Los Alamo: STAR Cryoelectronics, LLC.

[34] Chakravarti, A., Ranganathan, R., and Roy, S. B. (1992). Competing Interactions and Spin-glass-like Features in the UCu_2Ge_2 System. *Physical Review B*, 46 (10): 6236.

[35] Mukherjee, S., Ranganathan, R., and Roy, S. B. (1994). Linear and Nonlinear AC Susceptibility of the Canted-spin System: $Ce(Fe_{0.96}A_{0.04})_2$. *Physical Review B*, 50 (2): 1084.

[36] Roy, S. B., Kumar, S., Chaddah, P., et al. (1992). Magnetic Field Dependence of the Harmonic Generation in Sintered Pellets of YBaCuO: the History Effects. *Physica C: Superconductivity*, 198 (3–4): 383–388.

[37] Roy, S. B., Kumar, S., Pradhan, A. K., et al. (1993). Study of Minor Hysteresis Loops of YBaCuO: DC and AC Hysteresis Effects in Granular Samples. *Physica C: Superconductivity*, 218 (3–4): 476–84.

[38] Francavilla, T. L., Hein, R. A. and Liebenberg, D. H., eds. (2013). *Magnetic Susceptibility of Superconductors and Other Spin Systems*. Springer Science & Business Media.

[39] Bajpai, A., and Banerjee, A. (1997). An Automated Susceptometer for the Measurement of Linear and Nonlinear Magnetic AC Susceptibility. *Review of Scientific Instruments*, 68 (11): 4075–4079.

[40] Dutta, B., Kumar, K., Ghodke, N., and Banerjee, A. (2020). An Automated Setup to Measure the Linear and Nonlinear Magnetic AC-Susceptibility Down to 4K with Higher Accuracy. *Review of Scientific Instruments*, 91 (12): 123905.

[41] Pal, S., Kumar, K., Banerjee, A., Roy, S. B., and Nigam, A. K. (2020). Magnetic Response of Canonical Spin Glass and Magnetic Glass. *Journal of Physics: Condensed Matter*, 33 (2): 025801.

7

Magnetic Resonance and Relaxation

Electromagnetic (EM) radiation consists of coupled electric and magnetic fields oscillating in directions perpendicular to each other and the direction of propagation of radiation. EM radiation can be an interesting probe to study materials' properties. It is the electric field component of EM radiation that interacts with molecules and solids in most cases. Two conditions need to be fulfilled for the absorption of EM radiation during such interaction: (i) the energy of a quantum of EM radiation must be equal to the separation between energy levels in the atom/molecule, (ii) the oscillating electric field component must be able to stimulate an oscillating electric dipole in the atom/molecule. EM radiation in the microwave region of the EM spectrum can interact with molecules having a permanent electric dipole moment created by molecular rotation. On the other hand, infrared radiation would interact with molecules in vibrational modes giving rise to a change in the electric dipole moment.

Similarly, a solid or molecule containing magnetic dipoles is expected to interact with the magnetic component of EM radiation. EM irradiation of a molecule over a wide range of spectral frequencies does not normally result in absorption attributable to magnetic interaction. The absorption of EM radiation attributable to magnetic dipole transitions may, however, occur at one or more characteristic frequencies if the material of interest is additionally subjected to a static magnetic field. The application of a magnetic field $\vec{B}$ can cause precession of a magnetic moment at an angular frequency of $|\gamma\vec{B}|$, where γ is the gyromagnetic ratio. A material with magnetic moments placed in a magnetic field can absorb energy at this frequency. It is thus possible to observe a resonant absorption of energy from an EM wave tuned to an appropriate frequency. This phenomenon is known as "magnetic resonance", and can be studied with a number of different experimental techniques, depending upon the type of magnetic moment involved in the resonance.

The presence of a static magnetic field is a crucial requirement for magnetic dipolar transitions. If there is no static magnetic field, the energy levels will be coincident. The permanent magnetic moments in a material are associated either with electrons or with nuclei. The magnetic dipoles arise from net electronic or nuclear angular momentum. The fundamental phenomenon involved here is the quantization of angular momentum. The appropriate frequency of the EM wave is determined by both the type of magnetic moment and the size of the applied magnetic field.

The local environment of the magnetic moment in a solid material is determined by the crystal field. In addition, spin–orbit coupling and hyperfine interactions also play an important role. Magnetic resonance experiments are well suited to investigate such effects. In the following sections, we shall discuss some of these experimental techniques.

7.1 Nuclear Magnetic Resonance

The most widely used magnetic resonance technique is nuclear magnetic resonance (NMR). The NMR technique is based on the interaction of the spin of a nucleus with external magnetic fields, neighbouring nuclei, the surrounding electrons of the nucleus itself and also the electrons of neighbouring atoms [1]. The excitation of a nuclear spin by an alternating or pulsed magnetic field can be described within a quantum mechanical framework as a resonant absorption of photons. The externally applied magnetic field is the determining factor for the effective field at a nucleus of paramagnetic and diamagnetic materials. In ferromagnetic materials, the most prominent contribution to the effective field arises from the hyperfine field, which refers to the interaction of the nuclear magnetic moment with magnetic fields originating in the spin and orbital moments of the surrounding electrons [1]. NMR experiments probe the local structure by measuring the resonance frequencies and thus the local hyperfine fields.

The essential requirement of an NMR experiment is a nucleus with non-zero spin. In a simple configuration of this experiment, a sample is inserted in a coil placed between the pole pieces of an electromagnet. A magnetic field $\vec{B}$ is generated by this magnet along a particular direction, say z-direction. The unit of nuclear magnetism is expressed as:

$$\mu_N = \frac{e\hbar}{2m_p} = 5.0508 \times 10^{-27} Am^2 \qquad (7.1)$$

Here m_p is the mass of the proton. The nuclear spin quantum number I can take one of these values: $0, \frac{1}{2}, 1, \frac{3}{2}, 2....$ The z-component of the angular momentum of the nucleus m_I takes only integral values between $-I$ and I. The magnetic moment associated with a nucleus resolved along the z-direction of the spin state with the largest value of m_I is expressed as:

$$\mu = g_N \mu_n I \quad (7.2)$$

Here g_N is the nuclear g-factor that reflects the detailed structure of the nucleus. This nuclear g-factor is a number of the order of unity. In general in an external field $\vec{B}_0$ the energy of nucleus can be expressed as [2]:

$$E = -\vec{\mu}.\vec{B}_0 = -g_N \mu_n m_I |\vec{B}_0| \quad (7.3)$$

This energy corresponds to a ladder with $2I+1$ equispaced levels of separation $g_N \mu_n B_0$. An externally applied radio frequency (RF) field $\vec{B}_1$ can excite transitions between pairs of levels in the ladder, and this phenomenon forms the basis of the NMR.

If the RF field $\vec{B}_1$ is applied along the x-direction, it would lead to a perturbation to the system proportional to $B_1 \hat{I}_x$. The matrix element of this perturbation is proportional to $\langle \acute{m}_I | \hat{I}_x | m_I \rangle$ [2]. This matrix element is zero unless $\acute{m}_I = m_I \pm 1$. The following selection rule describes the allowed transition:

$$\Delta m_I = \pm 1 \quad (7.4)$$

This selection rule indicates that transitions only between the adjacent levels are allowed.

Experimental details of NMR

Fig. 7.1 presents a schematic diagram of an NMR experimental setup. The sample is placed within an RF transmitter coil, which produces an oscillatory magnetic field $\vec{B}_1$. A highly homogeneous magnet provides the static field $\vec{B}_0$. This static field $\vec{B}_0$ is perpendicular to the direction of the oscillatory field $\vec{B}_1$. In the most common arrangement of the nuclear experiment, the RF coil not only induces the excitations, but also forms a part of a tuned circuit with a large quality factor (Q). The applied RF field $\vec{B}_1$ excites transition in the nucleus and the energy is transferred between the sample and the RF circuit. This causes a small change in the Q factor of the circuit. There are two ways to perform the NMR experiment.

1. The frequency of the RF field is kept constant and the magnetic field is varied.
2. The magnetic field is kept constant and the frequency of the RF field is varied.

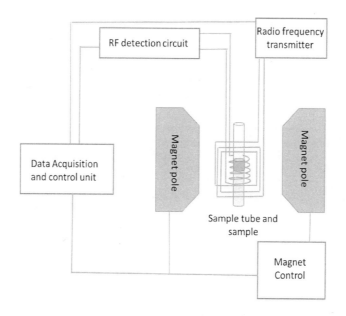

FIGURE 7.1 Schematic diagram of a nuclear magnetic resonance (NMR) experimental setup

A very crucial requirement of an NMR experiment is to have a highly homogeneous magnet, which produces a constant field $\vec{B}_0$. If the magnet is not homogeneous, the different parts of the sample will experience a slightly different field during field sweep. This leads to the broad spatial distribution of the energy gap between excited states throughout the sample (see Eqn. 7.3). As a result, the measured resonance will be very broad and even can get washed out totally.

In the case of a proton $I = \frac{1}{2}$ and m_I can take values of only $\pm\frac{1}{2}$. These two states are separated by an energy $\Delta E = g_N \mu_N B_0$. In a usually available magnetic field $\tilde{B}_0 \approx$ 1 T this means energy splitting for a proton $\approx 10^{-7}$ eV. This energy corresponds to a really small temperature $\approx$ 1 mK. At room temperature, the nuclei therefore only show a small tendency to line up with a magnetic field of this magnitude. This is because on average the thermal randomizing energy will dominate vastly over the weak alignment energy of the magnetic field. As a result, unless a resonant technique is employed it is practically impossible to detect any effect due to the magnetism of the nuclei.

The question now is why do we need radio frequency (RF) in the NMR experiment. We can excite transition across the energy splitting ΔE in the NMR experiment when the system is perturbed with exactly the right angular frequency ω given by the expression:

$$\hbar\omega = \Delta E \qquad (7.5)$$

For an applied magnetic field $\tilde{B}_0 = 1$ T, the frequency of the electromagnetic wave ν is given by:

$$\nu = \frac{\omega}{2\pi} = \frac{g_n \mu_n B_0}{2\pi \hbar} = 42.58 MHz \qquad (7.6)$$

So we can see that the required frequencies in the NMR experiment lie in the RF region of the electromagnetic spectrum, and hence oscillators and coils can be used in the experimental setup. The system will absorb energy only at the right frequency of RF excitations. This is a resonance between the frequency of RF excitations and the energy separation of the nuclear levels, and is hence called "nuclear magnetic resonance". Modern NMR spectrometers in the laboratory nowadays use 12–15 T magnets. The corresponding frequency is in the range of $\approx$ 500–650 MHz, and they are still in the RF regime of the electromagnetic spectrum.

In a material, the nuclei are slightly shielded from the applied magnetic field by the electrons orbiting the nucleus. Thus, depending on the chemical environment of the nucleus, the NMR resonance frequency at a given applied magnetic field can be slightly shifted from the exact resonance frequency. This is known as chemical shift, which is typically a few parts per million. The amount by which a nucleus is shielded in a given chemical environment is known, hence a given molecule can be fingerprinted through this chemical shift. The NMR lines can also be split due to magnetic coupling between neighbouring nuclei. This is known as spin–spin coupling, and this takes place via an indirect contact hyperfine interaction mediated by electrons [2]. This also can provide information about the environment of the nucleus.

In a static magnetic field $\vec{B}_0$ applied to the sample along the z direction, a very weak magnetization, say with equilibrium value $\vec{M}_0$, will exist. A radio frequency excitation $\vec{B}_1$ along, say, x direction will progressively destroy this magnetization. If the magnetization is reduced to zero, the spin system will take some time to recover its equilibrium value if $\vec{B}_1$ is switched off. This time is called spin–lattice relaxation time T_1, which measures the time constant of the interaction between the spin system and its surroundings. Thus, it is expected that [2]:

$$\frac{dM_Z}{dt} = \frac{M_0 - M_Z}{T_1} \qquad (7.7)$$

This spin–lattice relaxation time T_1 involves interactions with the lattice as energy must be exchanged with it. The M_X and M_Y components are expected to be zero. If they are not zero, they will relax back to zero in a time T_2 such that:

$$\frac{dM_X}{dt} = -\frac{M_x}{T_2}, \frac{dM_Y}{dt} = -\frac{M_Y}{T_2} \qquad (7.8)$$

T_2 is known as spin–spin relaxation time. It characterizes the time for dephasing and it corresponds to the interaction between different parts of the spin system.

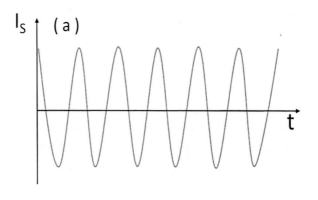
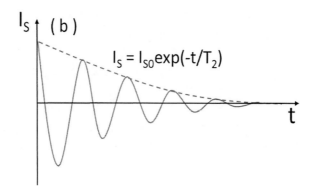

FIGURE 7.2 (a) The intensity of the signal in an NMR coil in the absence of any transverse RF field. (b) Free induction decay signal in the presence of transverse RF field due to spin–spin relaxation.

This can also arise from the inhomogeneity (if any) of the static magnetic field $\vec{B}_0$. It leads to the differences in precession frequency due to the interactions between the observed spin and the spins in the neighbourhood [2].

The applied magnetic field also causes precession of spin, and the respecive equations for M_x, M_y and M_z are [2]:

$$\frac{dM_x}{dt} = \gamma(M \times B)_x - \frac{M_x}{T_1} \tag{7.9}$$

$$\frac{dM_y}{dt} = \gamma(M \times B)_y - \frac{M_y}{T_1} \tag{7.10}$$

$$\frac{dM_z}{dt} = \gamma(M \times B)_z + \frac{M_0 - M_z}{T_1} \tag{7.11}$$

These are known as Bloch equations. They are useful in a number of cases, especially in the rotating reference frame method where the coordinates rotate in the x–y plane at the resonance frequency.

Experimental techniques to measure spin–spin relaxation time

In thermal equilibrium in the absence of the RF field B_1 there is a weak magnetization M_z parallel to the z direction. A short pulse of RF signal with amplitude $\vec{B}_1$ and of duration t_p is applied to the system with its frequency set close to the resonance frequency of the spins. This RF pulse causes the spin to rotate by an angle $\gamma B_1 t_p$, and this angle is adjusted by changing the duration of pulse t_p. If $t_p = \pi/2\gamma B_1$, the RF pulse is called a 90^0 pulse. The amplitude and time of the RF pulse are chosen in such a manner that the magnetization just precesses for a short time and rotates into the x–y plane. In the absence of RF field $\vec{B}_1$ the magnetization now precesses in the static field $\vec{B}_0$, with a steady rotation in the x–y plane at a frequency γB_0, and that results in an induced voltage in the coil. The spin–spin relaxation causes this relaxation to die away with time, and the resulting decay voltage in the coil is known as free induction decay [2]. Fig. 7.2(a) shows the intensity of the signal in the absence of a transverse field. Fig. 7.2(b) shows the decay of the intensity due to spin–spin relaxation. The oscillations die away, and the intensity decays as I = I$_0$exp(-t/T$_2$). This enables the measuring of spin–spin relaxation time T_2.

One problem with this method is that it also measures the relaxation arising out of the inhomogeneity of the magnet. Different parts of the sample give slightly different precession frequencies because of this inhomogeneity. As a result, the interference pattern in the free induction decay signal, in reality, is not as uniform as shown in Fig. 7.2(b). This is because the nuclei may experience slightly different magnetic fields due to their positions in the crystal solid. The real spin–spin relaxation can be separated from the relaxation arising out of the magnet inhomogeneity using the spin-echo technique. Blundell [2] has described the working principle of this experimental technique in an interesting way, which we reproduce here. This technique employs a 90^0 RF pulse, which tips the spins in the x–y plane, and then a 180^0 RF pulse which rotates the spin by 180^0. The effect is akin to a marathon race where all athletes run steadily but some athletes run a bit faster than others. At the beginning of the race, say at the 90^0 pulse, all athletes start their run in a pack. After a while, the fitter athletes move ahead while the others gradually trail behind. Then suddenly at time τ the rule of the game changes with the 180^0 pulse, and runners are instructed to turn around and head towards the starting point of the race. The slower runners now will have an advantage since being closer to the starting point they have a shorter distance to cover. However, it is easy to see by symmetry that all the runners would arrive at the starting point simultaneously at time 2τ should they stick to their original speeds. (There is a basic assumption here that the individual speeds of the runners remained the same during this time 2τ.) Analogously, all the spins in the system realign after time 2τ regardless of their individual dephasing. This is the concept of spin-echo. This

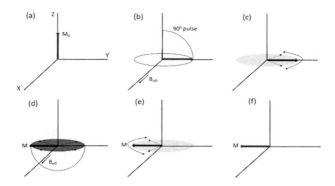

FIGURE 7.3 Schematic representation of the spin-echo experiment in the rotating frame of reference frame. (a) Initial situation, (b) 90^0 pulse, (c) dephasing and free-induction decay, (d) 180^0 pulse, (e) rephasing and (f) spin echo. (Adapted from reference [1] with permission from IOP Publishing.)

technique can be used to remove the problem associated with the inhomogeneity of the magnetic field and also the spread in fields due to chemical shifts. The effect of spin–spin interaction which arises from the time-dependent fluctuating random magnetic fields due to the neighbouring nuclei is not, however, influenced by this experimental process. The spin-echo sequence uses two pulses with a time interval τ in between, which produces a spin-echo at a time 2τ after the first pulse. Fig. 7.3 shows a schematic presentation of the spin-echo experiment in the rotating frame formalism [1]. The first pulse termed as a 90^0 pulse (Fig. 7.3(a)) causes a rotation of the spins from the z direction into the x–y plane (Fig. 7.3(b)). The spins start to dephase at this stage leading to free-induction decay (Fig. 7.3(c)). In the next step of the sequence a second pulse is applied after the time τ, which rotates the nuclear moments by 180^0 (Fig. 7.3(d)). This leads to a rephasing of the moments. The spins precess around the z-axis in the laboratory framework, and this gives rise to an inductive signal after the time interval 2τ. This is the so-called spin echo. The intensity of the spin-echo signal follows $I(2\tau) = I_0 e^{-2\tau/T_2}$, and this makes possible a true assessment of spin–spin relaxation time T_2. In practice, a scan of the RF frequency is used to measure the distribution of effective magnetic fields in a material [1].

The effective field experienced by a nucleus in ferromagnetic materials can be expressed as [1]:

$$\vec{B}_{Effective} = \vec{B}_{Applied} + \vec{B}_{Dipolar} + \vec{B}_{HF} \tag{7.12}$$

Here $\vec{B}_{Applied}$ denotes the external applied magnetic field, $\vec{B}_{Dipolar}$ the dipolar field, and $\vec{B}_{HF}$ the hyperfine field. The dipolar magnetic field arises due to the combined effects of the fields of all electronic moments from all other atoms in the material. It can be represented as a sum of the demagnetization field $\vec{B}_{Demag}$, the Lorentz cavity term $\vec{B}_{Lorentz}$ and the local field term $\vec{B}_{Local}$ [1]:

$$\vec{B}_{Dipolar} = \vec{B}_{Demag} + \vec{B}_{Lorentz} + \vec{B}_{Local} \qquad (7.13)$$

The demagnetization field $\vec{B}_{Demag}$ is determined by the macroscopic shape of the sample, and it is anisotropic except for spherical samples (see Appendix C). The Lorentz cavity field term $\vec{B}_{Lorentz}$ is given by $\vec{B}_{Lorentz} = \frac{4}{3}\pi m$, with m being the saturation value of the electron magnetization. The local field term $\vec{B}_{Local}$ originates from all moments in the Lorentz sphere except the central moment [1]. The sum of the demagnetization and Lorentz cavity field term $\vec{B}_{Demag} + \vec{B}_{Lorentz}$ is zero, whereas the local field term $\vec{B}_{Local}$ goes to zero in the case of a sample with cubic symmetry.

The dominating contribution to the effective field in a ferromagnetic material originates from the hyperfine field $\vec{B}_{HF}$. This represents (hyperfine) interaction of the nuclear magnetic moment with magnetic fields originating from the spin and orbital moments of the surrounding electrons, and can be expressed by a sum of three terms [1]:

$$\vec{B}_{HF} = \vec{B}_{HF,Dipolar} + \vec{B}_{HF,Orbital} + \vec{B}_{HF,CF} \qquad (7.14)$$

The dipolar field $\vec{B}_{HF,Dipolar}$ originates from the dipolar interaction between the nuclear magnetic moment and the spins outside the nucleus. It is dominated by the interaction of the nuclear magnetic moment with the spin density of the d shells at the same atom. The orbital field $\vec{B}_{HF,Orbital}$ is related to the non-quenched orbital moments of the valence electrons. In $3d$ metals, the most important term is the Fermi-contact field $\vec{B}_{HF,CF}$, which arises from the spin polarization of all electrons with a certain probability density at the nucleus. These electrons are mainly s electrons, which can be further categorized as the closed-shell s electrons and valence s electrons. The Fermi-contact interaction term consists of three terms [1]:

$$\vec{B}_{HF,CF} = \vec{B}_{HF,Core} + \vec{B}_{HF,Cond} + \vec{B}_{HF,Transferred} \qquad (7.15)$$

The term $\vec{B}_{HF,Core}$ represents the core polarization due to exchange interactions of the inner s electrons with the on-site magnetic moments of the $3d$ electrons. The term $\vec{B}_{HF,Cond}$ relates to the spin polarization of the conduction s electrons with the on-site magnetic moment of the nucleus itself. The polarized d orbitals of the nearest neighbour atoms hybridize with the valence s orbitals, which causes polarization of the valence s electrons. This polarization contributes to the hyperfine field, and this is termed as the transferred hyperfine field $\vec{B}_{HF,Transferred}$. The magnitude of the first two terms in Eqn. 7.15 is proportional to the on-site magnetic moment, whereas the contribution of the number and the magnetic moments of the nearest neighbour atoms determine the hyperfine field $\vec{B}_{HF,Transferred}$.

The NMR spin-echo signals are usually much larger for ferromagnetic materials in comparison to diamagnetic and paramagnetic materials. There are two types of enhancement in ferromagnetic materials, the transmitting and the receiving enhancement [1]. The RF field induces oscillations in the electronic moments of the 3d metals. Such oscillations affect the hyperfine field due to their coupling via the hyperfine interaction and result in an RF component of the hyperfine field. The actual RF field experienced by the nucleus $\vec{B}_{1,eff}$ is the sum of the applied RF field $\vec{B}_{RF,Applied}$ and the RF field induced by the oscillating electronic moments. This latter field causes transmitting enhancement, and the enhancement factor η is expressed as [1]:

$$\eta = \frac{B_{1,eff}}{B_{RF,Applied}} \qquad (7.16)$$

The receiving enhancement represents the amplification of the NMR signal by the hyperfine interaction. The precession of a resonant nucleus causes a corresponding coherent motion of the electronic moments. This gives rise to a much larger signal in the probe coil than the nuclear induction itself. This enhancement is proportional to the magnetic permeability of the electronic origin, and the enhancement can be expressed as the ratio $\vec{B}_{HF}/\vec{B}_{Restoring}$ [1]. Here $\vec{B}_{HF}$ is the hyperfine field and $\vec{B}_{Restoring}$ is the effective restoring field that pins the electronic moment along their original rest orientation. The maximum sensitivity for ferromagnetic materials in NMR measurements is obtained without the application of an external magnetic field. Thus, zero-field NMR measurements are preferred in studies of ferromagnetic materials. Furthermore, a measurement of the enhancement factor at a local scale can provide information on the magnetic stiffness and the restoring field. We note here that standard magnetometry measures the average magnetic anisotropy or stiffness.

In general, in a ferromagnetic material, the magnitude of the enhancement is determined by the local electronic moment stiffness related to anisotropy fields, coercive fields, exchange bias fields and in the case of a layered structure interlayer coupling fields. In an externally applied field, η is expressed as [1]:

$$\eta = \frac{B_{1,eff}}{B_{RF,Applied} + B_{RF,Anisotropy}} - 1 \qquad (7.17)$$

Here $B_{RF,Applied}$ is the externally applied field and $B_{RF,Anisotropy}$ is directly related to anisotropy, coercivity, and exchange coupling fields. Thus a local measurement of anisotropy, coercivity, and exchange coupling fields is possible by measuring η as a function of the external field strength and direction.

Experimental techniques to measure spin–lattice relaxation time

Spin–lattice relaxation time T_1 can be measured similarly, but this time a 180^0 pulse is applied first. This pulse will induce the magnetization initially aligned along the z direction (denoted by $\hat{z}$) to rotate to -z (denoted by $-\hat{z}$) direction. It will then relax back to $\hat{z}$ but without any precession with a time constant T_1. The magnetization $\vec{M}$ as a function of time τ after the application of the 180^0 pulse can be expressed as:

$$\vec{M}(\tau) = \hat{z} M_Z(\tau) = \hat{z}(1 - 2e^{-\tau/T_1}) \qquad (7.18)$$

A 90^0 pulse is then applied at a time τ after the 180^0 pulse to rotate the magnetization to x–y plane. The magnetization then undergoes a free induction decay with an initial amplitude $M_z(\tau)$. Thus the spin–lattice relaxation time T_1 can be obtained experimentally by measuring the decay of free induction signal as a function of the time delay between the 180^0 pulse and 90^0 pulse.

Information obtained from NMR experiments

There are different types of NMR measurements which include frequency or field scan measurements, measurements of the enhancement, and measurements of the relaxation rates. Varieties of information can be obtained from such measurements:

1. NMR spectrum obtained by the frequency or field sweep displays the distribution of hyperfine fields within a sample. From a study of the local hyperfine field distribution, information can be obtained about the crystallographic phases (for example, in a Co sample whether it is in fcc, bcc or hcp structure), degree of short-range order or disorder in alloys and compounds, the interface characteristics (interface roughness, topology and concentration profile), and strain in thin films.
2. Information on local magnetic anisotropies can be obtained by measuring the change in the enhancement factor while varying the strength and direction of an applied magnetic field.
3. The line shapes are quite complex unlike typical Gaussian line shapes. In the case of anisotropic contributions to the hyperfine field, for example, in hcp Co, quadrupole interactions with electric field gradients split the nuclear energy levels. This causes a broadening of the resonance line. Such splitting can be studied via the spin–spin relaxation rates, which will exhibit modulation of the exponential decay.

Fig. 7.4 shows ^{59}Co spin-echo spectra of Cu-covered Co thin films grown by molecular beam epitaxy on Cu(111) single-crystal substrates [1,3]. These thin Co films are in the thickness ranging between $6 \mathring{A}$ and $50 \mathring{A}$, corresponding to 3 monolayers (ML) and 25 ML, respectively. Neither pure face centred cubic (fcc) nor pure hexagonal close packed (hcp) layers were found in the whole Co thickness range under investigation [1]. An fcc stacking is favoured in the initial growth of Co on Cu(111), whereas a predominantly hcp stacking is observed for larger Co thicknesses.

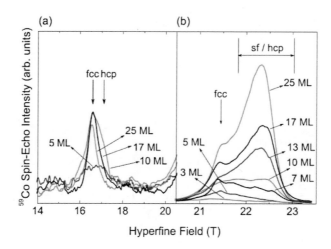

FIGURE 7.4 ^{59}Co spin-echo spectra of Cu-covered Co thin films grown by molecular beam epitaxy on Cu(111) single-crystal substrates. (a) Interface part (× 10); (b) bulk part. The terms sf refers to stacking fault environments, while hcp and fcc refer to hexagonal-closed-packed and face-centred-cubic Co environments, respectively. (Adapted from reference [3] with permission from Springer Nature.)

A clear NMR resonance is observed in the hyperfine field range of 16.0–17.5 T in films thicker than 5 ML. This corresponds to Co in an environment with three nearest Cu neighbours, which indicates the existence of a well-defined Co/Cu(111) interface. These spectra were measured in a zero applied field at T = 1.5K.

7.2 Electron Paramagnetic Resonance

Electron paramagnetic resonance (EPR), sometimes also called electron spin resonance (ESR), is an analogous experiment to NMR where one deals with electron spin instead of nuclear spin. The presence magnetic moment of an unpaired electron in magnetic solids forms the basis of EPR techniques. In this technique, the hierarchy of interactions of this magnetic moment with its immediate and the more distant crystal environment is exploited to gather information on a material under study. The EPR technique is thought to be an extension of the well-known Stern-Gerlach experiment. Zavoisky observed the first peak in the electron paramagnetic resonance in early 1940 in magnetic solids like $MnSO_4$ and $CuCl_2 2H_2O$. The term electron paramagnetic resonance (EPR) was introduced by H. E. Weaver of Varian Associates as a term that would account for contributions from orbital as well as spin angular momentum, and this "EPR" designation is now widely used [4].

The precession frequencies associated with electrons are much higher. This is because the magnetic moment of the electron is much higher than that of nuclei. The electromagnetic radiation in EPR experiments lies in the microwave region.

Principles of EPR and experimental details

The technique of the EPR experiment can be understood in a simple way in terms of a paramagnetic centre with an electron spin of S =1/2. Examples of such paramagnetic centres in materials can be ions of transition metals, free radicals, trapped electrons, and atoms, etc. In an external magnetic field $\vec{B}$ the interaction with the electron depends on the electron magnetic moment $\vec{\mu}_e$ associated with the spin and the direction of the spin:

$$E = -\vec{\mu}_e.\vec{B} = g\vec{\mu}_B\vec{S}.\vec{B} \tag{7.19}$$

where g is electron g-factor, and μ_B the Bohr magneton number. If the magnetic field is directed along the z direction, then this equation becomes:

$$E = g\mu_B B S_Z \tag{7.20}$$

Now for an electron with spin S = 1/2 there are two possible energy states often labelled as up(↑) or 1/2 and down(↓) or -1/2, which designate the orientation of S_Z parallel and antiparallel with respect to the applied magnetic field $\vec{B}$.

$$E_\pm = \frac{1}{2}g\mu_B B \tag{7.21}$$

The magnetic field $\vec{B}$ thus provides magnetic potential energy and splits the electron energy state by an amount proportional to the magnitude of $\vec{B}$ (Fig. 7.5). These are sometimes termed the Zeeman energies. If the material under study is now subjected to an additional oscillating magnetic field $\vec{B}_1$ (say, in the form of electromagnetic radiation) perpendicular to $\vec{B}$, a transition from one spin state to the other is possible at an appropriate frequency known as resonance frequency ν_0. The energy associated

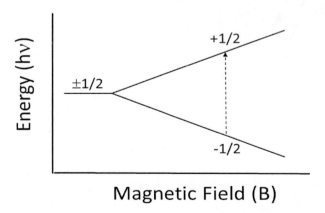

FIGURE 7.5 Schematic presentation showing the split of the electron energy state in varying magnetic field.

with this transition can be expressed in terms of the external magnetic field $\vec{B}$, the electron g-factor g, and the Bohr magneton μ_B, and the condition of resonance is expressed as [5]:

$$\Delta E = E_+ - E_- = h\nu_0 = \mu_B g B \tag{7.22}$$

Thus, if a paramagnetic sample is shined with appropriate electromagnetic radiation, the absorption of electromagnetic radiation will cause a transition from the low energy state to the higher energy state of the electron at the resonance frequency $\nu = \Delta E/h$. In a magnetic field of about 0.35 T readily available in standard laboratories, the resonance frequency ν lies in the so-called X-band between 9 and 10 GHz. The resonance condition is determined by linearly varying the magnetic field over a small interval in the vicinity of the resonance value B_0. The change of the absorbed energy with the external magnetic field is recorded as the signal of a so-called EPR spectrometer.

The samples of real materials contain a large number of spins, which in the constant magnetic field and thermal equilibrium would be distributed between the two allowed energy states. The Boltzmann distribution describes the relative population of two levels [5]:

$$N_+/N_- = exp(\Delta E/k_B T) \tag{7.23}$$

where k_B is Boltzmann constant and T is the temperature of the sample. The population of the lower energy state, however, is larger than that of the upper one. This leads to a larger number of transitions to the upper level, and, in turn, to energy absorption by the sample. This phenomenon forms the basis for the most widely used experimental technique, the so-called continuous wave EPR experiment. Here the paramagnetic sample is irradiated by microwaves with a frequency that is fixed by the microwave source, and the external magnetic field is varied to get the resonance condition for the microwave absorption. The variation of microwave absorption observed in such an experiment as a function of the external magnetic field produces the EPR spectrum.

Fig. 7.6 is a schematic presentation of a setup for the EPR spectroscopy experiment. An EPR spectrometer consists of four basic parts: (i) source system with typically used microwave frequency ranges are L-(4 GHz), X-(9 GHz), and Q-(35 GHz) bands, (ii) microwave cavity system, (iii) magnet system, and (iv) microwave detector.

The sample is housed in a resonance microwave cavity with a quality factor typically varying between 2000 and 20000. The high-quality factor of the microwave cavity enhances the detection capability of the weak EPR signal. This cavity is linked to a microwave bridge composed of a microwave source, an attenuator, and a detector. The magnet system typically supplies fields up to 2 T. The frequency

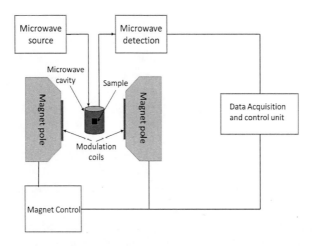

FIGURE 7.6 Schematic presentation of an EPR experimental setup.

of microwave radiation is usually kept fixed and the applied magnetic field is varied to observe resonance absorption. In addition to the static magnetic field provided by an electromagnet or superconducting magnet, a set of modulation coils is also added to provide a small oscillatory field at the sample. This superposed oscillatory field enables synchronous detection of the EPR signal. A phase-sensitive detection technique is often employed to improve the experiment further. In this technique, the oscillatory magnetic field is driven by a signal produced by a lock-in amplifier. The same lock-in amplifier also receives the microwave signal, and any part of the signal with a frequency other than that of the oscillatory magnetic field is rejected. In this way, the noise signals in the EPR experiments can be reduced significantly.

EPR lineshape and relaxation phenomena

The concept of the continuous wave (CW) EPR experiment can be further simplified with the introduction of a magnetization vector $\vec{M}$. In a sample under consideration, a collection of electron spins has a magnetic moment $\vec{M}_0$ in an applied magnetic field $\vec{B}_0$, and that is proportional to $\vec{B}_0$. In thermal equilibrium, the magnetization is parallel to the direction of the applied magnetic field, and it represents the difference in population among the spin states, whose degeneracy has been lifted by the applied magnetic field. There are only two spin states in the case of spin $S = 1/2$. With the external filed $\vec{B}_0$ applied along the z direction, the corresponding equilibrium magnetization is expressed as [5]:

$$M_z = \frac{Ng^2\mu_B^2 B_0}{4k_B T} \tag{7.24}$$

Here $N = N_+ + N_-$ is the total number of spins in the sample.

In a sample subjected to a microwave field, the spin-flipped population relaxes back to the thermal equilibrium state after excitation by microwave absorption. The characteristic time to return to the equilibrium magnetization is defined as the relaxation time. This characteristic time is described in terms of a single exponential decay.

The relaxation time with the magnetization along the direction of the applied magnetic field $\vec{B}_0$ is known as the longitudinal relaxation time T_1. This longitudinal relaxation in a sample is accompanied by a change of the energy of the spin system. In this relaxation process, the thermal motion is the source and sink of energy exchange. Thermal motion in solids is usually described by phonons, which are quanta of lattice vibrations. Longitudinal relaxation takes place in a solid due to absorption or stimulated emission of phonons. This longitudinal relaxation is often called spin–lattice relaxation because the coupling between the spin system and the lattice is required. While the phonon spectrum spans a wide range of frequencies, only those thermal fluctuations with frequencies that match the EPR frequency are important here. There are several mechanisms in solid phases involving the interaction of the spin system with phonons, namely direct, Raman, and Orbach processes. Since the electron moment is more strongly coupled to the crystal lattice than the nuclei, the spin–lattice relaxation times in EPR are quite short compared to those in NMR.

There is another characteristic time, namely the transverse relaxation time T_2. This is the relaxation time in the plane perpendicular to the static magnetic field $\vec{B}_0$. In the presence of the microwaves with a magnetic field amplitude $\vec{B}_1$ perpendicular to the field $\vec{B}_0$ there exist components of magnetization M_x and M_y that are otherwise zero in the absence of a microwave field. In contrast to longitudinal relaxation, there is no exchange of energy with the lattice in the transverse relaxation. Spin flips caused by mutual interactions within the ensemble of spins in the sample play the lead role in the transverse relaxation and thus is often termed as spin–spin relaxation.

The longitudinal and transverse relaxation processes with times T_1 and T_2 determine the lineshape of the spectrum of the two-level system. This phenomenon is elaborated with the help of Bloch equations that describe the changes in the spin ensemble with time under the influence of magnetic fields $\vec{B}_0$ and $\vec{B}_1$. We define a magnetization vector $\vec{M}$ in a coordinate system, which is rotating about the z-axis with an angular velocity $w_0 = \gamma_e B_0$ to match with the resonance frequency. Let the microwave field $\vec{B}_1$ be applied along the x-axis in the rotating coordinate system. This leads to [5]:

$$\frac{d\hat{M}_x}{dt} = (w_0 - w)\hat{M}_y - \frac{\hat{M}_x}{T_2} \tag{7.25}$$

$$\frac{d\hat{M}_y}{dt} = (w_0 - w)\hat{M}_x - \frac{\hat{M}_y}{T_2} + w_1 M_z \quad (7.26)$$

$$\frac{dM_z}{dt} = -w_1 \hat{M}_y - \frac{M_z - M_0}{T_1} \quad (7.27)$$

Here $\hat{M}_z = M_z$ since transverse magnetization does not depend on the frame, and M_0 is equilibrium magnetization. The spin precession will reach a steady state once the microwave field $\vec{B}_1$ is on for a sufficient time. In this situation, all derivatives are equal to zero and the stationary solution for the constants w_0 and w_1 is given by [5]:

$$\hat{M}_x = \frac{(w_0 - w)w_1 T_2^2}{1 - (w_0 - w)^2 T_2^2 + w_1^2 T_1 T_2} M_0 \quad (7.28)$$

$$\hat{M}_y = \frac{w_1 T_2}{1 - (w_0 - w)^2 T_2^2 + w_1^2 T_1 T_2} M_0 \quad (7.29)$$

$$M_z = \frac{1 + (w_0 - w)^2 T_2^2}{1 - (w_0 - w)^2 T_2^2 + w_1^2 T_1 T_2} M_0 \quad (7.30)$$

The M_x component of transverse magnetization is in phase with the microwave field $\vec{B}_1$ and is proportional to the dispersion mode of the EPR signal. The M_y component of transverse magnetization is 90° out of phase with the microwave field $\vec{B}_1$ and is proportional to power absorption. Apart from the phases, Eqns. 7.28–7.30 also describe the shape of the signal in a CW-EPR experiment. Fig. 7.7 shows the characteristic line shapes of dispersion and absorption modes in a CW-EPR experiment as a function of the frequency w.

When the microwave field $\vec{B}_1$ has a small value, the quantity $w_1^2 T_1 T_2 \ll 1$ and the energy absorption can be represented by a Lorentzian lineshape:

$$g(w) = \frac{T_2}{\pi} \frac{1}{1 + T_2^2(w - w_0)^2} \quad (7.31)$$

This line shape has a width $2/T_2$ at the half-height, and this means the width is influenced by the transverse relaxation time. With the increase in microwave field $\vec{B}_1$, at a certain power level of microwave irradiation the spin–lattice relaxation may not be fast enough for sustaining the population difference necessary for measuring the full amplitude of the resonance signal. As a result, the signal level gets reduced, and this phenomenon is known as saturation.

The highest intensity of the absorption mode at the resonance, i.e., $w = w_0$ reduces as $(1 + w_1^2 T_1 T_2)^{-1}$. This effect, however, is smaller at the flanks of the absorption line. The saturation influences the steady-state resonance lineshape by suppressing the central part relative to the wings. This results in an apparent broadening of the spectrum. Since the relaxation time T_1 is there in the equation for power absorption, the experimental measurement of EPR saturation in CW-EPR experiments can be used for the determination of T_1.

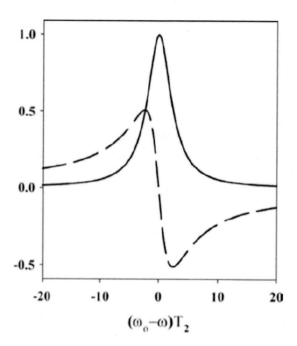

FIGURE 7.7 Schematic presentation of characteristic line shapes of dispersion and absorption modes in CW-EPR experiment as a function of the frequency w. Dashed (solid) line presents the dispersion (absorption) mode of the magnetic resonance transition. (From reference [5] with permission from Springer Nature)

Spin-Hamiltonian in EPR experiments

The spin-Hamiltonian operator in EPR experiments is written as:

$$\hat{H} = g\mu_B B \hat{S}_Z \qquad (7.32)$$

Here μ_B is Bohr magneton and g is having a free electron g-value is 2.00232. The magnetic component of the microwave electromagnetic field induces magnetic dipole transitions between the energy levels in the material under study, when the energy of the microwave photon matches the separation of the energy level $\Delta E = \Delta m_J g \mu_B B$. Here Δm_J represents the selection rule in an EPR transition and is given by [2]:

$$\Delta m_J = \pm 1 \qquad (7.33)$$

This is the dipolar selection rule since magnetic dipolar transitions are being induced here. The g-factor considered above is, in general, a tensor,

If the interaction of an electron with a magnetic field is the only effect present in a material, then the EPR spectra of a liquid material would consist of a single EPR line and the spectra of "powder" solid three lines corresponding to the principal directions of the generalized g-tensor. The only useful information that can thus be obtained from these spectra is the g-factor, which is rather limited information. However, there are fortunately other effects in a material that can produce spectra rich in line components. One such effect is the interaction of the electron spin magnetic dipole with the nuclei in its vicinity. We have already studied earlier that many nuclei have an intrinsic spin angular momentum and the spin quantum number I of such nuclei has one of the values $1/2, 1, 3/2, 2, ...$, etc. There is a corresponding multiplicity of nuclear spin states given by $(2I+1)$. As in the case of the electron, there will be a magnetic moment associated with the nuclear spin.

The interaction of an unpaired electron and a magnetic nucleus in a material is known as nuclear hyperfine interaction. The intrinsic spin angular momentum $I\hbar$ produces a hyperfine field that adds on B_0 and that gives rise to a new resonance condition. The hyperfine field arises due to the non-zero probability of an unpaired electron being found at the nucleus. Such a probability is more for electrons in s orbitals with isotropic wave functions. The mutual magnetic interaction between the nuclear dipoles and the unpaired electron gives rise to the anisotropy of the hyperfine interaction. The selection rule in EPR experiments is given by a combination of Eqn. 7.33 and the following one:

$$\Delta m_I = 0 \tag{7.34}$$

The last selection rule arises because the microwave frequencies cannot induce a transition between the nuclear levels.

The hyperfine coupling between electron and nucleus in a solid introduces a term $A\tilde{I}.\tilde{J}$ in the Hamiltonian. The term A is much smaller in comparison with $g\mu_B B$ for magnetic fields $\vec{B}$ available in the laboratory in an EPR experiment. So only the components $\vec{J}$ parallel to $\vec{B}$ are important, and the energy levels in a magnetic field are expressed as:

$$E = g\mu_B B m_J + A m_I m_J \tag{7.35}$$

Here the energy of the nuclear spin in the applied magnetic field is also ignored, since that is even smaller than the hyperfine energy. It can be shown using the selection rule that the resonances occur at [2]:

$$h\nu = [g\mu_B B(m_J+1) + Am_I(m_J+1)] - [g\mu_B B m_J + A m_I m_J] = g\mu_B B + A m_I \tag{7.36}$$

Each EPR line splits up into $2I+1$ hyperfine line. For a nucleus such as ^{1}H, $I = 1/2$, and the nuclear hyperfine interaction splits each of the electron Zeeman energy levels into two levels.

Crystal field splitting often takes place in GHz frequency range, hence EPR experiments can probe electronic spins in a useful range. Thus EPR experiments

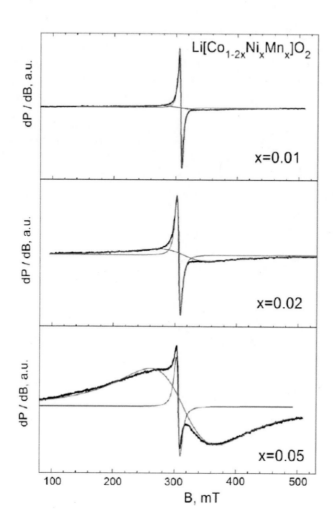

FIGURE 7.8 The X-band continuous wave (CW) and HF-EPR spectra of the Mn and Ni ions in $LiCo_{1-2x}Ni_xMn_xO_2$ measured at 103 K. The bold black line represents the experimental EPR spectra. The thin black lines correspond to the simulated signal due to low-spin Ni^{3+} and non-isolated Mn^{4+} ions. (From reference [6] with permission from ACS Publishing.)

can provide detailed information on the crystal field effects in paramagnetic salts. The crystal field effect is often anisotropic and the EPR line moves with the rotation of the magnetic field with respect to the crystal axis.

Fig. 7.8 shows the X-band continuous wave (CW) and HF-EPR spectra of the Mn and Ni ions in $LiCo_{1-2x}Ni_xMn_xO_2$ compound measured at 103 K [6]. The bold black line represents the experimental EPR spectra. The thin black lines correspond to the simulated signal due to low-spin Ni^{3+} and non-isolated Mn^{4+} ions. The resonances in EPR can be very sensitive to the atomic positions in a molecule, and the technique is very useful for the study of chemical reactions and particularly of free radicals.

The term "free radical" represents atoms or fragments of molecules with an unpaired electron, and they are often very reactive. Free radicals are of great interest from the environmental and biological points of view.

7.3 Ferromagnetic Resonance

Ferromagnetic resonance (FMR) uses the same equipment as EPR to study interactions between magnetic moments or spins at lattice sites via exchange interactions, which characterize crystalline ferromagnetic materials. In FMR experiments the frequency of the resonant mode depends on the direction and value of the applied magnetic field and also on the interactions internal to the ferromagnetic material. These internal interactions are the energy of the demagnetizing field (Appendix C) and the magnetocrystalline anisotropy energy (see Chapter 3.7). Similar to EPR experiments, to detect the resonance an FMR spectrum is recorded by varying the magnetic field while keeping the microwave frequency fixed. The parameters characterizing the magnetocrystalline anisotropy of a ferromagnetic sample can be determined by investigating the variation of resonance fields with the change in direction of the external field in different crystallographic planes of the sample. On the other hand, the analysis of the line widths can be used to determine the characteristic times for relaxation of the magnetization. The inherent assumption here is the existence of a uniform resonance mode in which the precession of all the spins in the system is in phase. However, there can also be excitation of non-uniform modes, namely spin waves, and their study would provide other important information including the value of the exchange parameter.

Principles of FMR and experimental details

In principle, FMR is similar to EPR. However, the parameters determining the positions and widths of the resonance lines can differ significantly between the two techniques because of the characteristic properties of ferromagnetic and ferrimagnetic materials. FMR occurs at relatively high field values in comparison to EPR. The exchange interaction energy between unpaired electron spins in ferromagnetic materials gives rise to the line narrowing. Before proceeding further we briefly recapitulate the characteristic properties of a ferromagnetic material that are important in the present context:

1. The strong exchange interaction between unpaired electrons located at the lattice sites of a ferromagnetic material tends to align their spins or magnetic moments. This results in magnetization much stronger than that in paramagnetic materials.

2. The presence of a demagnetization field $\vec{B}_D$ is dependent on the shape of the ferromagnetic sample. This field originates from the magnetic fields created by all the magnetic dipoles in the ferromagnetic materials and has a direction opposite to the direction of the magnetization $\vec{M}$ inside the sample (Appendix C). This demagnetization field is uniform in a uniformly magnetized ferromagnetic sample with an ellipsoidal shape, and can be expressed as [7]:

$$\vec{B}_D = -\tilde{\vec{N}}.\vec{M} \qquad (7.37)$$

Here $\tilde{\vec{N}}$ is the tensor of the coefficients of the demagnetizing field and is also known as the shape factors. The principle axes $\{x, y, z\}$ of the tensor $\tilde{\vec{N}}$ are the axes of the ellipsoid, and its trace is equal to 1. In this case, the energy density associated with the dipolar interactions is given by [7]:

$$F_D = \frac{1}{2}\mu_0 \vec{M}.\tilde{\vec{N}}.\vec{M} \qquad (7.38)$$

This energy is minimum when the magnetization lies in the direction of the largest dimension of the sample.

3. The combination of spin–orbit coupling and Coulomb interaction generates magnetocrystalline anisotropy energy in ferromagnetic materials. This energy can be described phenomenologically with the assumption that it depends only on the direction of $\vec{M}$ relative to the crystal axes. This direction can be identified by the angles (θ, ϕ) in a system of spherical coordinates or by the components $(\alpha_1, \alpha_2, \alpha_3)$ of the unit vector $\vec{m}$ in the direction of $\vec{M}$. In the latter case, the magnetocrystalline anisotropy energy is described by a polynomial in $\{\alpha_i\}$ only involving even powers [7]. The magnetocrystalline anisotropy energy density for a ferromagnetic material with cubic symmetry is expressed in terms of anisotropy constants Ks as [7]:

$$F_K = K_4(\alpha_1^2\alpha_2^2 + \alpha_2^2\alpha_3^2 + \alpha_3^2\alpha_1^2) + \acute{K}_4\alpha_1^2\alpha_2^2\alpha_3^2 \qquad (7.39)$$

If the sample is also subjected to uniaxial deformation in the z direction, then a term with the form $K_2 sin^2\theta$ is added to the above equation, where θ is the angle between $\vec{M}$ and z direction.

The minimization of the total energy, which is the sum of the exchange, demagnetization, and magnetocrystalline anisotropy energies, would result in the formation of magnetic domains in a ferromagnetic material in the absence of an external magnetic field. The domains have their magnetization oriented in different directions, and this domain structure of a ferromagnet is such that the total magnetization of the sample is generally very weak or zero. The configuration of the domains changes with the application of an external field. The total magnetization increases with the increase in the external field and tends to reach a saturation value when the sample attains a single domain structure.

We now consider a single domain ferromagnetic sample and assume a uniform resonant mode. All the magnetic moments in the sample rotate in phase while conserving the amplitude of the magnetization. We further assume that the energy of the demagnetization field and of the magnetocrystalline anisotropy is negligible in comparison to the exchange energy. In the absence of any dissipation or relaxation, the rate of variation of the magnetization $\vec{M}$ in the sample in a uniform external field $\vec{B}$ can be expressed as [7]:

$$\frac{1}{\gamma}\frac{d\vec{M}}{dt} = \vec{M} \times \vec{B} \qquad (7.40)$$

Here $\gamma = e/2m_e$ is the gyro-magnetic ratio.

The magnetization $\vec{M}$ precesses around $\vec{B}$ with an angular velocity $\omega_0 = g\mu_B B/\hbar$. The angle between $\vec{M}$ and $\vec{B}$ can be changed by the application of a second field $\vec{B}_1$ perpendicular to $\vec{B}$ and rotating with an angular velocity $\omega_0 = \gamma\mu_B B$. In an FMR experiment, this second field $\vec{B}_1$ is a microwave field with linear polarization. This field, however, can be considered as a sum of two circularly polarized components rotating in opposite directions, with the efficient component being the one that rotates in the same direction as $\vec{M}$. As in the case of the EPR experiment the resonance frequency is $\omega_0/2\pi$. If the frequency used in the FMR spectrometer is w, then the resonance field can be expressed as [7]:

$$B_{res} = \frac{\omega}{\gamma\mu_B} \qquad (7.41)$$

In the presence of demagnetizing field and magnetocrystalline anisotropy, the precession of the magnetization takes place around a direction that is not necessarily the direction of $\vec{B}$. The volumetric energy density of the sample F now needs to be considered to determine the angular velocity ω, which is not the same as ω_0. The volumetric energy density F depends on the directions of $\vec{B}$ and $\vec{M}$ relative to the axes of the crystal, which are identified by the angles (θ_B, ϕ_B) and (θ, ϕ) (Fig. 7.9). The following notations are used to designate the crystallographic directions and planes [7]: a direction parallel to a vector with components (l, m, n) is denoted by $[l, m, n]$, and a plane perpendicular to this vector by (l, m, n). The values of θ_{eq} and ϕ_{eq} defining the direction of equilibrium magnetization can be determined by minimizing $F(\theta, \phi)$. If the angle of precession around the direction of equilibrium magnetization is small, then the angular precession frequency ω can be expressed

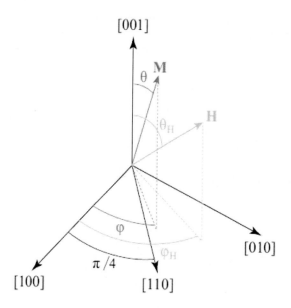

FIGURE 7.9 Schematic presentation of the directions of magnetization and magnetic field vectors with respect to the crystallographic axes. (From reference [7] with permission from Springer Nature.)

as [7]):

$$\omega^2 = \frac{\hbar^2}{g^2 \mu_B^2 M^2 \sin^2 \theta_{eq}} \left[\frac{\partial^2 F}{\partial \theta^2} \frac{\partial^2 F}{\partial \phi^2} - \left(\frac{\partial^2 F}{\partial \theta \partial \phi} \right)^2 \right] \quad (7.42)$$

The partial derivatives are calculated for $\theta = \theta_{eq}$, $\phi = \phi_{eq}$.

We now consider a ferromagnetic sample with cubic symmetry and having the shape of an ellipsoid of revolution and which is subjected to a uniaxial stress in the [001] direction. The energy density F for this sample can be written as [7]:

$$F = -\vec{M}.\vec{B} + \frac{1}{2}\mu_0 \vec{M}.\tilde{N}.\vec{M} + K_4(\alpha_1^2 \alpha_2^2 + \alpha_2^2 \alpha_3^2 + \alpha_3^2 \alpha_1^2) + K_2 \sin^2 \theta \quad (7.43)$$

The first term in the above Eqn. 7.43 represents the energy of the interaction with the external field $\vec{B}$, the second term represents the energy of the demagnetization field (Eqn. 7.38), the third term represents the anisotropy energies due to the cubic symmetry of the sample (Eqn. 7.39), and the last term is due to the energy of uniaxial constraint. Components of magnetization (α_1, α_2, α_3) can be expressed as a function of (θ, ϕ). The expression for the resonance frequency and resonance field can now be obtained with the help of Eqns. 7.42 and 7.43 for a given direction of the magnetic field. If the $\{x, y, z\}$ axes of the ellipsoid sample are parallel to the crystal axes, then for the magnetic field applied along z direction, the following equation is obtained [7]:

$$\omega^2 = \gamma^2 \left[B_{res} + \mu_0(N_x - N_z)M + 2\mu_0 \frac{K_2}{M} + \mu_0 \rho \frac{K_4}{M} \right]$$
$$\times \left[B_{res} + \mu_0(N_y - N_z)M + 2\mu_0 \frac{K_2}{M} + \mu_0 \rho \frac{K_4}{M} \right] \quad (7.44)$$

Here B_{res} stands for the resonance field and $\gamma = g\mu_B/\hbar$. The parameter ρ is equal to -4/3 when z is the crystallographic direction [111], and 2 when it is in the direction [100]. In general, the ellipsoid and crystallographic axes do not coincide, and the calculation of the resonance field and the resonance frequency is much more involved. In the case of spherical samples, $N_x = N_y = N_z = \frac{1}{3}$, and from Eqn. 7.44 we can see that there will be no effect of the demagnetization field on the resonance field and resonance frequency.

The basic components of an FMR spectrometer are very similar to that of an EPR spectrometer. But there is a major constraint on the sample in FMR spectroscopy, where excitation of the sample must be homogeneous throughout its volume. This is particularly an issue with metallic samples, where the skin effect does not allow microwaves to penetrate the sample beyond a certain depth. However, this depth is about a micrometer (μm) for microwaves at a frequency of 10 GHz, and hence generally sub-micrometre film samples are quite suitable for FMR spectroscopy. In the standard configuration of an FMR spectrometer, the sample of a few nm thickness and an area of 1–10 mm^2 is placed in the centre of a resonant cavity with a quality factor varying between 2000 and 20000. This cavity is coupled to a microwave bridge consisting of a microwave source, an attenuator, and a detector. Magnetic fields usually up to 2 T can be supplied by an electromagnet. A sinusoidal magnetic field with a frequency around 100 kHz and an amplitude of around 1 mT is superposed on the quasi-static magnetic field. This allows synchronous detection of the signal, and, consequently, the resonance spectrum takes the shape of a Lorentzian derivative. It is evident from Eqn. 7.44 that the resonance field in FMR experiments is a function of the magnetization of the sample. Hence, the magnetization M must be known and is measured independently using various possible magnetometric techniques discussed in Chapter 6. In an FMR experiment itself, the value of the spectrometer frequency is known, and the variation in the resonance field is generally monitored as a function of the direction of $\vec{B}$ in certain crystallographic planes. Using Eqn. 7.44 it is then possible to determine the easy and hard axes of magnetization and anisotropic constants associated with the magnetocrystalline anisotropy.

Apart from the classical FMR spectrometers with resonant cavities, inductive techniques can also be used when the sample produces high-intensity FMR [7]. There are several specific measurement types including those based on strip lines, vector network analyzers (VNA), and pulsed inductive microwave magnetometers (PIMM) [8]. The most widely used electrical FMR instruments are those based on a VNA. The sample is placed on a co-planar waveguide connected to the VNA, and the reflection and transmission coefficients are then measured by the VNA at either fixed frequency while sweeping the magnetic field or sweeping the frequency by keeping the magnetic field fixed. This procedure has the advantage of allowing continuous variation of the microwave frequency over a wide range.

FMR lineshape and relaxation phenomena

In the FMR experiment, the magnetization $\vec{M}$ of a ferromagnetic sample would relax towards its equilibrium direction. The relaxation phenomena do not influence the value of the resonance field but result in the broadening of the resonance line for the uniform mode. To study this relaxation phenomenon, additional terms must be added to Eqn. 7.40 expressing the rate of variation of the magnetization $\vec{M}$. These additional terms represent underlying phenomena of creation, annihilation, and scattering of magnetic excitations, namely the spin waves or magnons. There are several approaches to study these phenomena on a purely phenomenological basis, and amongst these the Gilbert equation is used most widely [7].

$$\frac{d\vec{M}}{dt} = -\gamma \vec{M} \times \vec{B} + \frac{\alpha}{M_S} \vec{M} \times \frac{d\vec{M}}{dt} \qquad (7.45)$$

The dimensionless parameter α is known as the damping factor and M_S is the saturation magnetization. Eqn. 7.45 is used to investigate the relaxation phenomena by correlating them to an effective field proportional to $\frac{d\vec{M}}{dt}$. There is another parameter G, known as the Gilbert coefficient, which is also sometimes used in the literature. It is linked to damping coefficient α by [7]:

$$\alpha = \frac{G}{\gamma \mu_0 M} \qquad (7.46)$$

The Gilbert coefficient G is determined by studying variation of the width of the resonance line in the uniform mode as a function of the microwave frequency. The peak-to-peak width $\Delta H_{PP}(w)$ can be shown to be the sum of two terms:

$$\Delta H_{PP}(w) = \Delta H_{inhom} + \Delta H_{hom} = \Delta H_{inhom} + \frac{2G}{\sqrt{3}\gamma^2 M} w \qquad (7.47)$$

Here γ stands for the gyromagnetic ratio. The first term of the Eqn. 7.47 represents the broadening of the line associated with the magnetic inhomogeneity of the thin-film sample under investigation. This term is strongly influenced by the microstructure of the thin films, but it is not dependent on the microwave frequency.

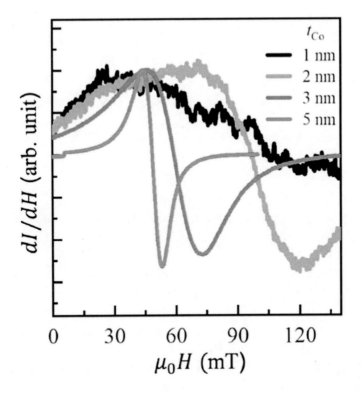

FIGURE 7.10 FMR spectra of the Co thin films of 1, 2, 3 and 5 nm thickness grown onto an SiO_2 (300 nm)/Si substrate. (From reference [9] with Creative Commons Attribution License.)

The second term is a linear function of the microwave frequency and contains the Gilbert coefficient.

Fig. 7.10 shows the FMR spectra of the Co thin films of 1, 2, 3 and 5 nm thickness grown onto an SiO_2 (300 nm)/Si substrate. The resonance field and the linewidth of the FMR spectra changes drastically with the Co thickness, and the FMR signal is barely visible when the thickness of the Co is less than 2 nm.

7.4 Muon Spin Rotation

In the muon spin rotation (μSR) technique, muon spins probe the local magnetic fields in their close vicinity, typically at the atomic scale. The muon is a spin-$\frac{1}{2}$ particle with charge $\pm e$ and mass intermediate between that of the electron and the proton. It carries a magnetic moment, the value of which along with the gyromagnetic ratio also lies between that of electron and proton. A natural source of muons is cosmic rays, and roughly one muon arrives vertically on each square centimetre of the Earth's surface every minute [10]. As we shall see here muons can be quite useful to study the magnetic properties of materials. However, for such experiments a greater intensity of muons is needed than that available from cosmic rays. That necessitated the development of sources involving charge particles. In such muon sources high energy proton beams are fired into a target to produce pions π through the following process:

$$p + p \rightarrow \pi^+ + p + n \qquad (7.48)$$

And then pions decay into muons:

$$\pi^+ \rightarrow \mu^+ + \nu_\mu \qquad (7.49)$$

Here ν_μ is a muon-neutrino. Considering the case of pions, which are at rest in the laboratory frame, to conserve momentum the muon and the neutrino must possess equal and opposite momentum. With the spin of a pion being zero, the spin of a muon must be opposite to the neutrino spin. The spin of a neutrino is aligned antiparallel with its momentum, and this, in turn, implies that the muon spin is aligned similarly. Thus, a beam of 100% spin-polarized muons can be produced by selecting pions, which stop in the target and which are therefore at rest when they decay. This is the method used most commonly for producing muon beams for condensed matter physics research [10, 11].

In a magnetic field $\vec{B}$, the muon spin precesses with angular frequency $w_\mu = \gamma_\mu B$, where $\gamma_\mu = ge/m_\mu$ is the gyromagnetic ratio of the muon. This is known as the Larmor precession, and the field-dependent precession frequencies for the muon lie in the frequency regime intermediate between the microwave frequencies suitable for ESR and radio frequencies suitable for NMR. However, unlike in those resonance techniques, no electromagnetic field is necessary for the muon experiments since the precessing muon can be followed directly. The experiment thus is known as muon spin rotation (μSR).

In an experiment, muons are implanted into a sample under investigation. The muons decay after a time t with a probability proportional to $e^{-t/\tau}$ where $\tau = 2.2$ μs is the lifetime of the muon. It is to be noted here that the muons themselves never emerge from the sample. It is the positrons into which the muons decay that come out from the sample and are detected for obtaining the useful information in μSR experiment. The muon decay is a three-body process [10]:

$$\mu^+ = e^+ + \nu_e + \bar{\nu}_\mu \qquad (7.50)$$

The initial momentum of the stopped muon is zero, and from conservation of momentum, this will be the total vector momentum of the positron e^+ and two neutrinos ν_e and $\bar{\nu}_\mu$. As a result, the energy of the resultant positron will vary depending on how momentum is distributed between the three particles. The decay of muon involves weak interaction and does not conserve parity [10]. This phenomenon causes the emitted positron to emerge predominantly along the

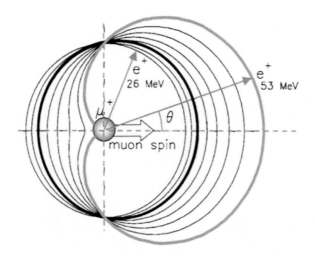

FIGURE 7.11 The angular distribution of the probability of positron emission during muon decay with respect to the initial muon spin direction. (From reference [12] Creative Commons Attribution License.)

direction of the muon spin when it decays. The angular distribution of the probability of positron emissions is shown in Fig. 7.11. The positron energy may vary between 0 and 53 MeV, the mass-energy of muon, and most of the positrons possess enough kinetic energy to reach the detectors outside the sample [12]. The higher the energy of the emitted positron, the more asymmetric will be the positron emission.

In condensed matter physics, one is essentially dealing with various aspects of electrons rather than nuclei, hence the best place for the test particle to be placed is in the electron cloud. The implanted positive muons μ^+ sits away from nuclei in regions of large electron density. It is thus well suited for the investigation of various interesting phenomena in condensed matter physicists, such as magnetism, superconductivity, etc. The negative muons μ^- are in general much less sensitive to experiments in condensed matter physics, since those are implanted close to atomic nuclei.

Muon spin rotation (μSR) experiment

A μSR experimental setup is shown schematically in Fig. 7.12. A muon polarized antiparallel to its momentum is implanted in the sample under investigation. After the decay of the muon, positrons are detected either in the forward detector or in the backward detector. The muon will precess in the presence of a transverse field. The muon will not have time to precess if it decays immediately, and a positron will be emitted preferentially into the backward detector. If the muon lives a little longer, then it will have time to precess. If the muon lives for half a revolution, the resultant positron will be preferentially emitted into the forward detector [10]. The

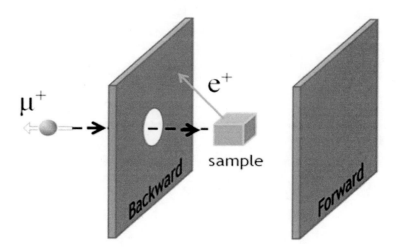

FIGURE 7.12 Schematic diagram of a μSR experimental setup. (From reference [12] Creative Commons Attribution License.)

positron beam associated with an ensemble of precessing muons is like the beam of light from a lighthouse. The number of positrons detected as a function of time in the forward and backward detectors can be represented by the functions $N_F(t)$ and $N_B(t)$, respectively, and these are shown in Fig. 7.13(a). The decay of muons is a radioactive process and these terms $N_F(t)$ and $N_B(t)$ sum up to exponential decay. The time evolution of the muon polarization can thus be studied by examining the normalized difference of these two functions via the asymmetry function $A(t)$, expressed as [10]:

$$A(t) = \frac{N_B(t) - N_F(t)}{N_B(t) + N_F(t)} \quad (7.51)$$

The asymmetry function $A(t)$ is shown in Fig. 7.13(b). This function can be obtained experimentally and has a calculable maximum value A_{max} for a particular experimental configuration, which depends on the initial beam polarization, the intrinsic asymmetry of the weak decay, and the efficiency of the detectors for positrons of different energies, and usually turns out to be around $A_{max} \approx 0.25$ [10]. The function can be normalized to 1, and in that case, it expresses the spin autocorrelation function of the muon, $G(t) = A(t)/A_{max}$, which represents the time-dependent spin polarization of the muon.

It is possible to perform muon spin rotation experiments in two different ways depending on the time structure of the muon beam. The muons arrive at the sample intermittently if the muon beam is continuous (Fig. 7.14(a)). When a muon enters the experiment, it needs to be detected to start a clock. This clock is stopped when the positron is detected in either the forward or backward detector. However, in the event of a second muon arriving before the first one has decayed, then there is

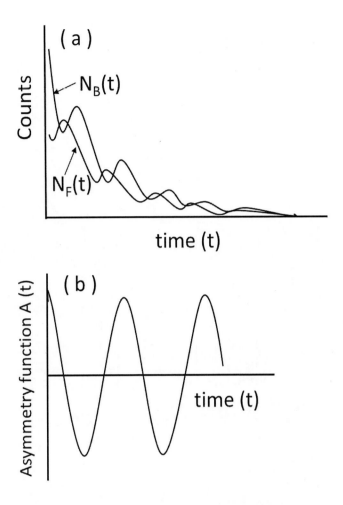

FIGURE 7.13 Schematic presentation of (a) number of positrons collected in backward and forward detectors as a function of the time elapsed since the implantation of muon; (b) asymmetry function A(t) as a function of time.

no way of knowing whether a subsequently emitted positron came from the first or second muon. Hence, this event must be disregarded. This gives rise to the need for sophisticated high-speed electronics and a low-incident muon arrival rate. Alternatively, an electrostatic deflector triggered by the detectors can be used to ensure no muons enter the experiment until the first implanted muon decays [10].

The above experimental complications can be circumvented with a pulsed muon beam of width τ_w (Fig. 7.14(b)). In this arrangement, a large number of muons arrive in a very intense pulse, and there is no need to detect the arrival of each muon. Instead, the detection of positrons is made and each event is timed with respect to the arrival of the pulse. There are several million detected positrons

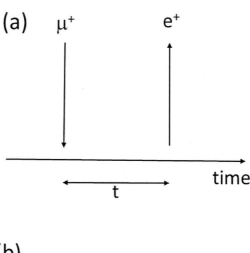

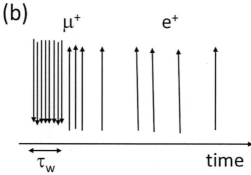

FIGURE 7.14 Schematic presentation of (a) continuous wave (CW) muon beam and (b) pulsed muon beam.

in a typical dataset so that an appreciable number of muons live for 20 μs or longer. The difficulties of detecting long-lived muons with CW muon beams are thus circumvented by using a pulsed source, and those can be detected accurately. The pulsed muon source also has certain drawbacks originating from the finite width τ_w of muon pulse. That causes a slight uncertainty in all the timing measurements, leading to an upper limit on the measured precession frequencies.

CW muon beams are available at the Paul Scherrer Institute in Switzerland and TRIUMF in Canada. On the other hand, KEK in Japan and ISIS, the spallation source at the Rutherford Appleton Laboratory, UK, host pulsed muon sources.

Muons and the study of magnetism

Muons offer an interesting probe to investigate the magnetic properties of materials. In studying a ferromagnet with muons, even a magnetic field does not need to be

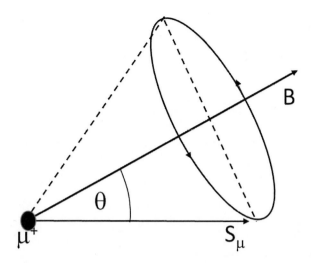

FIGURE 7.15 Schematic presentation of precession of muon spin $\vec{S}_\mu$ with a magnetic field $\vec{B}$ applied at an angle θ.

applied. Implanted muons precess in the internal magnetic field of the ferromagnet and directly yield signals proportional to that magnetic field. The muon behaves like a microscopic magnetometer. Muons are very sensitive to extremely small magnetic fields even down to $\sim 10^{-5}$ T, and hence can be very useful in studying small moment magnetism. With the increase in temperature in the ferromagnetic sample towards its Curie temperature, the frequency of oscillations decreases as the internal field decreases and goes to zero and vanishes at the Curie temperature. No oscillations can be observed above the Curie temperature, except a weak spin relaxation arising from spin fluctuations in the paramagnetic state. μSR is also very useful for studying materials with random or very short-range magnetic order. Muons stop uniformly throughout a sample. As a result, each signal appears in the experimental spectrum with a strength proportional to its volume fraction, and thus μSR is useful for multiphase samples or samples with partial magnetic ordering.

In a μSR experiment, the possible sources of the internal field at the muon site of the sample are the following:

1. The externally applied magnetic field (if it is used).
2. The dipolar and demagnetization fields, which can be calculated from the magnetization.
3. The hyperfine field induced by the applied field.

There exist a small set of possible interstitial sites that the muon usually can occupy. In favourable circumstances, only one of those will be consistent with the observed data. It is necessary to know such a muon site for obtaining quantitative

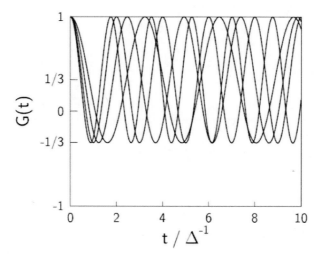

FIGURE 7.16 Schematic presentation of the time evolution of the muon-spin polarization with different values of the magnitude of the local field $|\vec{B}|$ (From reference [10] with permission from Taylor & Francis.)

information from μSR experiments. It is also helpful to look further into some aspects of spin precession to appreciate the ability of the muon in studying randomness and dynamics in magnetism. Let us assume that the local magnetic field at the site of the implanted muon is at an angle θ to the initial direction of muon spin $\vec{S}_\mu$ at the moment of implantation as shown in Fig. 7.15. The muon spin will then precess around the end of a cone of semi-angle θ about the magnetic field (Fig. 7.15). The normalized decay positron asymmetry will be expressed by [10]:

$$G(t) = cos^2\theta + sin^2\theta cos(\gamma_\mu Bt) \tag{7.52}$$

In the event of the local magnetic field being entirely random, then averaging over all directions would result in:

$$G(t) = \frac{1}{3} + \frac{2}{3}cos(\gamma_\mu Bt) \tag{7.53}$$

If the strength of the local magnetic field is represented by a Gaussian distribution of width Δ/γ_μ centred around zero, then a straightforward averaging over this distribution leads to [10]:

$$G(t) = \frac{1}{3} + \frac{2}{3}e^{-\Delta^2 t^2/2}(1 - \Delta^2 t^2) \tag{7.54}$$

This result was first due to an entirely theoretical exercise by Kubo and Toyabe [13]. Fig. 7.16 shows several curves of equation 7.54 for different values of the internal field $\vec{B}$ where the time is measured in units of Δ^{-1}. Initially, all the curves fall from 1 to a minimum value and then increase, but after a short time, they dephase with respect

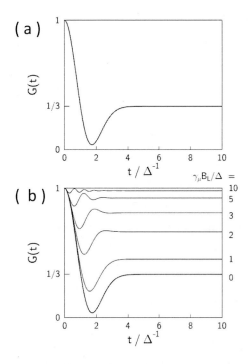

FIGURE 7.17 (a) The Kubo–Toyabe relaxation function as presented in Eqn. 7.54 with its characteristic dip and the subsequent recovery to a value of 1/3. (b) Schematic representation of the effect of a longitudinal magnetic field B_L on the Kubo–Toyabe relaxation function. (From reference [10] with permission from Taylor & Francis.)

to each other. The average of these curves, the Kubo and Toyabe relaxation function (Eqn. 7.54), is then expected to fall from unity to a minimum and ultimately recover to an average value of one-third (Fig. 7.17(a)).

The Kubo and Toyabe relaxation function is often observed experimentally, say, in copper at 50 K with zero applied field [14]. The muon implanted in the copper sample is stationary at this temperature, and it precesses in the field due to the neighbouring nuclear dipoles in copper atoms, which are randomly orientated with respect to each other. This gives rise to a field distribution, and each component of this field has Gaussian distribution of about zero. This, however, is not seen at higher temperatures because of thermally activated hopping of muons, and also at lower temperatures due to the possibility of quantum diffusion. The form of the observed muon-spin time evolution will be affected whenever there is a deviation from such distribution of internal fields. The internal fields, for example, in a magnetic material with a spin-density wave incommensurate with the crystal lattice, will be sinusoidally modulated. The muons will randomly sample this internal field, and in this case, the muon-spin relaxation would follow a Bessel function [10, 15]. This effect is observed, for example, in classic spin-density wave material chromium [16].

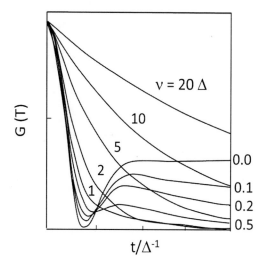

FIGURE 7.18 Schematic presentation of the the relaxation function for a muon for different hopping rates ν. (From reference [10] with permission from Taylor & Francis.)

If there is a small variation from site to site in an otherwise almost uniform static internal field in a sample, different muons will precess at slightly different frequencies and become progressively dephased so that the oscillations in the data will be damped [10]. No oscillations will be observed, however, if the site-to-site variation in the internal field is very significant and thus leading to very large damping. Such effect can also arise due to fluctuations either of the internal fields because of some intrinsic property of the sample, of the position of the muon, or because of muon diffusion. One possible way to distinguish between these different possibilities is to do the measurement in an external magnetic field B_L applied in the longitudinal direction parallel to the initial muon-spin direction. This will modify the Kubo–Toyabe relaxation function as shown in Fig. 7.17(b). There is an increase in the "$\frac{1}{3}$-tail" because the muons now precess in both the internal field of the sample and the applied magnetic field B_L. In Fig. 7.17(b) the longitudinal field values B_L are shown in units of Δ/γ_ν and the time in the units of Δ^{-1}. In the limit of the large applied magnetic field, the muon spin is held constant and does not relax from a value near unity. This method is quite useful to distinguish the effects of inhomogeneous line broadening and fluctuations since they lead to very different behaviours in a longitudinal field.

The effects of muon hopping on the relaxation for different hopping rates ν are shown in Fig. 7.18. The value of the internal field after each hop is taken from a Gaussian distribution about zero with width Δ/γ_ν. The curve for the hopping rate $\nu = 0$ in Fig. 7.18 corresponds to the Kubo–Toyabe relaxation function (expressed in Eqn. 7.54) for zero field. The relaxation of the muon spin is dominated by the hopping process in the case of fast hopping, and the relaxation is exponential. The relaxation rate decreases as the dynamics become faster; this is known as motional narrowing. A very small effect is observed at very short times for slow hopping. However, a large sensitivity to weak hopping is observed in the $\frac{1}{3}$-tail that is observed at longer times. This sensitivity to slow dynamics via the behaviour of the tail of the relaxation function observed at long times enables measurement of dynamics across a very large range of time [10].

A longitudinal magnetic field has a large effect on the muon relaxation if the dynamics are weak, but the effect is much less if the dynamics are fast. A careful combination of zero-field and longitudinal-field experiments thus enables procurement of information on the nature of the internal field distribution. This kind of analysis has been carried out in many materials where both the dynamics and/or magnetic order have their interesting manifestations. A prominent example of such a system is canonical spin-glasses, where dilute magnetic impurities like Mn (or Fe) in a non-metallic matrix like Cu (or Au) couple via an RKKY exchange interaction which alternates in sign as a function of distance. The magnetic impurities are distributed randomly, hence spin-glasses do not show long-range magnetic order. They, however, have built-in frustration due to competition between positive and negative exchange interactions. With decreasing temperature, a slowing down of Mn spin fluctuations and a divergence in the correlation time between Mn spin fluctuations are observed at the characteristic temperature – spin-glass transition temperature. A static component of the local field is observed with muons below the spin-glass transition temperature. This corresponds to some degree of spin freezing, with each Mn spin having its preferred orientation, but at the same time fluctuating around this orientation [10, 17].

Fig. 7.19 shows zero-field muon-spin relaxation function observed in CuMn(1.1%) alloy which undergoes a spin-glass transition below $T_{SG} = 10.8$ K [17]. In the paramagnetic region above the spin-glass transition temperature, the Kubo–Toyabe relaxation function for the nuclear dipolar field of Cu is observed. However, the slowing down of the fluctuation of Mn-moments causes a rapid depolarization of muon spins with the decrease in temperature. In the spin-glass regime below T_{SG} the rate of quick initial damping of $G_Z(T)$ increases gradually with a decrease in temperature. In the spin-glass regime, the nuclear field is decoupled by the larger static field from Mn atoms. Muon-spin relaxation measurements can also be used to follow the spin-glass dynamics above the spin-glass transition temperature and obtain directly the form of the autocorrelation function of the spins.

In an antiferromagnetically ordered lattice if there is one muon site close to each spin, in the absence of hopping there will be a single precession frequency. If there is hopping, the sign of the internal field reverses every time the muon hops. The muon will precess one way and then after hopping precess back again. The oscillations are

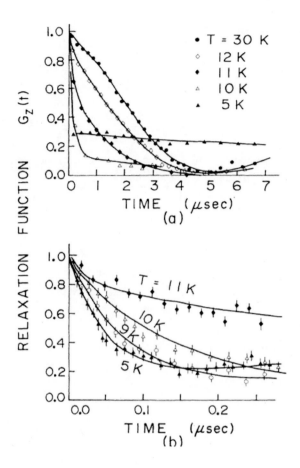

FIGURE 7.19 Zero-field muon-spin relaxation function observed in CuMn(1.1%) alloy spin-glass with transition temperature $T_{SG} = 10.8$ K: (a) Kubo–Toyabe function for nuclear dipolar fields from Cu is observed at $T \gg T_{SG}$, (b) the rate of quick initial damping of $G_Z(T)$ gradually increases with decrease of temperature below T_{SG}. (From reference [17] with permission from APS.)

gradually obliterated with the increase in the hopping rate until no relaxation is possible.

7.5 Mössbauer Spectroscopy

The Mössbauer effect discovered by Rudolf Mössbauer forms the basis of this spectroscopic technique. In this experimental technique, a source contains ^{57}Co nuclei, which decays by electron capture into the excited state of ^{57}Fe nuclei. Subsequently, the majority of the ^{57}Fe nuclei decay rapidly to the $I = \frac{3}{2}$ state, which ultimately decays to $I = \frac{1}{2}$ ground state. This decay releases a 14.4 KeV gamma (γ) ray photon corresponding to a frequency $\nu = 3.5 \times 10^{18}$ Hz, and this γ-ray photon

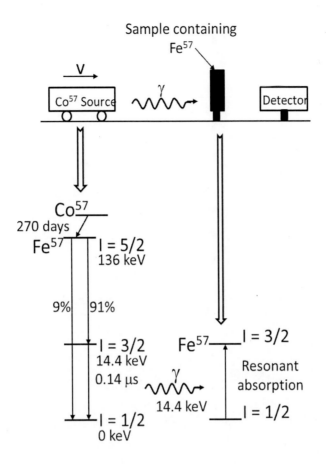

FIGURE 7.20 Schematic representation of the principles of Mössbauer spectroscopy technique

can excite a transition in the ^{57}Fe nuclei of the sample under investigation if it is absorbed in the sample resonantly. However, for this event to take place the energy of the γ-ray photon needs to be matched with the energy gap in the sample.

The Mössbauer effect is very sensitive to small changes in the chemical environment of certain nuclei. Three types of nuclear interactions are possible: (i) isomer shift due to differences in nearby electron densities, (ii) quadrupole splitting due to atomic-scale electric field gradients, (iii) and magnetic Zeeman splitting due to magnetic fields of non-nuclear origin. High energy and extremely narrow line widths of nuclear gamma rays make Mössbauer spectroscopy a highly sensitive technique in terms of energy resolution, capable of detecting changes of just a few parts in 10^{11}.

Fig. 7.20 shows the schematic representation of the principles of Mössbauer spectroscopy technique. The ^{57}Co source in the experiment is moved with a speed

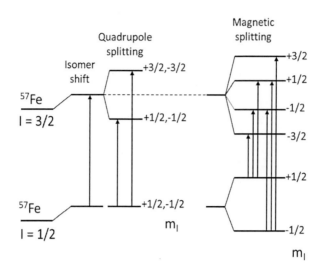

FIGURE 7.21 Schematic representation of the effects of chemical shift, quadrupole splitting, and magnetic splitting on the nuclear energy levels of ^{57}Fe

v relative to the sample that must contain some ^{57}Fe nuclei. By varying the speed v of the source it is thus possible to adjust slightly the frequency of γ-ray photons to different values through the Doppler effect. The resultant Doppler shifts can be quite significant as the frequency of γ-ray photons is quite high. The function of the detector is to measure the transmission of the γ ray photons through the sample, and the record is then used to deduce the absorption in the sample. It is thus possible to study any splitting in the source or absorber nuclei in the sample that might result from interactions including magnetic ones, which are known as hyperfine interactions.

It may be noted here that the Mössbauer spectroscopy technique would have been pretty useless if the γ-ray photons emitted from the excited ^{57}Fe nuclei did not have a well-defined frequency. The relatively large 0.2 μs half-life of ^{57}Fe nuclei leads to an uncertainty of 2 MHz in the frequency of emitted γ-ray photons; the relatively slow decay of ^{57}Fe nucleus is very important for the experiment. The second important feature of the experiment is that ^{57}Fe nuclei are embedded in the sample under study, and hence they exist in the solid state. A free Fe atom would have attained a recoil velocity of the order of $h\nu/m_{Fe}c \approx 80$ ms^{-1} to conserve momentum, and this would spoil the experiment. But the ^{57}Fe atom here is held rigidly within the solid sample, and the acquired momentum is transmitted to the entire crystal lattice of the sample. The probability of the excitation of the lattice vibrations or emission of phonons becomes very small in the case of recoil energy below a certain limit. In this case, the γ-ray is emitted without any loss of recoil energy. The basic

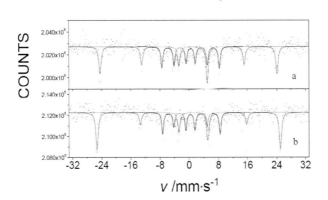

FIGURE 7.22 The ^{57}Fe Mössbauer spectra of single-molecule magnet Fe(L1)$_2$ system measured at 5 K: (a) in zero magnetic field (b) in an external magnetic field of 5 T applied parallel to the gamma rays. (From reference [18] with Creative Commons Attribution License.)

ingredients of Mössbauer spectroscopy techniques are this recoil-free emission and resonant-absorption of γ-rays. The conditions discussed here would indicate that the effect will be optimum for low energy γ-rays emitted from nuclei strongly embedded in a crystal lattice at relatively low temperatures. The ^{57}Fe isotope is the most used isotope but not the only one suitable for Mössbauer spectroscopy experiments. The other suitable isotopes include ^{119}Fe, ^{127}I, ^{151}Eu, ^{119}Sn, and ^{197}Au.

The output data from a Mössbauer spectroscopy experiment is a plot of γ-ray counts in the detector or the relative absorption in the sample against the velocity of the source with respect to the absorber. Now, let us see what we can learn from the results of the experiment. The results contain information on the effects of chemical shift, quadrupole splitting, and magnetic splitting of the nuclear energy levels of ^{57}Fe isotope. The resonant absorption may not take place exactly where one expects it to happen (i.e., when the source is stationary) but it may be shifted slightly (i.e., when the source is moving at a particular velocity) [2]. A slight change takes place over the nuclear volume in the Coulomb interaction between the nuclear and electronic charge distributions due to the slight increase in the size of ^{57}Fe nucleus in the $I = \frac{3}{2}$ state. The resultant shift in the resonant absorption is known as the Isomer shift. Then, there need not necessarily be just one absorption line as a function of source velocity. There can be several lines due to quadrupole splitting or magnetic splitting. In the ground state of ^{57}Fe with $I = \frac{1}{2}$ there is no electric quadrupole moment, but the excited state of of ^{57}Fe with $I = \frac{3}{2}$ has non-zero electric quadrupole moment. If the ^{57}Fe isotope embedded in a sample is subjected to an electric field gradient due to the crystal environment of that sample, then the interaction between the nuclear quadrupole moment and the crystal electric field gradient lifts the degeneracy of the excited state partially. This results in two excited states and two lines in the Mössbauer spectrum.

Magnetic splitting occurs due to the interaction between the magnetic moment of ^{57}Fe nucleus and the local magnetic field in the sample. This effect splits the $I = \frac{1}{2}$ ground state into a doublet and the $I = \frac{3}{2}$ state into a quadruplet. This leads to six possible lines in the Mössbauer spectrum with selection rule $\Delta m_I = 0, \pm 1$ [2]. This magnetic splitting is used to study magnetic exchange interactions and local magnetic ordering. All these shifts and splitting are shown schematically in Fig. 7.21. It is important to note that the figure is not drawn to scale. The observed splitting are in the range of 10^7–10^8 Hz, and these are 11 or 12 orders of magnitude smaller than the energy gap between the ground ($I = \frac{1}{2}$) and the excited ($I = \frac{3}{2}$) state of the ^{57}Fe nucleus [2].

It must be appreciated that in Mössbauer spectroscopy experiments, energy splitting of less than 1 μeV is measured with 14.4 KeV photon. This has been made possible due to the remarkable combination of the Mössbauer effect, emission of such γ-rays, and recoil-free resonance absorption. Figure 7.22 shows the ^{57}Fe Mössbauer spectra of single-molecule magnet $Fe^{II}[C(SiMe_3)_3]_2(Fe(L1)_2)$ measured at 5 K, with and without an applied magnetic field [18]. The external field of 5 T is applied parallel to the γ-rays. There are two sextets in both spectra, which provide evidence for the presence of two phases in the sample. One of the sextets S1 is the fingerprint of the iron atom linearly coordinated by the two carbon atoms in $Fe(L1)_2$, while the other sextet S2 belongs to the ferrihydrite ($Fe_2O_3.0.5H_2O$) present in the sample as a second phase. The applied magnetic field enhances the internal magnetic field manifested by the S1 sextet (Fig. 7.22(b)), and this indicates the presence of ferromagnetic exchange interaction in $Fe(L1)_2$ at low temperatures.

7.6 Summary

In this chapter we have studied various experimental techniques involving the core principle of *magnetic resonance*, and also muon spin rotation and Mössbauer spectroscopy. These techniques provide complementary tools to bulk thermodynamic measurements like magnetometry and obtain deeper information on the magnetic properties of materials. We see that apart from obtaining magnetic information, magnetic resonance experiments are also useful for investigating the local environment of the magnetic moment in a solid material, along with spin–orbit coupling and hyperfine interactions. In NMR study of the local hyperfine field distribution, information can be obtained on the crystallographic phases, degree of short-range order or disorder in alloys and compounds, the interface characteristics (interface roughness, topology, and concentration profile), and strain in thin films. Additional information on local magnetic anisotropies can also be obtained by varying the strength and direction of an applied magnetic field. EPR experiments probe electronic spins and can provide detailed information on the crystal field effects in paramagnetic materials. The crystal field effect is often anisotropic and the EPR line moves with the rotation of the magnetic field with respect to the crystal axis. FMR experiments study interactions between magnetic moments or spins at the lattice sites via exchange interactions in ferromagnetic materials and can determine parameters characterizing the magnetocrystalline anisotropy of a ferromagnetic sample. The study of magnetic hyperfine splitting in Mössbauer spectroscopy provides information on magnetic ordering in materials. In μSR experiment the muon behaves like a microscopic magnetometer and can be very sensitive to extremely small magnetic fields even down to $\sim 10^{-5}$ T. This technique is very useful in studying materials with small moment magnetism, and also materials with random or very short-range magnetic order.

Bibliography

[1] Wurmehl, S., and Kohlhepp, J. T. (2008). Nuclear Magnetic Resonance Studies of Materials for Spintronic Applications. *Journal of Physics D: Applied Physics*, 41 (17): p. 173002.

[2] Blundell, S. (2001). Magnetism in Condensed Matter: Oxford Master Series. *Condensed Matter Physics* (Oxford Series Publications, 2001), 29.

[3] Strijkers, G. J., Kohlhepp, J. T., Swagten, H. J. M., and de Jonge, W. J. M. (2000). Nuclear Magnetic Resonance Study of Epitaxial Co-layers on Single-crystal Substrates. *Applied Magnetic Resonance*, 19 (3): 461–469.

[4] Wertz, J. E., and Bolton, J. R., (1986). Experimental Methods; Spectrometer Performance. In *Electron Spin Resonance: Elementary Theory and Practical Applications*, New York: Chapman and Hall, pp. 450–467.

[5] Dikanov, S. A., and Crofts, A. R. (2006). Electron Paramagnetic Resonance Spectroscopy. In *Handbook of Applied Solid State Spectroscopy:* 97–149. Springer, Boston, MA. Paramagnetic Resonance Spectroscopy. In *Handbook of Applied Solid State Spectroscopy*, Boston: Springer, pp. 97–149.

[6] Stoyanova, R., Barra, A. L., Yoncheva, M., et al. (2010). High-Frequency Electron Paramagnetic Resonance Analysis of the Oxidation State and Local Structure of Ni and Mn Ions in Ni, Mn-Codoped LiCoO2. *Inorganic Chemistry*, 49 (4): 1932–1941.

[7] Bertrand, P., ed. (2020). *Electron Paramagnetic Resonance Spectroscopy*. Switzerland: Springer Nature AG.

[8] Kalarickal, S. S., Krivosik, P., Wu, M., et al. (2006). Ferromagnetic Resonance Linewidth in Metallic Thin Films: Comparison of Measurement Methods. *Journal of Applied Physics*, 99 (9): 093909.

[9] Yoshii, S., Ohshima, R., Ando, Y., et al. (2020). Detection of Ferromagnetic Resonance from 1 nm-thick Co. *Scientific Reports*, 10 (1): 1–7. https://doi.org/10.1038/s41598-020-72760-7

[10] Blundell, S. J. (1999). Spin-polarized Muons in Condensed Matter Physics. *Contemporary Physics*, 40 (3): 175–192. DOI:10.1080/001075199181521.

[11] Brewer, J. H. (1994). Muon Spin Rotation/Relaxation/Resonance. In *Encyclopedia of Applied Physics*. New York: VCH Publishers, Vol. 11, pp. 23–53.

[12] Bert, F. (2014). Probes of Magnetism, NMR and μSR: A Short Introduction. *École thématique de la Société Française de la Neutronique*, 13: 03001.

[13] Kubo, R., and Toyabe, T. (1967). A Stochastic Model for Low Field Resonance and Relaxation. In *Magnetic Resonance and Relaxation*. Amsterdam: North Holland, pp. 810–823.

[14] Luke, G. M., Brewer, J. H., Kreitzman, S. R., et al. (1991). Muon Diffusion and Spin Dynamics in Copper. *Physical Review B*, 43 (4): 3284.

[15] Amato, A. (1997). Heavy-Fermion Systems Studied by ?SR Technique. *Reviews of Modern Physics*, 69 (4): 1119.

[16] Herlach, D., Majer, G., Major, J., et al. (1991). Magnetic Flux Distribution in the Bulk of the Pure Type-II Superconductor Niobium Measured with Positive Muons. *Hyperfine Interactions*, 63 (1): 41–8.

[17] Uemura, Y. J., Yamazaki, I. T., Harshman, N. D., et al. (1985). Muon-spin Relaxation in AuFe and CuMn Spin Glasses. *Physical Review B*, 31 (1): 546.

[18] Kuzmann, E., Homonnay, Z., Klencsár, Z., and Szalay, R. (2021). ^{57}Fe Mössbauer Spectroscopy as a Tool for Study of Spin States and Magnetic Interactions in Inorganic Chemistry. *Molecules*, 26 (4): 1062. https://doi.org/10.3390/molecules26041062.

8

Optical Methods

8.1 Magneto-optical Effects

The magneto-optical effect arises in general as a result of an interaction of electromagnetic radiation with a material having either spontaneous magnetization or magnetization induced by the presence of an external magnetic field. Michael Faraday in 1846 demonstrated that in the presence of a magnetic field the linear polarization of the light with angular frequency w was rotated after passing through a glass rod. This rotation is now termed as Faraday rotation, and it is proportional to the applied magnetic field $\vec{B}$. The angle of rotation $\theta(w)$ can be expressed as [1].

$$\theta(w) = V(w)|\vec{B}|l \tag{8.1}$$

Here $V(w)$ is a constant called the Verdet constant, which depends on the material and also on the frequency w of the incident light; $|\vec{B}|$ is the magnitude of the applied magnetic field, and l thickness of the sample. The Faraday effect is observed in non-magnetic as well as magnetic samples. For example, the Verdet constant of SiO_2 crystal is 3.25×10^{-4} (deg/cm Oe) at the frequency $w = 18300 cm^{-1}$ [1]. This implies that a Faraday rotation of only a few degrees can be observed in a sample of thickness 1 cm in a magnetic field of 10 kOe. A much larger Faraday rotation can, however, be observed in the ferromagnetic materials in the visible wavelength region under a magnetic field less than 10 kOe.

In 1877 John Kerr showed that the polarization state of light could be modified by a magnetized metallic iron mirror. This magneto-optical effect in the reflection of light is now known as the magneto-optical Kerr effect (MOKE), and it is proportional to the magnetization $\vec{M}$ of the light reflecting sample. Today MOKE is a popular and widely used technique to study the magnetic state in ferromagnetic and ferrimagnetic samples. With MOKE it is possible to probe samples to a depth, which is the penetration depth of light. This penetration depth can be about 20 nm in the case of metallic multilayer structures. In comparison to the conventional magnetometers like vibrating sample magnetometer and SQUID magnetometer which measure the bulk magnetization of a sample, MOKE is rather a surface-sensitive technique. On the other hand, in comparison to magnetic microscopy techniques using electrons, which are only sensitive to a few atomic layers at or near the sample surface, MOKE is a depth-sensitive magnetic technique allowing one to study buried ferromagnetic layers. We summarize below some characteristic features of the MOKE technique:

1. MOKE is a very sensitive technique, and has an advantage over SQUID and VSM in the investigation of the magnetism of ultrathin films. With MOKE it is possible to detect the magnetization of a fraction of the atomic layer of the ferromagnetic material.
2. The short duration of the light/matter interaction makes the MOKE technique a very fast method and also allows time-resolved measurements of the magnetization. Time resolution down to 100 fs can be achieved using a femtosecond pulsed laser.
3. A good lateral resolution down to 0.2 μm is possible with MOKE. Spatial resolution in MOKE is usually limited by wavelength limit. This is approximately 300 nm for visible light, but sub-wavelength resolution is also possible. It is thus possible to observe magnetic domains or to study the spatial distribution of magnetization in various kinds of samples in ferromagnetic nanowires, patterned magnetic arrays, self-organized magnetic structures, wedge-shaped samples, etc.
4. In a MOKE experiment it is possible to investigate samples located far from the light source and the detector. This makes MOKE a useful and popular technique to study thin-film magnetism inside vacuum chambers or under extreme conditions like a magnetic field, temperature, pressure, etc.

8.1.1 Principles of magneto-optical effects

The first quantum mechanical description of the magneto-optical effect was provided by Hulme [2, 3] in 1932. He introduced the idea that the spin-dependent dielectric constant is a consequence of the spin–orbit interaction. The spin–orbit interaction couples the electron spin with its motion, and, in turn, connects the magnetic and optical properties of a magnetic sample. To some extent, spin–orbit interaction can be thought of as an effective magnetic vector potential acting on the motion of the electron. This spin–orbit coupling needs to be present in the ground state or the excited states of the material in order to have a magneto-optical effect.

A complete derivation of the magneto-optical effect in a magnetic sample needs a rigorous perturbation theory [4]. Here we shall rather try to understand the general principles in a simpler way. A plane-polarized light beam may be considered as a sum of left and right circularly polarized components. If a beam of plane-polarized light shines on the surface of a metallic sample, the reflected light will also be plane-polarized. However, if in addition there is a magnetic field present, the refractive index for the right circularly polarized light will be different from that of the left circularly polarized light. This is because right-handed and left-handed circularly polarized light will cause the charges in a material to rotate in opposite senses. Thus each polarization will lead to a contribution to the orbital angular momentum having an opposite sign. In the presence of a magnetic field, there will be spin-polarization along the direction of the magnetic field. The spin–orbit interaction will then cause an energy contribution for the two circular polarization with the same magnitude but with opposite signs. This, in turn, will mean that right-handed and left-handed polarization will have different refractive indices in a material, and they will propagate with different speeds. When these two light beams re-emerge after being reflected from a metallic surface, they will recombine but the phase lag between them implies that the emerging beam will have a rotated plane of polarization. This effect is generally known as the magneto-optic Kerr effect (MOKE). In a ferromagnetic sample, the internal field in the ferromagnetic state will play the role of an external magnetic field.

In an isotropic medium the dielectric tensor can be expressed as [3]:

$$\tilde{\epsilon} = \epsilon \begin{pmatrix} 1 & iQ_z & -iQ_y \\ -iQ_z & 1 & iQ_x \\ iQ_y & -iQ_x & 1 \end{pmatrix} \quad (8.2)$$

Here $\vec{Q} = (Q_x, Q_y, Q_z)$ is called Voigt factor, which is aligned with the magnetic field and with magnitude depending on the material. The off-diagonal terms in the dielectric tensor contribute to the magneto-optical effects. It actually leads to the two circularly polarized normal modes with dielectric constants $\epsilon_\pm = \epsilon(1 \pm \vec{Q} \cdot \vec{k})$, where $\vec{k}$ is the direction of propagation of light [3]. These circular modes travel in the material with different velocities and attenuate differently. These different attenuations cause slight elliptical polarization in the recombined light emerging from the material.

MOKE is used in the study of metallic samples rather than the Faraday effect. This is because metals absorb light, and hence cannot be studied in transmission unless the metallic sample is very thin. A plane-polarized light incident on a metallic

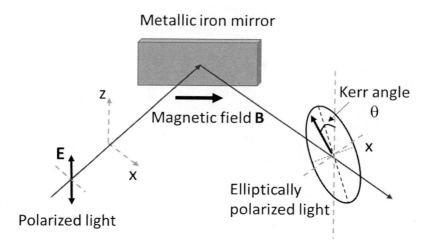

FIGURE 8.1 Scchematic representation of the change of polarization of light after reflection from a magnetized metallic iron mirror.

surface will become elliptically polarized light after reflection with a rotated plane of polarization. Fig. 8.1 shows schematically the state of polarization of light after reflection from a magnetized metallic iron mirror. The Kerr effect is particularly useful in the study of magnetic thin films and surfaces, and the technique is sometimes known as *surface magneto-optical Kerr effect* (SMOKE).

Fig. 8.2 shows three common geometries of MOKE depending on the direction of the magnetization relative to the plane of incidence of light and sample planes [3, 5]. If the direction of the magnetization is parallel to the surface normal, the MOKE is called the polar Kerr effect. The MOKE is called the longitudinal Kerr effect where the direction of the magnetization is parallel to the surface and in the plane of incidence of the light. On the other hand, MOKE is called the transverse or equatorial Kerr effect when the direction of the magnetization is parallel to the surface but perpendicular to the plane of incidence of light. The longitudinal and polar MOKE are both sensitive to the polarization of the incoming light parallel (p) and perpendicular (s) to the plane of incidence. This transverse MOKE is possible only when the light is polarized in the plane of incidence. The Kerr signal in polar Kerr effect is usually the largest, whereas in the other two effects it is usually an order of magnitude less than the polar effect.

8.1.2 Experimental methods

A typical experimental setup for MOKE measurements is shown in Fig. 8.3. This consists of a stabilized low noise light source (usually a continuous wave laser), a linear polarizer defining polarization of the incident light, the sample stage located in a variable magnetic field, an analyzer, and a photodetector (usually a photodiode).

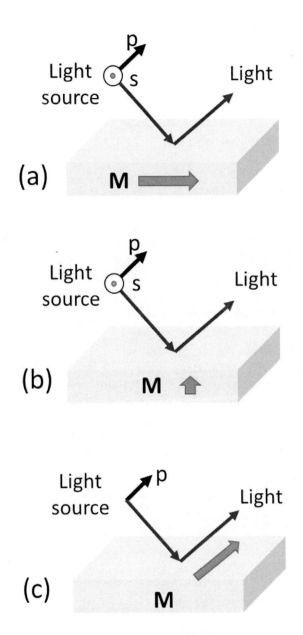

FIGURE 8.2 Geometries of magneto-optical Kerr effect: (a) longitudinal, (b) polar, and (c) transverse.

If the aim is to extract spectroscopic magneto-optic properties, then the laser is replaced by a lamp and monochromator to generate wavelength dependencies. The magnet consists of two pairs of split-coil solenoids. With either of these pairs, it is possible to generate a field either in the plane of the sample for longitudinal

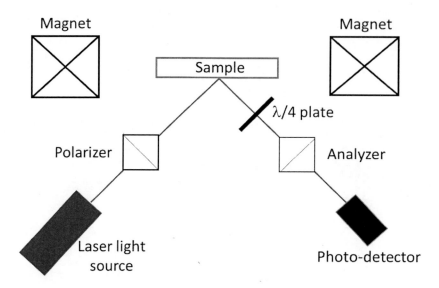

FIGURE 8.3 Schematic representation of a general MOKE experimental setup.

measurements or perpendicular to the plane for polar measurements. Provision is made in many experimental setups for rotation of the sample stage as well. As we saw earlier, MOKE is a useful and popular technique to study thin-film magnetism in situ inside ultra-high vacuum (UHV) chambers. However, in such measurements, the UHV viewport window usually introduces a birefringence, which prevents the extinction condition to be realized. In this case, a quarter wave plate is used before the analyzing polarizer to nullify the effect of birefringence of the UHV viewport window. The Kerr intensity is measured by a photodiode after the light has passed through an analyzing polarizer, and this provides the basis for quantitative analysis. The analyzing polarizer is set at an angle δ from extinction, and the Kerr intensity can be expressed as [3]:

$$I = I_0 \left(1 + \frac{2\theta}{\delta}\right) \quad (8.3)$$

Here θ is the Kerr rotation and I_0 is the intensity at zero Kerr rotation. It is possible to determine the maximum Kerr rotation θ_{max} from the relative change of the Kerr intensity $\Delta I/I_0$ obtained by reversing the saturation magnetization of the sample [3]:

$$\theta_{max} = \frac{\delta}{4} \frac{\Delta I}{I_0} \quad (8.4)$$

The total Kerr rotation is linearly dependent upon the magnetization. Using this relationship an output plot of photo-detector voltage versus the applied magnetic field can be interpreted as the relative magnetization versus applied field hysteresis loop. While the absolute magnetization cannot be determined through such an experiment, the field for saturation magnetization and the coercive field can easily be obtained, and the nature of the hysteresis loop can be studied in a material. With the provision of sample rotation, the same measurements can be made at several angles of the sample with respect to the plane of light incidence. Thus, the easy and hard axes of magnetization in a material can be determined from the Kerr rotations. This experimental method uses only the DC signal from the photodetector to obtain a magnetization hysteresis loop and it is known as the DC-MOKE technique [3]. The Kerr rotation and ellipticity, however, cannot be separated in the DC-MOKE technique, and additional optical components and electronics are required for this purpose. Figure 8.4 shows the Kerr hysteresis loops for a cobalt (Co) thin film in the longitudinal Kerr geometry [6]. The magnetization is plotted in arbitrary units. The square hysteresis loop with a sharp jump in magnetization and the absence of hysteresis at 90^0 from that direction indicate the existence of an in-plane easy magnetization axis. The inset of Fig. 8.4(a) shows the angular dependence of the coercive field H_c (0 - 360^0) in the form of a polar graph. Fig. 8.4(b) shows the two in-plane components of magnetization M_l and M_t obtained simultaneously by the longitudinal MOKE and transverse MOKE at $\theta = 85^0$.

The AC-MOKE technique is capable of determining the values of the Kerr rotation and ellipticity in a given magnetic sample. In this technique, the laser beam is modulated at a frequency f by a photoelastic modulator, so that the Kerr ellipticity and rotation are measured at locked-in frequencies f and $2f$, respectively, and normalized by the quasi-continuous incoming light intensity [7]. Fig. 8.5 presents the schematic representation of an AC-MOKE experimental setup. The light beam reflected after interaction with the sample is deviated by a cube beam splitter. It then passes through a photoelastic modulator with a principal optical axis oriented along the optical axis polarizer and an analyzer whose axis makes a 45^0 angle with that of the polarizer. The laser beam then passes through an optical filter peaked at the laser wavelength to finally illuminate a photodiode. A more sophisticated version of this configuration has been used to build a high-resolution polar Kerr magnetometer for nanomagnetism studies [7]. The output signal of the photodetector contains a DC contribution I_0 and AC components at the fundamental f and harmonic $2f$ frequencies. The normalized magnitudes of these AC components, I_f/I_0 and I_{2f}/I_0, can be measured using a lock-in amplifier. The Kerr rotation θ and ellipticity ϵ for small values of rotation can be expressed as [7]:

$$\frac{I_f}{I_0} = -\frac{4\epsilon J_1(\phi_0)}{1 - 2J_0(\phi_0)\theta} \tag{8.5}$$

$$\frac{I_{2f}}{I_0} = -\frac{4\theta J_2(\phi_0)}{1 - 2J_0(\phi_0)\theta} \tag{8.6}$$

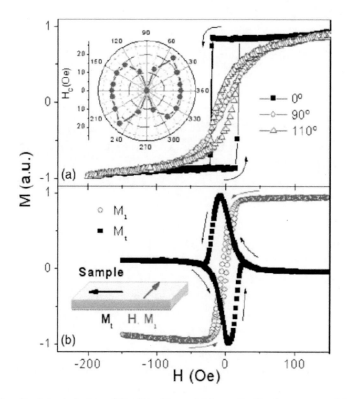

FIGURE 8.4 Kerr hysteresis loops of Co thin films: (a) Longitudinal component (M_L) for different orientations ($\theta = 0°$ (easy axis), $90°$ (hard axis), and $110°$) of the sample. The inset shows the angular dependence of the coercive field H_c. (b) In-plane components of magnetization M_l and M_t obtained simultanously by the longitudinal MOKE and transvers MOKE at $\theta = 85°$. (From reference [6] with permission from AIP Publishing.)

Here J_0, J_1, and J_2 stand for the zeroth, first, and second order Bessel functions, and ϕ_0 is the amplitude of polarization modulation. After adjustment of $J_0(\phi_0) = 0$, which is akin to symmetrization I_{2f}/I_0 with respect to θ, Eqns. 8.5 and 8.6 can be rewritten as [7]:

$$\frac{I_f}{I_0} = -4\epsilon J_1(\phi_0) \tag{8.7}$$

$$\frac{I_{2f}}{I_0} = -4\theta J_2(\phi_0) \tag{8.8}$$

This leads to the equations:

$$\epsilon = A\frac{I_f}{I_0} \tag{8.9}$$

$$\theta = B\frac{I_{2f}}{I_0} \tag{8.10}$$

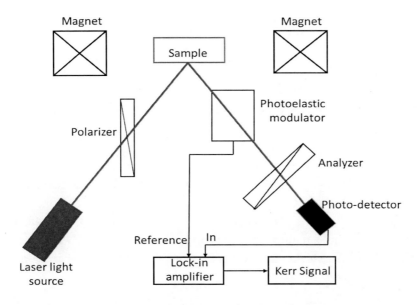

FIGURE 8.5 Schematic representation of an AC-MOKE experimental setup.

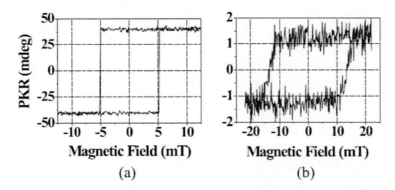

FIGURE 8.6 Polar Kerr rotation hysteresis loops of: (a) continuous Pt(4.5 nm) /Co(0.5 nm) / Pt(3.5 nm) layer with a perpendicular magnetic anisotropy, and (b) a single 130 nm wide magnetic nanodisk. (From reference [7] with permission from AIP Publishing.)

The coefficients A and B can be estimated by convenient calibration procedures. The Kerr rotation θ and ellipticity ϵ can then be deduced from the measured ratios of I_f/I_0 and I_{2f}/I_0. The laser source power fixes I_0. The measured magnitudes of the magneto-optical effect depend on the filling factor of laser spot area since I_f and I_{2f} are proportional to the probed area of the magnetic sample. Fig. 8.6 shows polar Kerr rotation hysteresis loops of a continuous Pt(4.5 nm) /Co(0.5 nm) / Pt(3.5 nm) layer with perpendicular magnetic anisotropy and a single 130 nm wide magnetic nanodisk patterned in this layer.

8.2 Scanning Near-field Optical Microscopy

Scanning near-field optical microscopy (SNOM) is a technique that combines the strength of optical contrast methods with the high resolving power of scanning probe microscopy. In this technique, an optical dipole radiator is scanned over a sample at a spacing smaller than the near-field zone of the dipole radiator. The far-field radiation characteristics of the dipole that are sufficiently modified by the optical features of the sample surface are detected to form a scanned image of the sample's surface. The characteristic length scale of near-field optics is set by the physical size of the dipole radiator and not the wavelength of light. This forms the basis of enhanced resolution in the SNOM technique. The change in the polarization of the incident light is required to be determined to study magneto-optical effects. This, in turn, leads to the requirement of polarization-sensitive SNOM, which would enable the determination of local magnetic properties of a sample and imaging magnetic domains with sub-micron resolution.

8.2.1 Principle of scanning near-field optical microscope

Any monochromatic light field can be represented by a set of waves of the form [8]:

$$E(\vec{r}) = \int_{-\infty}^{+\infty} E_k e^{i\vec{k}.\vec{r}} dk \qquad (8.11)$$

Here $\vec{r}$ represents spatial position. A wave vector $\vec{k}$ defines each wave in the set with a complex amplitude E_k. In a non-absorbing medium with refractive index n for a given far-field distribution wave vector $\vec{k}$ is real with an amplitude defined by:

$$k_0 = \sqrt{(k_x^2 + k_y^2 + k_z^2)} = 2\pi n/\lambda \qquad (8.12)$$

Here λ is the wavelength of light in vacuum. In an event where $(k_x^2 + k_y^2) > k_0^2 = w^2/c^2$, k_z must have a non-zero imaginary component. Any imaginary component within the far-field wave vector k in Eqn. 8.12 will generate an exponentially decaying amplitude E_k. A light wave characterized by such exponential decay of amplitude is called an evanescent wave. The light field near the originating surface at a distance $\vec{r}$ that is much less than the wavelength λ of the light involved is dominated by evanescent light. Fig. 8.7 shows the contrast between imaging in the far and near fields. The SNOM technique utilizes this near-field structure that contains information about the surface beyond the diffraction-limited maximum resolution of the far-field [8].

In a SNOM experiment, the probe needs to be positioned within the near-field, which is close to the sample surface at a distance r much less than the wavelength λ of the incident light. In practice SNOM is typically coupled to a scanning microscopy

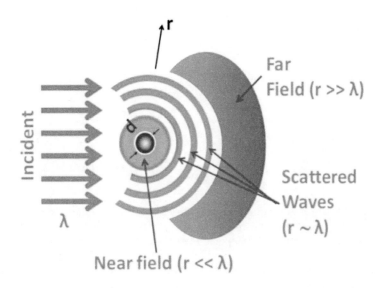

FIGURE 8.7 The far- and near-field configuration around a light emitting object. The evanescent light dominates the light field when the distance ($\vec{r}$) from the emitting object that is much less than the wavelength of light used. (From reference [8] Creative Commons Attribution License.)

system such as AFM [8], where the cantilever system allows the SNOM probe to be positioned very close to the surface at a distance of tens of nm ($r \ll 1$), to be within the near-field [8]. The near-field distribution is then mapped by scanning the sample in a raster pattern while a tip position very close to the surface is maintained. There are two distinct modes of SNOM, which are classified by the type of tip that is being used. Fig. 8.8(a) shows one mode known as the aperture SNOM or collection SNOM, which uses a hollow tip that allows light to shine from the back out of, or up into, the tip. Aperture SNOM works by generating near-field light by scattering far-field light from the nanoscale aperture at the end of the tip [8]. The sample surface then scatters the generated tip near-field light into the far-field, which is detected for the sought-after experimental information. The standard aperture tip consists of a circular opening with a typical diameter ranging from 80 to 250 nm. With advanced nanofabrication procedures, even a diameter as small as 10–20 nm can be achieved. The range of frequencies that can be coupled to or scattered by the sample is dictated by the tip composition as well as the size and shape of the tip and aperture. Fig. 8.8(b) shows apertureless-type SNOM mode, or scattering SNOM (s-SNOM) mode, where an apertureless tip scatters light from an external source. The tip acts as an antenna to produce a source of evanescent light near the nanostructured sample surface (Fig. 8.8(b)). The s-SNOM mode does not suffer from the fundamental size limitations of aperture SNOM and can achieve resolution below 10 nm. However, it has the drawback of a relatively intense far-field background

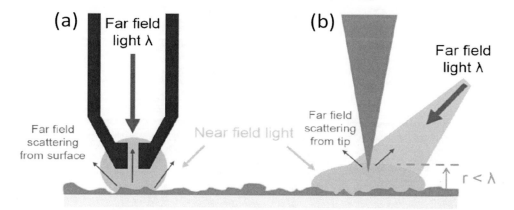

FIGURE 8.8 Schematic representation of (a) aperture SNOM mode where an evanescent light field created by a nanoscale opening at the end of a tip is scattered off the surface. Light scattered from the surface is detected in the far-field (b) apertureless SNOM mode utilizing near-field light emitted from the surface due to external illumination that is scattered off a sharp tip into the far-field to be detected. (From reference [8] Creative Commons Attribution License.)

composed of light reflected or scattered directly from the sample surface [8]. This latter effect creates a large background signal over a spatial area much larger than the near-field domain of the tip.

8.2.2 Magneto-optical measurement using scanning near-field optical microscope

In the case of magneto-optical measurements with SNOM, the polarization state of the signal needs to be analyzed for achieving magneto-optical contrast [9]. This technique can also perform a simultaneous measurement of the topography of the sample and enables the elimination of the topographic artifacts from the magneto-optical results. Fig. 8.9 shows a schematic representation of a typical magneto-optical SNOM setup. A quartz tuning-fork-based shear-force feedback can be used to ensure that the probe is in near-field proximity to the sample surface throughout the scan [9]. The sample is illuminated by an He-Ne laser at a wavelength of 633 nm through a system of mirrors, which controls the angle of incidence onto a sample in both transmission and reflection modes. A half-wave plate controls the polarization of the illuminating light. A sharpened optical fibre probe collects the light transmitted through or reflected from the sample. After decoupling, the light is sent through a polarizer onto the detector. To avoid the depolarization effects associated with metal-coated SNOM tips, chemically etched uncoated fibre tips are used. Uncoated SNOM fibre tips have a larger optical acceptance area than metal-coated tips. However, they still provide sub-wavelength resolution since they probe the evanescent fields present in the near-field region close to the sample

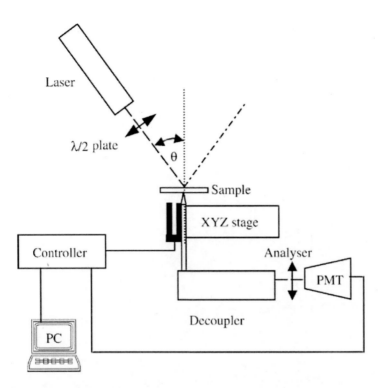

FIGURE 8.9 Schematic representation of a scanning near field optical microscope setup for magneto-optical measurement. (From reference [9] with permission from Elsevier.)

surface. Further, polarization analysis combined with SNOM measurements greatly improves the contrast and resolution obtained with uncoated fibre tips. These tips enable a resolution of the order of 100 nm in linear optical imaging and down to a few tens of nanometers in nonlinear SNOM imaging if the electromagnetic field enhancement is significant.

The capability of SNOM in magnetic imaging is shown in Fig. 8.10, which presents a magneto-optical image (along with a topographical image) of the magnetic domain structure of a 6 μm thick ferromagnetic film of yittrium-iron-garnet (YIG) of (111) orientation grown on a gadolinium-gallium-garnet (GGG) substrate [9]. The image was obtained in transmission mode configuration. The surface of the film was of sufficiently high quality and no topographic features were observed in the topographic image (Fig. 8.10(a)). The domain structure visible in Fig. 8.10(b) with bright and dark areas corresponding to magnetic domains with opposite magnetic orientation. The domain structure shown in Fig. 8.10(c) represents the same area as in Fig. 8.10(b) but with the application of an external magnetic field perpendicular to the sample surface. The magnitude of the applied magnetic field is close to that required for saturation of magnetization. No significant magneto-optical contrast is

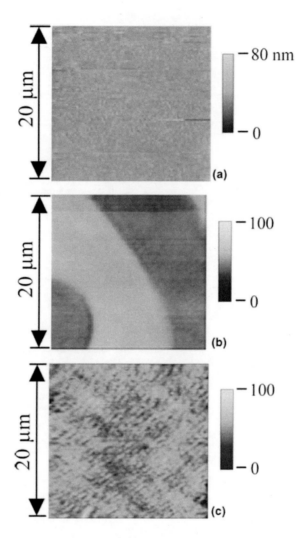

FIGURE 8.10 Topographical image (a) and magneto-optical images (b) and (c) of a YIG film obtained with SNOM. The plane of incidence perpendicular to the horizontal axes of the images and the angle of incidence is $0°$. Images (b) and (c) were obtained in the cross-polarization configuration. Image (c) was obtained after subjecting the sample to a magnetic field greater than required for saturation. (From reference [9] with permission from Elsevier.)

observed in Fig. 8.10(c), as no domain structure is expected in this kind of applied magnetic field.

8.3 Brillouin Light Scattering

The brillouin light scattering (BLS) technique is based on the phenomenon of inelastic light scattering in which the incident photons are scattered by spin waves in a magnetic sample. BLS provides access to spin waves without exciting them explicitly. BLS is particularly known for the following advantages over microwave spectroscopy [10]:

1. Ability to investigate spin waves with different absolute values and orientations of their wave vectors.
2. The possibility to detect both low-amplitude thermal spin waves and high-amplitude spin waves after excitation of the magnetic sample by an external microwave field.
3. A unique opportunity to investigate two-dimensional confinement effects with a high spatial resolution is achievable in BLS. This is determined by the size of the laser beam focus, which is typically 30–50 μm in diameter.

8.3.1 Principles of Brillouin light scattering

The BLS process can be understood within a quasi-classical framework as the inelastic scattering of light from the space- and time-dependent potential described by the dielectric tensor of a solid-state body [11]. In a non-magnetic material, acoustic waves are responsible for the temporal as well as spatial variation of the dielectric properties. In a magnetic material, a spin wave creates a phase grating due to magneto-optical effects, which travels with the phase velocity of the spin wave [10]. The light incident in the material is Bragg-reflected from the phase grating with its frequency being Doppler-shifted by the spin-wave frequency. The number of photons scattered into the solid angle dΩ in the frequency interval between w_s and $w_s + dw_s$ per unit incident flux density, i.e., the differential light scattering cross section $d^2\sigma/d\Omega dw_s$, can be expressed as [10]:

$$d^2\sigma/d\Omega dw_s \propto \langle \delta\epsilon^*(k_{in} - k_s)\delta\epsilon(k_{in} - k_s)\rangle_{w_{in}-w_s} \qquad (8.13)$$

Here $\delta\epsilon^*$ is the fluctuating term of the dielectric permittivity. This is caused by the spin waves due to magneto-optical effects and which gives rise to the scattering of photons. This fluctuating dielectric permittivity is proportional to the dynamic part of the magnetization $\vec{M}$ of the spin–wave. The correlation function is expressed as [10]:

$$\langle \delta\epsilon^*(k)\delta\epsilon(k)\rangle_w = \int d(t_2 - t_1)d^3(r_2 - r_1)exp[-iwt - ik(r_2 - r_1)]$$
$$\times \langle \delta\epsilon^*(r_1,t_1)\delta\epsilon(r_2,t_2)\rangle$$
$$\propto \int d(t_2 - t_1)d^3(r_2 - r_1)exp[-iwt - ik(r_2 - r_1)]$$
$$\times \langle M^*(r_1,t_1)\delta M(r_2,t_2)\rangle \qquad (8.14)$$

In the above equation ⟨...⟩ represents the statistical average. The spatial integration volume will encompass the entire space when the incident light is scattered from the propagating spin wave in an infinite medium. In this situation the correlation function in Eqn. 8.14 will be non-zero only if the relations $w = w_s - w_1$ and $k = k_s - k_{in}$ are fulfilled. This gives the conservation laws of energy and momentum. However, in a finite-size sample, for example, a thin-film sample, a spin-wave mode propagates in an integration volume, which is bounded by the two film surfaces. In this situation, the conservation conditions will be satisfied only for the two in-plane components of the wave vector $k_\parallel$. In the backscattering geometry $k_\parallel = 2k_1 \sin\phi$ when $k_s = -k_{in}$. Here ϕ is the angle of incidence of the probing light. The conservation laws, however, cannot be used to determine the component perpendicular to the film, because the system does not have the symmetry of translational invariance perpendicular to the film. The uncertainty in k_x is inversely proportional to ξ, which is determined by the thickness of the film, or the size of the confinement region of the mode, or the penetration depth of the probing light in the material. It can be neglected if $(k_s - k_{in})\xi \gg 2\pi$.

Within a quantum mechanical framework, the BLS process can be understood in terms of the interaction of light quanta or photons from a light source with energy $\hbar w_{in}$ and momentum $\hbar \vec{k}_{in}$ and the quanta of spin waves or magnons with energy $\hbar w_{SW}$ and momentum $\hbar \vec{k}_{SW}$. This principle of BLS is shown schematically in Fig. 8.11. In the BLS process if a magnon is annihilated, the scattered photon will gain in energy and momentum [10]:

$$\hbar w_{out} = \hbar(w_{in} + w_{SW})$$
$$\hbar \vec{k}_{out} = \hbar(\vec{k}_{in} + \vec{k}_{SW}) \tag{8.15}$$

It is evident from the above equation that the wave vector $\vec{k}_{out} - \vec{k}_{in}$ transferred in the scattering process is equal to the wave vector $\vec{k}_{SW}$ of the annihilated spin wave. This is known as the anti-Stokes process. In a similar way during the scattering process a magnon can also be created by energy and momentum transfer from the photon.

$$\hbar w_{out} = \hbar(w_{in} - w_{SW}) \tag{8.16}$$
$$\hbar \vec{k}_{out} = \hbar(\vec{k}_{in} - \vec{k}_{SW})$$

This is known as the Stokes process. Both the Stokes and the anti-Stokes process can take place with the same probability at finite temperatures (T$\gg \hbar w/k_B \approx 1$ K) [10].

8.3.2 Experimental method for Brillouin light scattering

The analysis of the frequency shift $|w_{out} - w_{in}|$ of the scattered photon provides information about the energy of the spin wave. This energy analysis of the spin-wave

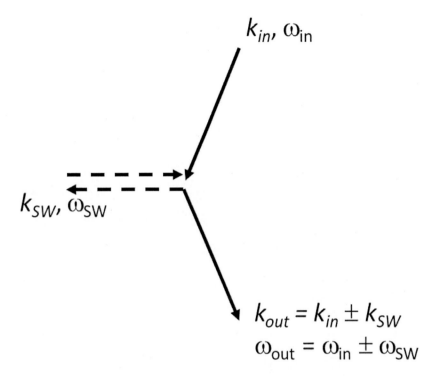

FIGURE 8.11 Schematic representation of scattering process of photons from spin-wave excitations.

signal forms the basis for all BLS measurements. The momentum of the detected spin wave can be estimated from the angle of incidence and the angle of detection of the light. Two essential components for building a BLS spectroscopy setup are: (i) a monochromatic light source and (ii) an instrument for high-contrast frequency analysis. Diode-pumped solid-state lasers with a wavelength of 532 nm are commonly used for providing the light that undergoes the inelastic scattering [11].

The spin-wave frequencies are in the range of a few gigahertz, which requires a spectrometer with a resolution of better than 0.05 cm^{-1}. Moreover, the BLS process has a small scattering cross section. This causes a weak output signal intensity in comparison to the fraction of light that is elastically scattered from the sample surface. The standard tool for the detection of BLS spectra is the tandem Fabry-Pérot interferometer (TFPI) because of its excellent performance in terms of its high contrast and dynamic spectral range [11]. Fig. 8.12(a) shows a schematic representation of a TFPI, which consists of a set of two Fabry-Pérot interferometers (FPI), i.e., etalons comprising of two highly reflective parallel mirrors. These are schematically shown as FPI1 and FPI2 in Fig. 8.12(a). An etalon allows transmission of light only when the mirror spacing is a multiple of half of the wavelength λ of light. The transmission of light is strongly suppressed for other mirror spacings. This also

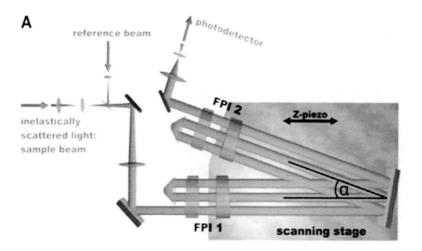

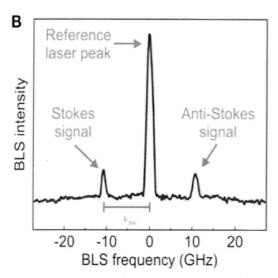

FIGURE 8.12 (A) Schematic representation of the principle scheme of a tandem Fabry-Pérot interferometer. (B) BLS spectrum showing the central reference laser peak as well as the Stokes and Anti-Stokes signal. (From [11]. Creative Commons Attribution License.)

implies that the transmission of light in an FPI is periodic with respect to the mirror distance. This periodicity is typically defined in the frequency space in terms of the free spectral range (FSR), which is the maximum frequency range accessible by the interferometer before higher transmission orders appear in the spectrum [11]. In the BLS spectroscopy experiment, the wavelength or the frequency of light is determined with the detection of the transmitted light intensity by scanning the distance of the interferometer mirrors. During a scan, one mirror of each FPI is fixed while the other mirror is placed on a common scanning stage. The transmission function of a single

FPI is periodic, hence the two interferometers are arranged under an angle α. This experimental process enables unambiguous identification of the spin-wave signals in the detected BLS spectrum. A comparison of the mirror spacing corresponding to light that is scattered from the sample with the position of a reference beam that directly enters the interferometer enables the determination of the absolute value of the frequency shift [11]. The zero position in a BL Spectrum is marked by the position of this reference beam. The inelastically scattered light passes each etalon three times (Fig. 8.12(a)) before it is finally detected in a photodetector. A high contrast better than 10^{10} that is needed for the detection of the small BLS signal intensities can be achieved in this experimental configuration.

Fig. 8.12(b) shows schematically a typical spectrum obtained in a BLS spectroscopy measurement, which exhibits three different peaks. The signal corresponding to the reference beam marks the zero position of the spectrum. In addition, the spectrum has one Stokes peak with a negative frequency shift with respect to the reference beam and one anti-Stokes peak with a positive frequency shift. The square of the amplitude of the dynamic magnetization is proportional to the BLS cross section; hence the spin-wave intensity can be obtained from the peak height in a BLS spectrum.

The momentum is conserved in the BLS process. This allows for a wave-vector-selective detection of spin waves. The modulus of the wave vector of the incident light $|k_{in}|$ is, however, fixed because of the use of a light source of constant wavelength. The direction of the light beam with respect to the sample surface is therefore used to control the momentum that can be transferred between photon and magnon, and thus to obtain wave-vector resolution. Fig. 8.13(a) shows schematically the experimental backscattering geometry with all relevant wave-vector components. This configuration uses a single objective for focusing the incident light as well as collecting the scattered light. The breaking of the translation symmetry at the sample surface dictates that the relevant momentum transfer is to be defined by the projection of the photon wave vector to the sample surface [11]. A change of the in-plane component of the incident light can easily be obtained by varying the angle of incidence ϕ if the objective has a large working distance. This situation can be realized experimentally by rotating the sample around the axis perpendicular to the plane of incidence while keeping the optical axis of the incident laser fixed. In the present backscattering geometry, the spin waves that can be probed need to satisfy the following relation for their wave vector:

$$\vec{k}_{SW,\parallel} = 2\frac{2\pi}{\lambda_L}sin\phi \tag{8.17}$$

Here λ_L is the wavelength of laser light and ϕ is the angle of incidence. If the wavelength of the laser used is $\lambda_L = 532$ nm, the maximum wave vector that can be detected is $\vec{k}_{SW,\parallel} = 23.6$ rad/μm. This corresponds to a minimum

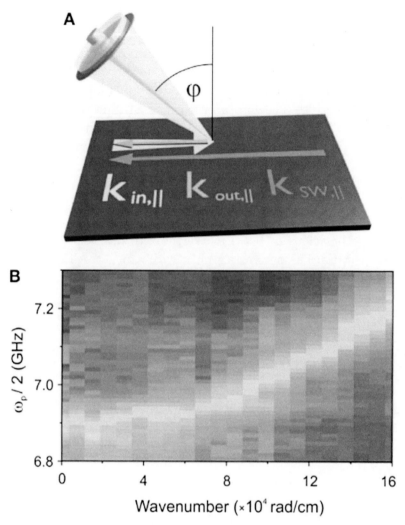

FIGURE 8.13 (A) Schematic representation the back scattering geometry. (B) BLS measurement of the spin-wave intensity as a function of the spin-wave wave vector in a 5 μm thick yttrium iron garnet film. (From [11] Creative Commons Attribution License.)

spin-wave wavelength $\lambda_{SW}^{min} = \frac{\lambda}{2} = 266$ nm. Fig. 8.13(b) shows the result of a measurement of the spin-wave intensity in a 5μ thick yttrium-iron-garnet film using a wave-vector-resolved BLS spectroscope in backscattering geometry [11]. Each column in this figure represents a BLS spectrum detected for a fixed wave vector. The signal intensity is colour-coded, with black (white) indicating maximum (minimum) intensity, respectively. Fig. 8.14 shows the BLS spectra in an $Ni_{80}Fe_{20}$ permalloy (Py)(25 nm)/Co(10 nm) exchange spring bilayer film obtained in the backscattering geometry using a single-mode solid-state laser operated at 532 nm and a TFPI in the presence of various bias magnetic field applied in the plane of the bilayer film

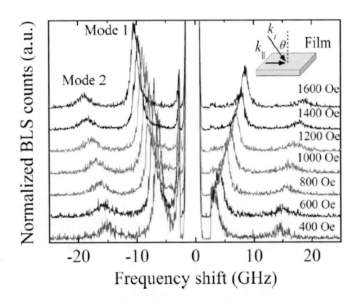

FIGURE 8.14 BLS spectra for $Ni_{80}Fe_{20}$ permalloy (Py)(25 nm)/Co(10 nm) film at different bias magnetic fields applied along the plane of the film. Inset shows the measurement geometry. (From [12] with permission from AIP Publishing.)

[12]. The angle of incidence ϕ was chosen as 45^0, which corresponds to a magnon wavenumber $\vec{k}_{SW,\parallel} = 1.67 \times 10^5 rad/cm$.

8.4 Summary

In this chapter we have studied experimental techniques employing interaction of electromagnetic waves with magnetic material without any resonant absorption. The magneto-optic Kerr effect (MOKE) can be used as a very sensitive technique, with an advantage over conventional magnetometry in the investigation of magnetism of ultrathin films. The short duration of light/matter interaction makes the MOKE technique a very fast method and also allows time-resolved measurements of magnetization. With MOKE, it is possible to observe magnetic domains or to study the spatial distribution of magnetization in various kinds of samples in ferromagnetic nanowires, patterned magnetic arrays, self-organized magnetic structures, wedge-shaped samples, etc. Scanning near-field optical microscopy (SNOM) technique combines the strength of optical contrast methods like MOKE with the high resolving power of scanning probe microscopy. With polarization-sensitive SNOM it is possible to determine local magnetic properties of a sample and imaging magnetic domains with sub-micron resolution. The Brillouin light scattering technique, which is based on the phenomenon of inelastic light scattering, has the ability to investigate spin waves in ferromagnetic materials with different absolute values and orientations of their wave vectors.

Bibliography

[1] Shinagawa, K. (2013). Faraday and Kerr Effects in Ferromagnets. In S. Sugano and N. Kojima, eds., *Magneto-Optics* (Vol. 128). Heidelberg: Springer Science & Business Media.

[2] Hulme, H. R. (1932). The Faraday Effect in Ferromagnetics. *Proceedings of the Royal Society of London. Series A, Containing Papers of a Mathematical and Physical Character*, 135 (826): 237–257.

[3] Qiu, Z. Q., and Bader, S. D. (1999). Surface Magneto-optic Kerr Effect (SMOKE). *Journal of Magnetism and Magnetic Materials*, 200 (1–3): 664–678.

[4] Argyres, P. N. (1955). Theory of the Faraday and Kerr Effects in Ferromagnetics. *Physical Review*, 97 (2): 334.

[5] You, C. Y., and Shin, S. C. (1996). Derivation of Simplified Analytic Formulae for Magneto-optical Kerr Effects. *Applied Physics Letters*, 69 (9): 1315–1317.

[6] Teixeira, J. M., Lusche, R., Ventura, J., et al. (2011). Versatile, High Sensitivity, and Automatized Angular Dependent Vectorial Kerr Magnetometer for the Analysis of Nanostructured Materials. *Review of Scientific Instruments*, 82 (4): 043902.

[7] Cormier, M., Ferré, J., Mougin, A., et al. (2008). High Resolution Polar Kerr Magneto-meter for Nanomagnetism and Nanospintronics. *Review of Scientific Instruments*, 79: 033706. https://doi.org/10.1063/1.2890839

[8] Bazylewski, P., Ezugwu, S., and Fanchini, G. (2017). A Review of Three-dimensional Scanning Near-field Optical Microscopy (3D-SNOM) and Its Applications in Nanoscale Light Management. *Applied Sciences*, 7 (10): 973.

[9] Dickson, W., Stashkevitch, A., Youssef, J. B. et al. (2005). SNOM Imaging of Thick Ferromagnetic Films: Image Formation Mechanisms and Limitation. *Optics Communications*, 250 (1–3): 126–136.

[10] Demokritova, S. O., Hillebrands, B., and Slavin, A. N. (2001). Brillouin Light Scattering Studies of Confined Spin Waves: Linear and Nonlinear Confinement. *Physics Reports*, 348 (6): 441–489.

[11] Sebastian, T., Schultheiss, K., Obry, B., et al. (2015). Micro-focused Brillouin Light Scattering: Imaging Spin Waves at the Nanoscale. *Frontiers in Physics*, 3: 35.

[12] Haldar, A., Banerjee, C., Laha, P., and Barman, A. (2014). Brillouin Light Scattering Study of Spin Waves in NiFe/Co Exchange Spring Bilayer Films. *Journal of Applied Physics*, 115 (13): 3901. https://doi.org/10.1063/1.4870053.

9

Neutron Scattering

A neutron is a nuclear particle, and it does not exist naturally in free form. Outside the nucleus, it decays into a proton, an electron, and an anti-neutrino. The scattering of low energy neutrons in solids forms the basis of a very powerful experimental technique for studying material properties. A neutron has a mass $m_n = 1.675 \times 10^{-27}$ kg, which is close to that of the proton and a lifetime $\tau = 881.5 \pm 1.5$ s. This lifetime is considerably longer than the time involved in a typical scattering experiment, which is expected to be hardly a fraction of a second.

A neutron has several special characteristics, which makes it an interesting tool for studying magnetic materials as well as engineering materials and biological systems. It is an electrically neutral, spin-1/2 particle that carries a magnetic dipole moment of $\mu = -1.913\ \mu_N$, where nuclear magneton $\mu_N = e\hbar/m_p = 5.051 \times 10^{-27}$ J/T. The zero charge of neutron implies that its interactions with matter are restricted to the short-ranged nuclear and magnetic interactions. This leads to the following important consequences:

1. The interaction probability is small, and hence the neutron can usually penetrate the bulk of a solid material.
2. Additionally, a neutron interacts through its magnetic moment with the electronic moments present in a magnetic material strong enough to get scattered measurably but without disturbing the magnetic system drastically. This magnetic neutron scattering has its origin in the interaction of the neutron spin with the unpaired electrons in the sample either through the spin of the electron or through the orbital motion of the electron. Thus, the magnetic scattering of neutrons in a solid can provide the most direct information on the arrangement of magnetic moments in a magnetic solid.

3. Energy and wavelength of a neutron matches with electronic, magnetic, and phonon excitations in materials and hence provide direct information on these excitations.

Neutrons behave predominantly as particles in neutron scattering experiments before the scattering events, and as waves when they are scattered. They return to their particle nature when they reach the detectors after the scattering events. In the scattering experiments, the wave nature of neutrons is often referred to in terms of wave vector of length $\vec{k} = \frac{2\pi}{\lambda}$ and with the same direction as the velocity v, where λ is the de-Broglie wavelength of neutron expressed as:

$$\lambda = \frac{2\pi\hbar}{m_n v} \tag{9.1}$$

The magnitude of wave vector $\vec{k}$ is expressed as:

$$k = \frac{m_n v}{\hbar} \tag{9.2}$$

In neutron scattering experiments the neutrons are considered as non-relativistic, and the neutron kinetic energy is expressed as:

$$E = \frac{1}{2}m_n v^2 = \frac{\hbar^2 k^2}{2m_n} \tag{9.3}$$

The kinetic energy of neutrons is measured in eV or meV, where 1 eV = 1.602×10^{-19} J.

The neutrons for condensed matter research are usually obtained by slowing down energetic neutrons after their production in neutron sources (nuclear reactors and spallation neutron sources) by some type of nuclear reaction. This moderation takes place using inelastic collisions within a moderator material containing light atoms. These kinds of neutrons are called thermal neutrons, and they have kinetic energies of the order of $k_B T$ where T is the temperature of the moderator. Using Eqn. 9.3 the wavelength of the neutron can be expressed as:

$$\lambda = \frac{h}{\sqrt{2mk_B T}} \tag{9.4}$$

This will lead to a value of neutron wavelength at $T \approx 300$ K to be $\lambda \approx 2$ Å, and this length will be comparable to the mean atomic separation in solid material and the typical length scale in a magnetic structure. Thus, the thermal neutrons are ideally suited for investigations of the atomic and magnetic structures of solids with neutron diffraction experiments. Furthermore, the kinetic energy of thermal neutrons is of the order of 25 meV, and that is the typical energy for collective excitations in materials including spin excitations in magnetic solids. Both wavelength and energy of thermal

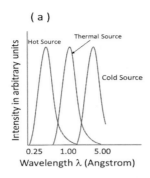 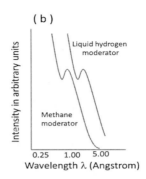

FIGURE 9.1 (a) Typical Maxwell wavelength distributions for neutrons originating from a reactor-based neutron source. The spectra shown here for a hot source, a thermal source, and a cold source are normalized so that the peaks of the Maxwell distributions are at unity. (b) Typical normalized neutron wavelength spectra from a spallation neutron source with liquid hydrogen moderator at 20 K and liquid methane moderator at 100 K. There is a high energy slowing down part along with a thermalized component with Maxwell distribution.

neutrons are thus very suitable for the studies of atomic and spin dynamics in condensed matter through inelastic neutron scattering experiments.

The spin of the neutron can be further utilized to get information on nuclear spins. In the event of a neutron being scattered by a nucleus with a non-zero spin, the strength of the interaction will also be influenced by the relative orientation of neutron and nuclear spins. So neutrons can also be used as a unique probe of nuclear spin correlation and magnetic ordering at low temperatures.

In the sections below we will provide a general introduction to the neutron scattering experiments and how can they be used to obtain useful information on the various magnetic properties of materials. There exist several classic books [1–6] and review articles [7–9, 14] on neutron scattering techniques, and interested readers are referred to those for more details on the subject.

9.1 Neutron Sources and Neutron Scattering Facilities

Neutrons used in neutron scattering experiments are produced in two different ways: (1) in a nuclear reactor by fission and (2) in proton-accelerator-based pulsed neutron sources by a process called spallation. These sources can provide neutron-flux densities required during neutron scattering experiments for the investigation of materials.

Reactor-based neutron source

Neutron scattering experiments in nuclear reactors were pioneered by Bertram Brockhouse and Clifford Shull in the early 1950s. Those pioneering works eventually led to them being awarded the Nobel Prize in physics in 1994. In those studies, a neutron beam of specific direction and wavelength was created from the neutrons coming out from a nuclear reactor by diffracting them through a crystal. The neutron beam was then aimed at a sample and the direction and energy of the neutrons were analyzed before and after the scattering events. Traditionally high-flux reactors at the Institute of Laue Langevein (ILL) at Grenoble, France, and Oak Ridge National Laboratory, USA, have provided facilities for neutron scattering, where a high continuous flux of neutrons is produced in the core of a conventional fission reactor. The high energy (MeV) neutrons are thermalized to meV energies by using a moderator and the neutron beam coming out from the moderator has a broad band of wavelengths. The energy distribution in the neutron beam can be changed to lower energies by allowing the beam to come into thermal equilibrium with a cold source such as liquid deuterium at 25 K, or to higher energies with a hot source, say, a self-heating graphite block at more than 2000 K. This results in a Maxwell distribution of energies or wavelengths in the neutron beam, which will have the characteristic temperatures of the moderators (Fig. 9.1(a)). In neutron scattering experiments a small band of wavelengths needs to be selected out of this continuous stream of electrons by a device called a monochromator at reactor sources (and choppers at spallation neutron sources).

The crystal spectrometers in reactor-based neutron sources are relatively inefficient because they only use a fraction of the neutrons scattered by the sample. This is mainly due to the narrow acceptance angles and wavelength bands associated with crystal diffraction, which means that as small as one-millionth of the energy and angle-dependent distribution of the scattered neutrons can be utilized. However, the crystal spectrometers are easily installed at research reactors, which were developed as a spin-off from the nuclear power projects. This is the reason why reactor-based neutron scattering facilities have been very popular and successful over many decades.

Spallation neutron source

Physical phenomena in many complex materials can only be understood with the exploration of the entire energy and angle distribution of scattered neutrons. In addition, one is also interested in higher energy scales. Over a period, advances in particle accelerators, mechanical engineering, and electronics have enabled the production of the intense pulsed neutron in principle of almost any energy. This intense beam of neutrons can be guided to a sample, and much of the scattered beam is then characterized and analyzed.

In an accelerator-based pulsed neutron source, neutrons are released by bombarding a target of heavy metal with high energy protons from a high-power accelerator. Neutrons are produced by the nuclear disintegration or "spallation" in the heavy-metal target. The protons are absorbed in the target atoms generating highly excited nuclei, which decay in part by the emission of neutrons. The resultant neutron beams have pulsed nature because the proton beam in the accelerator is pulsed. The spallation process releases typically about 30 MeV heat energy per neutron, which is significantly less than that of about 190 MeV in the fission process. This enables the pulsed neutron sources to deliver high neutron brightness with significantly less heat generation in the target. Rutherford Appleton Laboratory, UK, and Oak Ridge National Laboratory, USA, host such spallation neutron source.

As in the case of reactor-based sources, the neutrons produced in the spallation neutron sources have very high energies. These neutrons need to be slowed down by many orders of magnitude, say from MeV to meV. This energy reduction is achieved by hydrogen moderators, like water at room temperature, liquid methane at 100 K, and liquid hydrogen at 20 K. Such moderators surround the spallation target and they exploit the large inelastic-scattering cross section of hydrogen in slowing down the neutrons with repeated collisions with the hydrogen nuclei.

The neutron spectrum in a pulsed neutron source is different from that of a reactor source. It consists of two parts: (i) a slowing down region of hotter incompletely thermalized neutrons, and (ii) a Maxwellian distribution, which is characteristic of the moderator temperature (Fig. 9.1(b)). The neutrons of different energies interact in a complex manner with the moderator system, and hence the shape of the neutron pulse can be tailored and optimized for different types of experiments.

The characteristics of the neutrons produced by a pulsed source are quite different from those produced by a reactor. Even in the very powerful pulsed source, the time-averaged flux in neutrons per second per unit area is low in comparison with that of reactor sources. The use of time-of-flight techniques exploiting the high brightness of the neutron pulse can, however, compensate for this drawback.

The experiments performed at reactor-based and spallation neutron sources differ in details because of the different characteristics of these sources. At reactor sources, a single-wavelength neutron beam is normally used for experiments, which is produced by wavelength selection from a crystal monochromator or by velocity selection through a mechanical chopper. On the other hand, white beams containing neutrons of a wide range of wavelengths are employed for experiments in a spallation neutron source. Since the neutron beam in spallation neutron sources is pulsed in nature, the detectors perform an additional time-of-flight analysis. Energy analysis of the scattered neutron beam is performed by measuring the total time of flight (from source to detectors) or by Bragg scattering from an analyzer crystal. In the case of the former, the time a neutron takes to reach the sample from the sources needs to be determined as well. Long times of flights are required for high-resolution measurements. This can be achieved with the help of neutron guide tubes. They work on the principle of total reflection and are used to transport the neutrons (except the very high energy ones) without significant loss of intensity over long distances.

At present, there are neutron source facilities available in many parts of the world [14]: Australia: – ANSTO-HIFAR Reactor (Sydney); Canada: – Chalk River Laboratories Canadian Neutron Facility; Great Britain: – ISIS Spallation Source, Rutherford – Appleton Laboratory (Oxford); France: – ILL: Institut Laue Langevin LLB, Leon Brillouin Laboratory at CEA (Saclay); Germany: – FRM II Technical University (Garching), Hahn-Meitner Institute (Berlin); Hungary: – Budapest Neutron Centre (Budapest); India: – Bhabha Atomic Research Centre (Mumbai); Japan: – J-PARC-Japan Proton Accelerator Research Complex (Tokai); Netherlands: – Interfaculty Reactor Institute (Delft); Russia: – Joint Institute for Nuclear Research (Dubna); Switzerland: – Paul Scherrer Institute (Villigen); Sweden: European Spallation Source (ESS), Lund; USA: – Los Alamos Neutron Science Center (Los Alamos), NIST – Center for Neutron Research (Washington), ORNL – High Flux Isotope Reactor (Oak Ridge), and SNS – Spallation Source, Oak Ridge (Tennessee).

Normally, reactor-based neutron sources operate continuously and spallation neutron sources are pulsed. But this is not always the case. The reactor at the Franck Laboratory of Nuclear Physics (Dubna, Russia) consists of two pieces of sub-critical uranium, one of which is mounted on a rotating disc. As the two pieces come into proximity, the mass goes critical producing neutrons and as it goes sub-critical again, neutron production ceases [14]. The net result is that it acts like a pulsed neutron source. On the other hand, the SINQ facility at the Paul Scherrer Institute (Villigen, Switzerland) is a continuous spallation source.

In summary, both reactor and spallation neutron sources use a moderator to moderate high energy neutrons to "thermal" velocities and then transport the neutrons through neutron guide systems to the neutron scattering instruments. The wavelength of thermal neutrons is around 2 Å, which is similar to inter-atomic distances, and the energy is around 25 meV, which is similar to elementary excitations in solids. It is thus possible to obtain information simultaneously on the crystal structure and dynamics of solid using such thermal neutrons.

9.2 Basics of Neutron Scattering

A neutron scattering experiment is shown schematically in Fig. 9.2, in which a beam of neutrons characterized by a wave vector $\vec{k}_0$ and the spin σ_0 falls on the sample. The interaction probability of the neutron inside the sample is rather small in a typical experimental situation and most neutrons will be transmitted without any

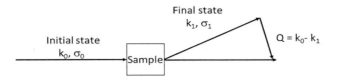

FIGURE 9.2 Schematic representation of a general neutron scattering scattering experiment.

interaction. Some neutrons, however, are scattered to a final state characterized by a wave vector $\vec{k}_1$ and the spin σ_1, which can be measured with a neutron detector placed in the direction $\vec{k}_1$. The vector $\vec{Q} = \vec{k}_0 - \vec{k}_1$ in Fig. 9.2 defines the scattering vector.

9.2.1 Neutron cross sections

The interaction of a neutron beam with materials can be described with the central concept of cross-sections. The flux of a neutron beam is defined as:

$$\Phi = \frac{n}{A} \tag{9.5}$$

where n stands for the number of neutrons impinging on a surface per second, and A for the surface area perpendicular to the neutron beam direction. The unit of Φ is in n/(cm²s).

The *neutron scattering cross section* σ of a system is defined by its ability to scatter neutrons:

$$\sigma = \frac{\acute{n}}{\Phi} \tag{9.6}$$

where $\acute{n}$ stands for the number of neutrons scattered per second. The *neutron scattering cross section* σ has units of area that for a single nucleus can be visualized as the effective area of the nucleus perpendicular to the neutron beam. The scattering cross section defined here is the total cross section, which depends on the volume of the sample.

Let the incident beam be characterized by a uniform flux Φ. If the sample has N identical atoms in the beam path, and the detector subtends a solid angle $\Delta\Omega$ and has efficiency η, one may expect the count rate C in the detector to be proportional to all these quantities. The constant of proportionality is called the "differential cross section" and is expressed as:

$$\frac{d\sigma}{d\Omega} = \frac{C}{\Phi N (\Delta\Omega) \eta} \tag{9.7}$$

The differential cross section is a function of the magnitude and direction of $\vec{k}_0$, and the direction of $\vec{k}_1$. It is also the characteristic of the material being studied, and it may also depend on the spin state of the incident neutron. The most important types of interaction for the study of solid materials are the nuclear interaction and the magnetic dipole interaction.

9.2.2 Conservation of energy and momentum

In Fig. 9.2 we see that during the neutron scattering experiment the neutron wave vector $\vec{k}_1$ changes direction. This indicates that there is a change in the neutron momentum, which must be exchanged with the sample. The momentum transferred to the sample is conventionally described in terms of the scattering wave vector $\vec{Q}$, and according to the law of momentum conservation:

$$\vec{k}_1 - \vec{k}_0 = \vec{Q} \qquad (9.8)$$

The scattering atom takes up the momentum $\hbar\vec{Q}$ and shares it with the rest of the sample. In certain cases, even the sample as a whole may recoil.

In general energy is also exchanged with the sample, and the law of conservation of energy can be expressed as:

$$\frac{\hbar^2 k_0^2}{2m} - \frac{\hbar^2 k_1^2}{2m} = E \qquad (9.9)$$

Here E denotes the energy transferred to the sample, and the two terms on the left hand of Eqn. 9.9 represent the energy of the incident and scattered neutron. Therefore, a set of values of $\vec{Q}$ and E is associated with a process in which the neutron is scattered from the momentum state $\vec{k}_0$ to $\vec{k}_1$. In a neutron scattering experiment, the intensity of the scattering as a function of these variables $(\vec{Q}, E)$ is a property of the sample being studied and the sample environment, i.e., temperature, pressure, magnetic field, etc.

9.2.3 Master formula for neutron scattering

We shall now provide a short introduction to the quantitative basis for the concepts introduced so far. This will not be a rigorous development of the subject, and for that, the interested readers are referred to many excellent works of literature including the classic textbook by Lovesey [3] and the review article by Price and Sköld [7].

With reference to the conceptual scattering experiment shown in Fig. 9.2 we now consider a change in state of the neutron-sample system from a state λ_0 to a state λ_1 and also a possible change in neutron spin from σ_0 to σ_1. The Eqn. 9.7 for the differential cross section can now be expressed as [7]:

$$\left(\frac{d\sigma}{d\Omega}\right)_{k_0,\sigma_0,\lambda_0 \to k_1,\sigma_1,\lambda_1} = \frac{1}{N\Phi\Delta\sigma} W_{k_0,\sigma_0,\lambda_0 \to k_1,\sigma_1,\lambda_1} \qquad (9.10)$$

Here $W_{k_0,\sigma_0,\lambda_0}$ is the of transitions per second of the combined neutron-sample system from the state $(k_0, \sigma_0, \lambda_0)$ to the state $(k_1, \sigma_1, \lambda_1)$.

The right-hand side of Eqn. 9.10 can be evaluated using Fermi's Golden Rule, in other words, the first-order perturbation theory [7].

$$W_{k_0,\sigma_0,\lambda_0 \to k_1,\sigma_1,\lambda_1} = \frac{2\pi}{\hbar} |<k_1,\sigma_1,\lambda_1|V|k_0,\sigma_0,\lambda_0>|^2 \rho_{k_1,\sigma_1}(E_1) \qquad (9.11)$$

Here V is the interaction potential between the sample and the neutron, and $\rho_{k_1,\sigma_1}(E_1)$ is the density of final scattering states per unit energy interval. The use of the first-order perturbation theory is valid here since the nuclear potential is short-range and only s-wave scattering is possible.

We now consider the neutron and the sample to be represented by a large box of volume V_0, and the incident and scattered wave-functions are $\frac{1}{\sqrt{V_0}} e^{i\vec{k}_0 \cdot \vec{r}} |\sigma_0>$ and $\frac{1}{\sqrt{V_1}} e^{i\vec{k}_1 \cdot \vec{r}} |\sigma_1>$, respectively. Then the number of states in scattered energy interval dE_1 is expressed as:

$$\rho_{k_1,\sigma_1}(E_1) dE_1 = \frac{V_0^3}{8\pi} k_1^2 dk_1 \Delta\Omega \qquad (9.12)$$

Now $dE_1 = \hbar^2 k_1 dk_1/m$; hence we can write [7]:

$$\rho_{k_1,\sigma_1}(E_1) = \frac{V_0}{(8\pi)^3} \frac{m k_1}{\hbar^2} \Delta\Omega \qquad (9.13)$$

The incident neutron beam flux can be written as the normalized number density times velocity v:

$$\Phi = \frac{v}{V_0} = \frac{\hbar k_0}{V_0 m} \qquad (9.14)$$

Using Eqns. 9.11, 9.13, and 9.14, we can rewrite Eqn. 9.10 as:

$$\left(\frac{d\sigma}{d\Omega}\right)_{k_0,\sigma_0,\lambda_0 \to k_1,\sigma_1,\lambda_1} = \frac{1}{N} \frac{k_1}{k_0} \left(\frac{m}{2\pi\hbar^2}\right)^2 |<\vec{k}_1,\sigma_1,\lambda_1|V|\vec{k}_0,\sigma_0,\lambda_0>|^2 \qquad (9.15)$$

Here $|\vec{k}_0>$ and $|\vec{k}_1>$ denote the plane waves $e^{i\vec{k}_0 \cdot \vec{r}}$ and $e^{i\vec{k}_0 \cdot \vec{r}}$, respectively.

We shall now invoke the energy conservation condition:

$$E = E_o - E_1 = E_{\lambda_1} - E_{\lambda_2} \tag{9.16}$$

This condition can be incorporated as a delta function in the scattering cross section, thus giving [7]:

$$\left(\frac{d^2\sigma}{d\Omega dE}\right)_{k_0,\sigma_0,\lambda_0 \to k_1,\sigma_1,\lambda_1} = \frac{1}{N}\frac{k_1}{k_0}\left(\frac{m}{2\pi\hbar^2}\right)^2$$

$$\times \left|<\vec{k}_1,\sigma_1,\lambda_1|V|\vec{k}_0,\sigma_0,\lambda_0>\right|^2 \delta(E + E_{\lambda_0} - E_{\lambda_1}) \tag{9.17}$$

We shall now sum over all final states of the sample λ_1, and final polarization states of the neutron σ_1, and average over all initial λ_0 of the sample, which occur with probability p_{λ_0}, and over the initial polarization states of the neutron, which occur with probability p_{σ_0}, to get an expression for the double differential cross section [7]:

$$\left(\frac{d^2\sigma}{d\Omega dE}\right)_{k_0 \to k_1} = \frac{1}{N}\frac{k_1}{k_0}\left(\frac{m}{2\pi\hbar^2}\right)^2 \sum_{\lambda_o,\sigma_0} p_{\lambda_0} p_{\sigma_0}$$

$$\times \sum_{\lambda_1,\sigma_1} \left|<\vec{k}_1,\sigma_1,\lambda_1|V|\vec{k}_0,\sigma_0,\lambda_0>\right|^2 \delta(E + E_{\lambda_0} - E_{\lambda_1}) \tag{9.18}$$

This Eqn. 9.18 is known as the "master formula" for neutron scattering, and it forms the basis of the interpretation of all neutron scattering experiments.

In the case of elastic scattering involving no energy changes between the initial and final states, the relation $|\vec{k}_0| = |\vec{k}_1|$ holds good. With an assumption a monochromatic but not polarized beam of neutrons, integrating out the energy dependence and after multiplication by the number of entities N causing the scattering, the master formula is simplified to [9]:

$$\left(\frac{d\sigma}{d\Omega}\right)_{k_0 \to k_1} = \left(\frac{m}{2\pi\hbar^2}\right)^2 |<\vec{k}_1,\sigma_1|V|\vec{k}_0,\sigma_0>|^2 \tag{9.19}$$

Here V is the interaction potential sensed by a neutron at coordinates defined by $\vec{r}$ in the solid material. It is useful to introduce a scattering amplitude operator $a(\vec{Q})$ expressed by the relation:

$$<\vec{k}_1,\sigma_1|a(\vec{Q})|\vec{k}_0,\sigma_0> = \left(\frac{m}{2\pi\hbar^2}\right)<\vec{k}_1,\sigma_1|V|\vec{k}_0,\sigma_0> \tag{9.20}$$

Here $\vec{Q} = \vec{k}_0 - \vec{k}_1$ is the scattering vector. We can then write the differential cross section for the scattering of an unpolarized neutron beam with no interference between nuclear and magnetic contributions as the sum of the nuclear and the magnetic parts [9]:

$$\frac{d\sigma}{d\Omega}(\vec{Q}) = \left(\frac{d\sigma}{d\Omega}\right)_N (\vec{Q}) + \left(\frac{d\sigma}{d\Omega}\right)_M (\vec{Q}) \tag{9.21}$$

Here nuclear and magnetic contribution has been indicated by subscripts N and M, respectively.

9.2.4 Nuclear scattering

The range of the strong nuclear force is in femtometres (fm), which is much smaller than the neutron wavelength (measured in Å) used normally in neutron scattering experiments. Hence, the nuclear scattering is isotropic, and is characterized by a single scalar parameter b, called "scattering length". The interaction between the incident neutron represented by a plane wave and a point-like nucleus can therefore be described using the Born approximation. It is then appropriate to describe it in the form of a delta function Fermi pseudopotential [9]:

$$V_N(\vec{r}) = \frac{2\pi\hbar^2}{m} b\delta(\vec{r} - \vec{R}) \tag{9.22}$$

Here b is called the scattering length. This in general will depend on the spin state of the neutron-nucleus system, and $\vec{r}$ and $\vec{R}$ represent the instantaneous positions of neutron and nucleus, respectively. At temperatures of few tens of a Kelvin where nuclear polarization is negligible, the state of the scattered neutron can be described by a spherical wave. The nuclear scattering amplitude for a single isotope nucleus becomes quite simple, insensitive to the scattering vector, and equal to the scattering length of the isotope b. The scattering length b is different not only for each atom but also for each isotope of the same nucleus. The sign and magnitude of b change irregularly with atomic number Z and atomic weight A. This is in contrast to X-ray scattering for which the atomic scattering length is a monotonically increasing function of Z as X-ray gets scattered from the electronic density. This phenomenon has some important consequences and makes neutrons sensitive to the presence of light atoms, particularly hydrogen, and also to the difference between atoms of similar atomic numbers. The variation in scattering length between different isotopes of the same element is often large, and this is utilized in experiments using isotope substitution.

The differential cross section for the nuclear scattering of an unpolarized neutron beam by a natural collection of nuclei without nuclear polarization can be expressed in terms of the coherent and incoherent parts:

$$\left(\frac{d\sigma}{d\Omega}\right)_N = \left(\frac{d\sigma}{d\Omega}\right)_{Coh} + \left(\frac{d\sigma}{d\Omega}\right)_{Incoh} \tag{9.23}$$

The coherent part contributes to the elastically measured Bragg reflections arising from the average nucleus with $\sigma_{Coh} = 4\pi b_C^2$. On the other hand, the incoherent scattering has two contributions, arising from (i) a random distribution of the deviations of the scattering lengths from the mean value, and (ii) the fluctuations of the nuclear spins. The value of b varies erratically as a function of the atomic number. Some elements like vanadium or aluminum have small cross sections; vanadium (dominated by an incoherent cross section) is usually used as containers for the samples in the neutron diffraction experiments, and aluminum (very small coherent and also incoherent cross sections) is used for cryostats and other auxiliary components for inelastic neutron experiments. Elements with large cross sections such as boron, cadmium, or gadolinium, on the other hand, are used for an effective shielding and lowering of the background signal.

The interaction potential $V(\tilde{r})$ between the neutron and the material sample can then be determined by summing over the atoms in the sample [7]:

$$V(\vec{r}) = \frac{2\pi\hbar^2}{m} \sum_i b_i \delta(\vec{r} - \vec{R}_i) \tag{9.24}$$

We now evaluate the average of Eqn. 9.20 over the neutron wave function and get:

$$<k_1|V|k_0> = \frac{2\pi\hbar^2}{m} \sum_i b_i \int dr e^{-i\vec{k}_1 \cdot \vec{r}} \delta(\vec{r} - \vec{R}_i) e^{i\vec{k}_0 \cdot \vec{r}} = \frac{2\pi\hbar^2}{m} \sum_i b_i e^{\vec{Q} \cdot \vec{R}_i} \tag{9.25}$$

where $\vec{Q} = \vec{k}_0 - \vec{k}_1$. The master formula in Eqn. 9.18 can now be rewritten as [7]:

$$\frac{d^2\sigma}{d\Omega dE} = \frac{1}{N}\frac{k_1}{k_0} \sum_{\lambda_0,\sigma_0} p_{\lambda_0} p_{\sigma_0} \left| \sum_{\lambda_1,\sigma_1} \sum_i b_i <\sigma_1 \lambda_1 | e^{i\vec{Q} \cdot \vec{R}_i} | \sigma_0 \lambda_0 > \right|^2 \delta(E + E_{\lambda_0} - E_{\lambda_1}) \tag{9.26}$$

Most of the neutron scattering experiments use unpolarized neutron beams where both spin states are equally probable. In such cases, the dependence on σ_0 and σ_1 can be ignored and the master formula gets simplified further [7]:

$$\frac{d^2\sigma}{d\Omega dE} = \frac{1}{N}\frac{k_1}{k_0} \sum_{\lambda_0} p_{\lambda_0} \left| \sum_{\lambda_1} \sum_i b_i <\lambda_1 | e^{i\vec{Q} \cdot \vec{R}_i} | \lambda_0 > \right|^2 \delta(E + E_{\lambda_0} - E_{\lambda_1}) \tag{9.27}$$

A sample of a solid material can be viewed as an array of nuclei arranged on a periodic lattice in three dimensions. The material in this sense is built by translations along unit cell vectors a, b, and c defining the lattice. This enables one to define the position of the n-th unit cell with respect to an arbitrary origin. Let $\vec{r}_j$ be the position of an individual atom with index j within such an n-th unit cell. Then the position of this atom within the crystal can be written as $\vec{R}_{nj} = \vec{R}_n + \vec{r}_j$. In the reciprocal space, the unit cell vectors are $a' = 2\pi b \times c/V$, $b' = 2\pi c \times a/V$, $c' = 2\pi a \times b/V$ where V is the volume of the unit cell in real space. The volume of the reciprocal unit cell as $(2\pi)^3/V$. The reciprocal lattice vector is defined as $\vec{Q} = h\vec{a}' + k\vec{b}' + l\vec{c}'$. The nuclear interaction potential expressed in Eqn. 9.24 can be rewritten as [9]:

$$V_N(\vec{r}) = \frac{2\pi\hbar^2}{m} \sum_{nj} b_j \delta(\vec{r} - \vec{R}_{nj}) \qquad (9.28)$$

The summation runs over all crystallographic unit cells and atoms within one unit cell. A material may contain different atoms and the index j contains the possibility of different scattering lengths. The resulting nuclear cross section from a material crystal reads as [9]:

$$\left(\frac{d\sigma}{d\Omega}\right)_N = \left|\sum_{nj} b_j e^{i\vec{Q}\cdot\vec{R}_{nj}}\right|^2 = \frac{(2\pi)^3}{V} N \sum_H |F_N(\vec{Q})|^2 \delta(\vec{Q} - \vec{H}) \qquad (9.29)$$

Here $\vec{H}$ denotes the reciprocal lattice vector where the diffraction occurs, and $F_N(\vec{Q})$ denotes the nuclear structure factor, which provides the structural information of atoms in a crystal at rest. At finite temperatures, all atoms vibrate around their equilibrium positions in the crystal lattice, and this effect leads to a decrease in the scattered intensities of neutrons, especially at higher $\vec{Q}$ values. This effect is represented by the Debye–Waller factor of a form e^{-W_j}, where W_j denotes how much the actual position of the atom deviates from its average position in the crystal lattice.

In a neutron scattering experiment, one detects how many neutrons are diffracted at a particular reciprocal crystal lattice point. In a crystal consisting of identical atoms (i.e., one nuclear species) and one atom per unit cell, the differential cross section is:

$$\left(\frac{d\sigma}{d\Omega}\right)_{coh} = b^2 \qquad (9.30)$$

The corresponding nuclear intensity is observed at 2θ angle for which the Bragg's condition $2d_{hkl}\sin\theta = \lambda$ is satisfied. These intensities are affected by several other factors apart from the Debye–Waller factor, such as the geometrical Lorentz factor, and need to be corrected further for absorption and extinction, which is different for different Bragg reflections. At the end of the experiment, one gets a set of nuclear structure factors that have to be compared with the calculated ones.

9.2.5 Magnetic scattering

In magnetic samples, neutrons are also scattered from unpaired electrons through magnetic interaction. Here the basic difference with respect to nuclear scattering is the fact that the magnetic interaction is not short-range and both the magnetic moments of the neutron and the unpaired electrons around the nucleus are vectors. The magnetic interaction potential of a neutron at position $\vec{r}$ according to standard results of quantum mechanics is expressed as [9]:

$$V_M(\vec{r}) = -\mu_n \sum_j \left(\nabla \times \left(\frac{\mu_j \times (\vec{r} - \vec{R}_j)}{|\vec{r} - \vec{R}_j|^3} \right) - \frac{2\mu_B p_j \times (\vec{r} - \vec{R}_j)}{\hbar |\vec{r} - \vec{R}_j|^3} \right) \quad (9.31)$$

Here μ_n, μ_j, and p_j are the neutron moment, moment due to the spin of the jth unpaired electron and its momentum, respectively. This magnetic scattering potential is not a short-ranged one, but it is weak and hence the usual conditions for the validity of the first-order perturbation theory apply here as well. The interaction potential has two components, the first one originating from the spins of unpaired electrons and the second one from their orbital motion around the nuclei. Inserting the magnetic interaction potential expressed in Eqn. 9.31 in the simplified master formula 9.18, one can write the magnetic scattering amplitude as:

$$a_M(\vec{Q}) = p\sigma |\hat{Q} \times \vec{M}(\vec{Q}) \times \hat{Q}| = p f(\vec{Q}) \mu_\perp . \sigma \quad (9.32)$$

Here $\hat{Q} = \vec{Q}/|\vec{Q}|$ is the unit scattering vector, $\vec{M}(\vec{Q})$ is the Fourier transform of the magnetization density in the direct space arising from both spin and orbital components of all unpaired electrons, and $p = 2.696$ fm is a constant. The function $f(\vec{Q})$ is termed as magnetic form factor, which is specific for each isotope of an atom in a particular valence state. The term $\hat{Q} \times \vec{M}(\vec{Q}) \times \hat{Q}$ is a projection of $\vec{M}(\vec{Q})$ on a plane normal to the scattering vector $\vec{Q}$. This enables simplification of the formula by introducing the perpendicular component of the magnetic moment $\mu_\perp$, and allows the determination of the magnetic moments involved in the magnetic ordering in a solid material. We can write the differential cross section of the magnetic ion for an unpolarized neutron beam as:

$$\left(\frac{d\sigma}{d\Omega} \right)_M (\vec{Q}) = p^2 f^2(\vec{Q}) \mu_\perp^2 \quad (9.33)$$

To study the scattering of the neutron by a magnetic solid with an expression analogous to nuclear diffraction, one has to, however, consider the magnetic moments as a new set of entities that are in principle allowed to order with a different periodicity. For example, the magnetic unit cell may be larger than the crystallographic counterpart. When the magnetic unit cell can be modelled by a simple multiplication of the crystallographic unit cell, then the magnetic structure is considered to be commensurate with the underlying crystal structure. In this situation, it is easy to investigate the mutual orientation of the moments involved in the ordering and the symmetry of the magnetic structures like a simple ferromagnet, antiferromagnet, and ferrimagnet. The determination of a magnetic structure factor here is analogous to the case of atomic structure, and the summation runs over the magnetic unit cell including only atoms carrying a magnetic moment. One has to take care, however, while evaluating the magnitude of the magnetic moment that the proper scaling is taken into account.

Magnetic structures can be also incommensurate with the crystallographic structure, and that leads to very different kinds of finite magnetic unit cells. In this case, a suitable approach to the determination of magnetic structures involves the concept of the reciprocal space and propagation vectors. This enables a general classification. Let us consider a magnetic crystal composed of periodically arranged magnetic moments $\vec{\mu}_{nj}$ with a long-range order, which is periodical in real space but the periodicity is different from that of the atoms. It is then possible to Fourier transform such a real-space function as:

$$\vec{\mu}_{nj} = \sum_k m_{jk} e^{-i\vec{k}\cdot\vec{R}_n} \tag{9.34}$$

Here m_{jk} is the Fourier component associated with the propagation vector $\tilde{k}$, and the summation runs over all existing propagation vectors. While magnetic moments are real quantity, both $\exp(-i\vec{k}\cdot\vec{R}_n)$ and m_{jk} components can in the general be complex. Thus, for every propagation vector $\vec{k}$ there is also an associated vector $-\vec{k}$ with $m_{j,-k} = m^*_{j,-k}$. The summation is performed only over k in special cases, when $\exp(-i\vec{k}\cdot\vec{R}_n)$ is real. Magnetic moments are associated with atoms, which are arranged periodically in the crystal lattice. Hence, the propagation vectors are periodic in reciprocal space, and they can be chosen within the first Brillouin zone for Bravais lattices. The differential cross section for the magnetic neutron scattering can be expressed as [9]:

$$\left(\frac{d\sigma}{d\Omega}\right)_M (\vec{Q}) = \frac{(2\pi)^3}{V} N \sum_H \sum_k p^2 \left|\sum_j m_{jk} f_j(Q) e^{i\vec{Q}\cdot\vec{r}_j}\right|^2 \delta(\vec{Q}-\vec{H}-\vec{k})$$

$$= \frac{(2\pi)^3}{V} N \sum_H \sum_k \left|F_M(\vec{Q})\right|^2 \delta(\vec{Q}-\vec{H}-\vec{k}) \tag{9.35}$$

Here the magnetic form factor $F_M(\vec{Q})$ is a vector quantity that can be either real or complex or even imaginary. Only $F_{M\perp}(\vec{Q})$ onto the plane perpendicular to the scattering vector $\vec{Q}$ can be used in the Eqn. 9.35. $F_{M\perp}(\vec{Q})$ is sometimes called magnetic interaction vector, and this leads to an expression similar to Eqn. 9.33. If all the magnetic moments contributing to a particular propagation vector $\vec{k}$ are parallel to the scattering vector $\vec{Q}$, the differential cross section becomes zero. There will be no neutrons scattered from such magnetic moments, and hence no contribution due to magnetic order to the Bragg reflection will be observed at places defined by $\vec{k}$.

9.2.6 Classification of magnetic structures

Simple ferromagnetic, antiferromagnetic, and ferromagnetic materials differ macroscopically by their response in an external magnetic field. On a microscopic scale, however, they differ in the manner how the individual magnetic moments are arranged. The propagation vectors associated with the magnetic structures can be commensurate with the crystal structure if magnetic unit cells are an integer multiple of the crystallographic unit cells. However, this may not be the case always. The concept of the propagation vector enables a systematic description of different magnetic structures, and this is achieved with the help of the modulus of $|\vec{k}|$. The magnetic structure is said to be commensurate with the crystal lattice if $|\vec{k}|$ is a rational number; otherwise, the magnetic structure is incommensurate. The number of propagation vectors that define the magnetic structure is another defining criterion. The magnetic structure is called single-$\vec{k}$ if only one propagation vector $\vec{k}$ is present; otherwise, multi-$\vec{k}$ structures are realized. Then there are collinear and non-collinear magnetic structures, which are distinguished with respect to the direction of moments.

Commensurate magnetic structures

We consider a Bravais lattice with only one magnetic atom in a two-dimensional crystallographic unit cell. A single propagation vector $\vec{k} = 0$ would indicate that the magnetic and crystallographic unit cells have equal size. If in one unit cell the magnetic moment points towards a particular direction, then this direction must be preserved in all other unit cells. This represents the case of a simple ferromagnet shown in Fig. 9.3. Here every magnetic atom in real space carries an identical magnetic moment pointing along the same direction as in all others. In the corresponding reciprocal space, the magnetic components appear at the same positions as the nuclear reflections (Fig. 9.3(b)). The magnetic contribution will, however, be zero at reciprocal points for which there is no perpendicular component of the magnetic moment (in Fig. 9.3(b) along the $\tilde{b}^*$). Hence, the only type of magnetic order possible for materials with a single magnetic moment in the unit cell and $\vec{k} = 0$ is ferromagnetic. The situation is more complicated in materials with two or more magnetic moment-carrying atoms in the unit cell. Here the magnetic

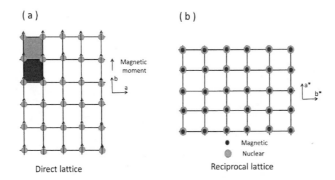

FIGURE 9.3 Schematic representation of a simple ferromagnet with moments residing on an orthorhombic lattice projected to the a–b plane: (a) the real space (left) and (b) the reciprocal space (right). Crystallographic unit cell (shown in gray colour) is of the same size as the magnetic unit cell (shown in black colour). Magnetic scattering intensities appear at the top of the nuclear reflections. (Adapted from reference [9] with permission from Elsevier.)

moments can point to different directions even in the case of $\vec{k} = 0$, and one needs to consider a non-collinear magnetic structure.

The magnetic and crystallographic unit cells are different if the propagation vector $\vec{k}$ has a non-zero value. If $\vec{k}$ is a rational number $W = m/n$, where m and n are integer numbers, then the magnetic unit cell is $W = n/m$ times larger than the nuclear or atomic unit cell. In these possibilities, magnetic structures with a single propagation vector $\tilde{k} = \tilde{H}/2$, which corresponds to a symmetry point of the Brillouin zone, represent a special case. A limited number of such symmetry points for each of the 14 Bravais lattices represent different antiferromagnetic structures. A simple antiferromagnetic structure in two dimensions is shown in Fig. 9.4. Each of the magnetic atoms carries a magnetic moment of equal size, but orientations of moment alternate between two exactly antiparallel directions as one moves along the direction of a-axis. The resultant magnetic structure can be visualized as comprising of two interpenetrating ferromagnetic sublattices, each having a unit cell twice as large as the crystallographic unit cell. The nodes corresponding to magnetic order in the reciprocal space do not coincide with the Fourier transform of the crystal structure lattice, but instead lie between them. The magnetic signal arises from different reciprocal lattice points than the crystalline one. It may be noted at this point that for $\tilde{k} = 0$ and $\tilde{k} = \tilde{H}/2$, the Fourier component m_{jk} is a real number and can be directly correlated to the magnetic moment μ_{nj}. A more general approach is needed for propagation vectors with other rational values and that will lead to the cases of incommensurate structures.

Summarizing the discussion (above) on the case of a single magnetic moment in a crystallographic unit cell, it is possible to make a simple correspondence between the type of the propagation vector $\vec{k}$ and the type of the magnetic ordering. A material

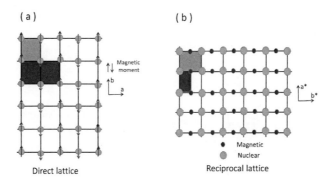

FIGURE 9.4 Schematic representation of a simple antiferromagnet with magnetic moments situated on an orthorhombic lattice projected to the a–b plane: (a) the real space (left) and (b) the reciprocal space (right). All magnetic moments are of the same size and point either along the b-axis direction or in the opposite direction. The magnetic unit cell for materials with one magnetic atom in the crystallographic unit cell (shown in black colour) in real space (reciprocal space) is larger (smaller) than the nuclear unit cell (shown in gray colour). Two crystallographic unit cells are necessary to repeat the pattern as one moves along the a-axis direction. Consequently, the propagation vector of the magnetic structure is $\vec{k} = (1/2, 0, 0)$, and the magnetic intensities appear at different reciprocal lattice spots than the nuclear reflections. (Adapted from reference [9] with permission from Elsevier.)

with Bravais lattice has to order ferromagnetically for $\vec{k} = 0$. The propagation vector $\vec{k}$ must be non-zero in the case of an antiferromagnet. This simple rule, however, is not good enough for two or more magnetic moments in the crystallographic unit cell, since they can get coupled within this cell either ferromagnetically or antiferromagnetically. In the antiferromagnetic case, if these moments are not equal in size, then one encounters a ferrimagnetic state as shown in Fig. 9.5. In this case, the moments in one unit cell point towards the same direction, and the resultant magnetic structure is collinear. It is clear, however, that different moments in one unit cell can have different orientations while preserving the $\vec{k}$ value. One thus realizes that the scheme introduced above involving Eqns. 9.31 and 9.32 describe the translational periodicity of magnetic moments and not how the moments couple within the magnetic unit cell. The information about the mutual orientation of the magnetic moments is contained in the values of the Fourier components. The number of possibilities increases with the number of moments involved, and the task of determining the coupling between them can be quite complex. However, these magnetic moments are often symmetry related, and that offers a possibility to simplify the task in some cases.

Incommensurate magnetic structures

One cannot define a magnetic unit cell for an incommensurate structure which would be an integer multiple of the crystallographic unit cell. Many types of magnetic

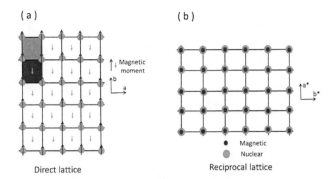

FIGURE 9.5 Schematic representation of a ferrimagnet with magnetic moments situated on an orthorhombic lattice projected to the a–b plane: (a) the real space (left) and (b) the reciprocal space (right). In a ferrimagnetic material, two different groups of magnetic moments lead to two sublattices. In one sublattice, all magnetic moments point along the same direction (in this case along the b-axis), which is antiparallel to the direction of moments within the other sublattice. The antiparallel coupling thus takes place within a single nuclear unit cell. The crystallographic unit cell (shown in gray colour) has the same size as the magnetic unit cell (shown in black colour), and the propagation vector of the magnetic structure is $\vec{k} = (0, 0, 0)$. The magnetic intensities appear in the same place as in the case of a ferromagnet. The intensities, however, are different. (Adapted from reference [9] with permission from Elsevier.)

structures fall into this category. The simplest example of an incommensurate magnetic structure is a sine wave-modulated structure as shown in Fig. 9.6(a). Here the magnetic moments are modulated in size in direct space as one moves along the $\vec{k}$ direction according to the relation:

$$\vec{\mu}_{nj} = \vec{\mu}_j \vec{u} cos(\vec{k}.\vec{R}_n + \phi_j) \tag{9.36}$$

Here the vector $\vec{u}$ represents the direction of the propagating magnetic moment and ϕ_j its phase. If $\vec{\mu}_j$ and $\vec{u}$ are parallel (perpendicular) to each other, a longitudinally (transversally) modulated structure is realized. Examples of such structures are shown in Fig. 9.6(b) and Fig. 9.6(c). The magnetic moment is a real quantity, and its spatial distribution consists of two components, one associated with $\vec{k}$ and the other with $-\vec{k}$. These two components are represented by $\vec{m}_{jk} = \vec{\mu}_j/\vec{u}\exp(-i\phi_j)$ and $\vec{m}_{j,-k} = \vec{\mu}_j/\vec{u}\exp(i\phi_j)$, respectively, which are conjugated complexes. In this case the magnetic structure factor in Eqn. 9.35 is expressed as:

$$F_M(\vec{Q}) = \frac{1}{2} p \sum_j f_j(\vec{Q}) \vec{\mu}_j.\vec{u} exp(-i\phi_j) exp(i\vec{Q}.\vec{r}_j) \tag{9.37}$$

Here the factor $1/2$ indicates the fact that the summation is over $\vec{k}$ and $-\vec{k}$. In the truly incommensurate magnetic structures, all the magnetic moments within the crystal are different either in magnitude or in direction.

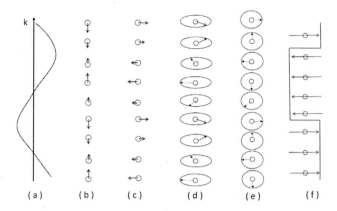

FIGURE 9.6 Schematic representation of: (a) a sine wave modulated structures, (b) longitudinally modulated collinear magnetic structure, (c) transversally modulated collinear magnetic structure, (d) a spiral magnetic structure, (e) a cycloidal magnetic structure, and (f) a commensurate collinear magnetic structure.

The collinear magnetic structures with sine wave modulation shown in Fig. 9.6(b) and Fig. 9.6(c) are a special case of a more general structure with non-collinear magnetic moments and have two or three non-zero Cartesian components. If there is only one propagation vector, they are simultaneously modulated along the same propagation vector. In the case of two Cartesian components (the third component is zero), the magnetic moments lie within one plane. For this structure, another degree of freedom being available is the phase shift between modulations of both components. The magnetic structure is again collinear if the phase shift is zero, and the moments remain aligned along a single direction within the plane defined by both Cartesian components. On the other hand, if the phase shift equals to $\pi/2$, the term $cos(\vec{k}.\vec{R}_n + \phi_j + \pi/2)$ in Eqn. 9.36 becomes $sin(\vec{k}.\vec{R}_n + \phi_j)$, and one gets a helical magnetic structure. Two possibilities arise here again, as the propagation vector $\vec{k}$ can be perpendicular to the plane in which the magnetic moments are modulated, i.e., spiral structure, or lies within this plane, i.e., cycloidal structure. Pure elements like Tb, Ho, and Dy display such spiral structures at low temperatures. A helix structure is either elliptically ($\vec{\mu}_{1j} \neq \vec{\mu}_{2j}$) or circularly modulated ($\vec{\mu}_{1j} = \vec{\mu}_{2j}$) depending on the magnitudes of the two Cartesian components. In real space in the more general elliptical case the magnetic moments are modulated as [9]:

$$\vec{\mu}_{nj} = \vec{\mu}_{1j}\vec{u}cos(\vec{k}.\vec{R}_n + \phi_j) + \vec{\mu}_{2j}\vec{v}sin(\vec{k}.\vec{R}_n + \phi_j) \qquad (9.38)$$

The corresponding magnetic form factor is expressed as [9]:

$$F_M(\vec{Q}) = \frac{1}{2}p\sum_j f_j(\vec{Q})[\vec{\mu}_{1j}.\vec{u} + \vec{\mu}_{2j}.\vec{v}_j]exp(-i\phi_j)exp(i\vec{Q}.\vec{r}_j) \qquad (9.39)$$

In the case of $\vec{\mu}_{2j}=0$ the formula in Eqn. 9.39 reduces to the case of a sine wave-modulated structure. Fig. 9.6(d) and Fig. 9.6(e) present simplified examples of such helical and cycloidal structures. A simple commensurate magnetic structure having a sequence of ++++-- is also shown in Fig. 9.6(f) for comparison.

The magnetic structures mentioned above are fully described by a single propagation vector $\vec{k}$ and the associated $-\vec{k}$ vector, and such magnetic structures are termed as single-$\vec{k}$ structures. However, there are systems where the magnetic moments in direct space are modulated in more directions at the same time. Description of such structures needs more than one $\vec{k}$ vector. The Fourier expansion of real space magnetic moment vector in Eqn. 9.34 contains several propagation vectors that might not be collinear. In this situation, the magnetic structures are termed as multi-$\vec{k}$ structures, and the coexistence of different ordering types is possible. There can be many combinations giving rise to various possibilities. For example, a canted arrangement of magnetic moments in real space comprising of a set of ferromagnetically aligned moments and moments alternating in opposite directions as one moves along the a-axis gives rise to a combination of a ferromagnetic vector $\vec{k}_F$ (0, 0, 0) with an antiferromagnetic vector (1/2, 0, 0) $\vec{k}_{AF}$. Appropriate summation of the two Fourier components according to Eqn. 9.34 leads to the correct description of the magnetic moment. Another interesting example is a conical type magnetic structure that combines a ferromagnetic or an antiferromagnetic component and a helical spiral one. This kind of magnetic structure is observed in several rare earth elements at low temperatures. In such a structure all moments rotate at a surface of a cone around the axis defined by the ferromagnetic or antiferromagnetic component.

9.3 Neutron Scattering Experiments

A neutron scattering experiment records the Bragg peak intensity $I(\tilde{Q})$ equal to the number of neutrons entering the detector when the diffraction condition according to Bragg's law $2d\sin\theta = \lambda$ is satisfied. This condition is achieved for a given d spacing and diffraction angle θ either by variation of the wavelength λ or, if the wavelength is fixed, then by a change in the diffraction angle. The first kind of approach is used in the Laue technique. In the second approach, a single crystal is rotated to get the Bragg reflection in a constructive reflection condition to a detector, which is positioned at an expected diffraction position. In the case of a powdered sample, a large-area detector is used to cover a larger range of diffraction angles. Then there is the time-of-flight (TOF) method, where the detection is coded using the time. One detector at one fixed angle covers a larger d-spacing range as the time of flight is proportional to wavelength and hence to d-spacing.

Equations 9.31–9.33, which relate magnetic moments to the differential magnetic cross sections, Fourier components, and the magnetic structure factors, are the key formulas for a magnetic structure determination. The recorded intensity $I(\tilde{Q})$ in a neutron scattering experiment corresponds to an integration of the differential cross sections and is proportional to the square of the relevant structure factors. In the case of magnetic neutron scattering, it is proportional to the square of the projected magnetic structure factor $|F_{M\perp}(\vec{Q})|^2$. The detected intensity $I(\tilde{Q})$, however, needs to be corrected for various geometrical factors. The final coefficient of proportionality between experimentally determined and calculated $|F_{M\perp}(\vec{Q})|^2$ provides a scaling factor and, in turn, offers the possibility of getting the magnitudes of magnetic moment determined on an absolute scale.

Identification of the signal of magnetic origin

The diffracted neutron signal has information on both nuclear and magnetic orders present in the materials under study. The recorded intensity $I(\tilde{Q})$ originating from the nuclear and magnetic processes can be comparable in magnitude, but their origins are very different. At the first sight, it may appear to be difficult to separate nuclear and magnetic signals, but at least in the case of elastic scattering in materials with large magnetic moments, it is a relatively straightforward exercise. This is of course subject to the condition that all other processes leading to variations in crystal structure and lattice distortions (like magnetostriction) can be ignored.

Nuclear scattering takes place at all temperatures and it does not depend on the scattering vector $I(\tilde{Q})$. In contrast, magnetic scattering appears only below the temperature of magnetic ordering of the material under investigation, and it decreases sharply as $\vec{Q}$ increases due to a reduction in magnetic form factor $f(\vec{Q})$ (Eqns. 9.32 and 9.33). An estimate of the magnetic scattering can thus be made by subtraction of neutron scattering data obtained at temperatures lower than the temperature (for example, T_{Curie} or T_{Neel}) of magnetic ordering from the high temperature ($T > T_{Curie}$ or T_{Neel}) data. Newer Bragg peaks appear at positions different from those of nuclear origin in the case of antiferromagnetic ordering, where the magnetic unit cell is larger than the crystallographic unit cell. The magnetic and nuclear Bragg peaks often coincide in systems with commensurate magnetic and crystallographic unit cells. In this case, the situation is relatively more difficult since the magnetic signals can be comparable with the statistical uncertainties of the experiment.

There are four specific steps towards fully determining the magnetic structure [9].

1. In the first step the temperature at which the moments develop some type of long-range magnetic order needs to be identified and compared with the phase transition temperature determined from another method. This is necessary because of the experimental uncertainties in the determination of the temperature. Critical phenomena occur near the magnetic ordering temperature, and knowledge of the exact value of the ordering temperature becomes indispensable.

2. The determination of the wave vector $I(\vec{k})$ of the propagation of the magnetic structure. This is possible either by mapping the reciprocal space using a single crystal sample or by indexing a diffraction pattern obtained from a powder sample using a suitable propagation vector or a combination of vectors to account for all the observed magnetic Bragg peaks. The first approach needs a large single crystal of a suitable quality to cover a large portion of reciprocal space. The powder neutron diffraction experiments do not require single-crystal samples; the mapping is relatively easy since the random orientation of crystallites enables the satisfaction of the reflection condition for all Bragg reflections in the available range. The problem here is that the signal-to-noise ratio is generally worse than in the case of a single crystal, more so in the case of weak ferromagnets. Despite such drawbacks, powder neutron diffraction is a popular first step in an attempt to identify the magnetic signal and to determine the propagation vector in new material. A trial-and-error method has been traditionally employed for this purpose, and automatic indexing routines are available nowadays to simplify the task. This approach, however, becomes difficult for samples having propagation vectors with non-rational components or if the magnetic structure is defined by two or more propagation vectors.
3. Determination of the coupling and direction of the Fourier component magnetic moments. This is normally achieved by inspecting systematic extinction rules and magnetic intensities. The magnetic scattering intensity becomes zero when the scattering vector $\vec{Q}$ is parallel to the magnetic moment. In a simple collinear magnetic structure, it is usually quite simple to find the series of (hkl) reflections that are absent because they arise from $\vec{Q}$s parallel to the magnetic moments. This simple approach, however, cannot be followed in materials with non-collinear and multi-$\vec{k}$ magnetic structures. Here the observed intensities (i.e., derived magnetic structure factors) need to be compared with the systematically calculated structure factors for a series of magnetic structure models. The aim is to theoretically generate all magnetic structure models and test those with the experimentally observed ones in the sample under investigation. It may seem formidable if not impossible to test all the combinations, especially for magnetic structures with large magnetic unit cells built from different magnetic moments in the presence of several propagation vectors. The difficulty of this task can be reduced significantly by taking into account symmetry constraints of the problem with the help of magnetic space group and group representation theory [9].
4. In this step the magnitude of the magnetic moments on the different sites is estimated from the determined Fourier components. The knowledge of

magnetic form factors is essential here, the standard values of which are tabulated in available literature [9].

There is, however, a fundamental limitation of neutron scattering techniques that leads to difficulty in determining the magnetic structure of materials. That is the impossibility of determining phase factors in Eqn. 9.36. The Fourier components $\vec{m}_{jk} = \vec{\mu}_j/\vec{u}\exp(-i\phi_j)$ and their conjugated complexes $\vec{m}_{j,-k} = \vec{\mu}_j/\vec{u}\exp(i\phi_j)$, which define the distribution of magnetic moments in the direct space, are in general complex vectors. In the magnetic diffraction experiments, the scattering intensity proportional to the square of the modulus of the magnetic structure factor (defined by Eqn. 9.37) is determined. The origin of the phase shift ϕ_j can be defined arbitrarily in Eqn. 9.37, and as a result the information regarding the ϕ_j is lost. In the case of a non-Bravais lattice, the only useful information retained is the phase difference between two distributions originating from moments residing at two different sites. It is possible to get the same neutron diffraction result from two different moment configurations given, which cannot be distinguished from this diffraction experiment. Hence, a careful consideration, taking into account macroscopic physical properties of the system and other physical arguments or techniques, is then needed to select one or the other magnetic structure from the existing possibilities.

9.3.1 Neutron powder diffraction

A powder sample of a material essentially consists of many single-crystal particles. Due to the random orientation of such single-crystal particles, in a powder diffraction experiment the nodes in reciprocal space form spheres with radii corresponding to the inverse separation of planes in real space. The same picture holds for "magnetic planes" made out of magnetic atoms. The "magnetic planes" are not necessarily the same as the "crystallographic planes" of the crystal structure of the material since the magnetic structure can be incommensurate with the crystal lattice. As a result, two sets of spheres can exist in the reciprocal space. One set corresponds to the crystal structure and the other corresponds to the magnetic structure. They can be detected in the diffraction experiment with Bragg reflections at diffraction angles θ given by the Bragg's law $2d\sin\theta = \lambda$. The relation between these angles and the modulus of the scattering vector $\vec{Q} = \vec{H} + \vec{k}$ can be obtained directly from the Bragg law:

$$|\vec{Q}| = |\vec{k}_0 - \vec{k}_1| = 4\pi\frac{\sin\theta}{\lambda} \tag{9.40}$$

Here $\vec{H}$ is a reciprocal lattice vector.

The diffracted intensity for powder samples lies on a cone. It is assumed that single-crystal grains in the powder are randomly oriented in space, and hence the information about the intensity on such a cone can be obtained by detecting only a portion of the cone. The detected intensity, therefore, originates from a representative statistical portion of the sample, and the signal strength is proportional to the covered cone part. Therefore, it is advantageous to cover as much spherical angle around the sample under study as possible. This can be achieved either with linear detectors covering a certain angular range around the horizontal plane or with position-sensitive detectors (PSDs). The PSDs have the advantage that it is possible to correct for the curvature of cones intersecting the active zone of the PSD. In the case of a linear detector, such curvature can give rise to an instrumental broadening of detected reflections, more so at low diffraction angles.

The neutron powder diffraction experiment produces a one-dimensional diffraction pattern, which contains the angular dependence of the detected scattering intensity. This dependence contains all the information about the materials under study – structural parameters (such as atomic positions, stoichiometry, and Debye-Waller factors), absorption, extinction, micro-absorption, and preferential orientation. But it is also affected by extraneous factors like instrumental conditions, for example, a systematic shift of reflections, peak shape and width, and background noise. If the material consists of a large number of randomly oriented crystals, then chances are high that all the coherent constructive signals will be detected. In the case of magnetic diffraction on a material with unknown propagation vector $\vec{k}$, it would then be a big advantage, because all possible orientations are covered for sure and the signal is unlikely to be missed.

There are, however, two problems encountered often in the powder neutron diffraction experiments. The first one is that the two Bragg reflections may accidentally have the same d-spacing and they appear at the same angle in the diffraction pattern. This problem of peak overlapping can be successfully circumvented by the Rietveld method [15]. This method involves least-square fitting of the experimental neutron diffraction pattern using fitting parameters defining the structure of the sample, shape, and width of Bragg reflections, background, and all other parameters defining the sample and experimental conditions. There are several excellent computer codes like Fullprof, GSAS, and Jana2006 accessible freely to the users, which make large numerical operations relatively easy. The users can judge the quality of the fit with the help of several sophisticated parameters and the integrated graphical interface where the observed and calculated profiles can be displayed.

The neutron scattering signal is not always strong enough to be detected although from a geometrical point of view the neutron has entered the detector. The second problem arises from this fact. There can be several reasons for this problem, namely that the magnetic moment involved in the sample is too small or the magnetic peak appears at the top of the nuclear Bragg peak with a relatively bad signal-to-noise ratio for the magnetic contribution. In the first case, larger samples may be of help but that might not always be possible. In the latter case, however, the problems remain as the proper determination of magnetic parameters may be inhibited by the nuclear structure parameters. In such a situation, only neutron single-crystal diffraction experiments, preferably combined with a polarized neutron beam, may be useful for the successful determination of the magnetic structure.

There are many powder neutron diffractometers available around the world, which may differ in detailed specifications. This includes flux at the sample position, available neutron wavelengths, collimation, angular and Q-range coverage, shielding, ability to accept different sample environments, detector system used, etc. All of these diffractometers, however, use more or less the same principles of detection of the diffracted neutrons, and the layout of the instrument is sort of generic. Typical geometry for a neutron powder diffraction experiment using a reactor source is shown in Fig. 9.7(a). The neutrons of one energy or wavelength are selected by a crystal monochromator or by velocity selection through a mechanical chopper, and those neutrons are then scattered by the sample. By contrast, in neutron diffraction experiments using a spallation neutron source white beams are used, which contain neutrons of a wide range of wavelengths. The typical geometry of such an experiment is shown in Fig. 9.7(b). The neutron beam in the spallation source is pulsed and the detectors perform an addition time of flight (TOF) analysis to determine the wavelength.

Magnetic Bragg scattering can be distinguished from nuclear scattering from the structure in several possible ways [10]. In a neutron powder diffraction, all

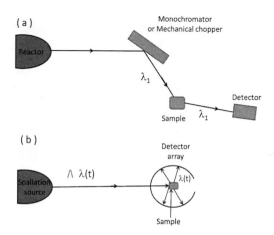

FIGURE 9.7 Schematic diagram showing the configuration of a neutron diffraction experient using: (a) recator source, (b) pulsed spallation source.

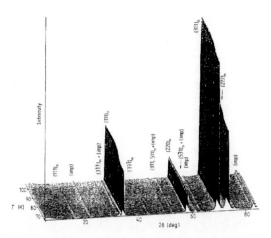

FIGURE 9.8 Neutron powder diffraction pattern from Ce(Fe$_{0.8}$Co$_{0.2}$)$_2$ between 65 and 105 K, showing ferromagnetic–antiferromagnetic phase transition around 85 K. Nuclear Bragg peaks are labelled with subscript n; antiferromagnetic Bragg peaks are labelled with subscript m; 1–2% Ce$_2$Fe$_{17}$ impurity peaks are labelled with subscript imp. (From reference [11] with permission from IOP.)

the Bragg peaks are nuclear in origin above the magnetic transition temperature. Magnetic Bragg peaks arise below the magnetic transition temperature and their intensities rapidly develop as the temperature decreases further. In the magnetic state, the nuclear and magnetic intensities simply add unpolarized neutrons. It is straightforward to distinguish magnetic scattering from nuclear scattering where magnetic Bragg peaks occur at positions different from the nuclear peaks. However, in the case of a ferromagnet or for some antiferromagnets that contain two or more magnetic atoms in the chemical unit cell, magnetic and nuclear Bragg peaks can appear at the same position [10]. Fig. 9.8 shows neutron powder diffraction pattern from a Ce(Fe$_{0.8}$Co$_{0.2}$)$_2$ sample between 65 and 105 K [11]. The sample undergoes a ferromagnetic to antiferromagnetic phase transition around 85 K and new magnetic Bragg peaks appear below this temperature. A standard technique for identifying the magnetic Bragg scattering is to make one diffraction measurement in the ordered state at the lowest temperature possible and another in the paramagnetic state well above the magnetic transition temperature, and then take the difference of the two sets of experimental data. This yields the magnetic Bragg peaks, on top of a deficit (negative) of scattering away from the Bragg peaks due to the disappearance of the diffuse paramagnetic scattering in the magnetically ordered state [10].

9.4 Single-Crystal Experiments

In this experiment using a constant-wavelength neutron source, it is necessary to bring a single-crystal sample in some reflection condition. This requires that at least one component changes its value or orientation. In practice, two basic designs are used to meet this condition. In one such design termed a four-cycle geometry, the single-crystal sample is mounted on a Eulerian cradle, which enables one to orient the crystal in any direction so that any scattering vector can be brought to the horizontal plane. In this arrangement, a large number of independent reflections can be measured with a reasonably short incident wavelength, and this is an advantage when details of the crystal/magnetic structure of the sample are investigated. However, the space available on a Eulerian cradle is rather limited, and as a result, only close cycle refrigerators and small furnaces can be used for the sample environment. The other design arrangement known as normal-beam geometry with a lifting detector is more preferable when a more involved sample environment such as helium- and nitrogen-cooled cryostats, dilution refrigerators, cryomagnets, pressure cells, and a combination of some of those are required. In this geometry the sample rotates around a pre-defined axis, usually the vertical axis of the diffractometer called ω. The detector can be positioned out of the horizontal scattering plane by rotating it by an angle ν around the sample position, but this imposes some constraints on the accessible reciprocal space. For example, all (00l) reflections would be inaccessible if the single-crystal sample is oriented with its c-axis is vertical. However, the fact that it is possible to inspect all accessible points of the reciprocal space individually would still result in an advantage over powder neutron diffraction. In other words, if all the propagation vectors in some k-star domain (where moments order with different propagation vectors) are reachable, it is then possible to collect all Fourier components belonging to such a propagation vector separately. Knowledge of the crystal's orientation with respect to the laboratory coordinate system is required here, which is obtained with the help of sufficiently known strong nuclear reflections. Since the magnetic intensity is proportional to the projection of the magnetization on the plane perpendicular to the corresponding scattering vector $\vec{k}$, the direction of the magnetic moments can be determined with the knowledge of the orientation of the crystal at each point of the reciprocal space.

9.5 Polarized Neutron Scattering

We know that the neutrons carry magnetic moments, and it is possible to extract neutrons with a particular orientation from an unpolarized beam with the help of polarizing monochromators. The polarization $\vec{P}$ of a neutron beam is the average of the polarization vectors of N individual neutrons:

$$\vec{P} = \frac{\sum_i \vec{P}_i}{N} \tag{9.41}$$

Polarization for a fully polarized beam $|\vec{P}| = 1$, and for unpolarized beam $|\vec{P}| = 0$. In reality, only a partially polarized beam is usually attainable, and for that $0 < |\vec{P}| < 1$.

A polarized neutron beam is brought to the sample using a neutron guide that keeps its orientation, and in the sample it undergoes interaction with both nuclei and unpaired electrons just like in the case of unpolarized neutron diffraction. The only difference with respect to the case of unpolarized neutrons is that now one has the interference between nuclear and magnetic structure factors, which also needs to be considered. The number of diffracted neutrons for different initial polarizations is not the same.

In the simplest polarized neutron experiment known as the flipping ratio method, the neutrons are detected irrespective of their spin without any analysis of the final spin state of the scattered beam. In this experiment the intensities I^+ and I^- of a Bragg reflection for an incident beam polarized parallel (superscript +) and antiparallel (superscript -) to the initial polarization are measured and the ratio between them is calculated. Both the intensities are measured at the same position and they are subject to the same corrections. The respective intensities are expressed as:

$$I^+ \approx |F_N + F_{M\perp}(\vec{Q})|^2 \tag{9.42}$$

and

$$I^- \approx |F_N - F_{M\perp}(\vec{Q})|^2 \tag{9.43}$$

In the case of unpolarized beam assuming a case where $F_N = 1$ $F_M = 0.2$, the detected intensity is $I \approx |F_N|^2 + |F_M|^2 = 1.04$. If instead a polarized neutron beam is used, then $I^+ \approx |F_N + F_M|^2 = 1.44$ or $I^- \approx |F_N - F_M|^2 = 0.64$. This highlights a significant improvement in the sensitivity in the magnetic measurements.

Fig. 9.9 shows the spin-up minus spin-down difference spectra obtained with polarized neutron diffraction measurements on polycrystalline samples of ferromagnets LuFe$_2$ and CeFe$_2$ at 5 K in the magnetic field of 5 T [12]. The difference signal at the (220) Bragg peak is negative, which corresponds to a flipping ratio of less than one.

9.6 Magnetic Small-Angle Neutron Scattering

Magnetic small-angle neutron scattering (SANS) is used for studying magnetic phenomena that occur over length scales that are large compared to atomic distances. This technique is particularly useful for exploring magnetic nanocomposites, magnetic domain structures, and long-wavelength oscillatory magnetic states, and magnetic flux-line lattice or vortex lattice in type-II superconductors. Magnetic fluctuations can also originate from dislocations in a metal as a result of magnetoelastic coupling between the elastic strain field of

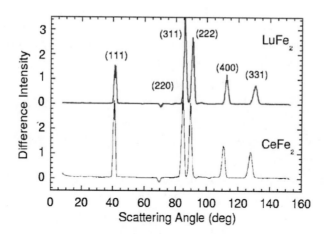

FIGURE 9.9 Difference spectra (spin-up minus spin-down) for CeFe$_2$ and LuFe$_2$ at 5 K in a magnetic field of 5 T. (From reference [12] with permission from Elsevier.)

the dislocations and the magnetization density. The resultant magnetic small-angle scattering can be significantly larger than the nuclear small-angle scattering caused by the dislocations.

In the SANS experiment, a monochromatized neutron beam is tightly collimated and directed onto the sample. The scattered beam is then registered in an area detector, and at the same time a beam stop is placed in front of the detector to absorb the direct beam that crosses the sample and that would saturate or even destroy the detector. One is dealing here with large (1–1000 nm) objects, which when compared with normal neutron diffraction (in the range of few Å), leads to a significantly different $\vec{Q}$ range. The typical $\vec{Q}$ range encountered in the SANS experiments is between 10^{-2} and 5 nm^{-1}, which corresponds in real space to values in the range of 0.5–500 nm.

The optical approximation can be applied in SANS measurements since one is not looking into the atomic details of the sample. A locally averaged scattering length density $\eta(\vec{r})$ can be used here because of the low spatial resolution required for the SANS cases. A nuclear scattering length density can be defined accordingly [9]:

$$\eta_N \propto \sum \frac{c_i b_i}{\Omega_i} \qquad (9.44)$$

Here b_i is the nuclear scattering length, c_i is the concentration, and Ω_i is the atomic volume of the specimen i. Similarly, the magnetic scattering amplitude encompassing the interaction between the neutron magnetic moment and a set of magnetic moments in the material under investigation can be expressed as [9]:

$$\eta_m \propto \sum \frac{c_i M_i^\perp}{\Omega_i} \qquad (9.45)$$

It is to be noted that magnetic scattering amplitude is a vector that depends on the orientation of the moment with respect to the scattering vector $\vec{Q}$.

SANS experiments can determine the form factor of a nano-object in a matrix, which is the difference of the Fourier transform of the scattering length density between the nano-object (η_p) and the surrounding matrix (η_{matrix}):

$$\Delta\eta = \eta_p - \eta_{matrix} \tag{9.46}$$

The magnetic form factor in a SANS experiment can suitably be conveniently defined as [9]:

$$F(\vec{Q}) = \int \Delta\eta e^{-i\vec{Q}\cdot\vec{r}_j} dr^3 \tag{9.47}$$

The interaction is limited again to the component of the magnetization perpendicular to the scattering vector $\vec{Q}$.

In a conventional SANS experiment involving an unpolarized monochromatic incoming neutron beam, the neutron spins are randomly distributed. The total scattering intensity is equal to the sum of the squared amplitudes from individual magnetic and nuclear contrasts. Let us now consider a sample of magnetic precipitates (nano-objects) in the presence of an applied magnetic field $\vec{H}$, which orients magnetic moments in these nano-objects or the nano-objects themselves along the field direction in a certain manner. The scattering cross section can be expressed as [9]:

$$\frac{d\sigma(\vec{Q})}{d\Omega} = A(\vec{Q}) + B(\vec{Q})\sin^2\alpha \tag{9.48}$$

This scattering cross section is anisotropic due to the vector nature of the magnetic form factor, with A and B representing the isotropic and anisotropic terms, respectively. The angle α is the azimuthal angle between the direction of the magnetic field $\vec{H}$ and the scattering vector $\vec{Q}$. It is possible to separate the contribution of the nuclear origin $A(\vec{Q})$ and the magnetic contribution $B(\vec{Q})$ in a SANS measurement if all magnetic moments in the sample are fully aligned along the applied magnetic field.

SANS is a powerful technique for measuring the magnetic contribution from magnetic nano-objects systems, but it has certain limitations:

1. The magnetic contrast is weak in comparison to the nuclear contribution.
2. It is not possible to obtain the composition, densities, and magnetization of the matrix and the particles on an absolute scale. This is because of the loss of the phase information between individual contributions.

Apart from magnetic nano-particles, other examples of the use of magnetic small-angle scattering techniques can be the study of magnetic domain structures

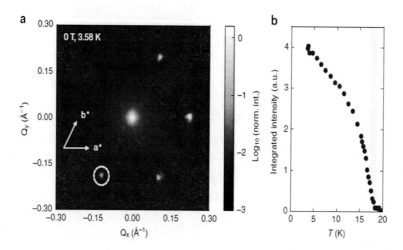

FIGURE 9.10 Results of small-angle neutron scattering experiments on $Gd_3Ru_4Al_{12}$ in zero magnetic field and at low temperature: (a) $Q_z = 0$ cut of scattering intensity in the SANS experiment. Scattered neutron intensity is shown on a logarithmic scale. (b) Integrated intensity of elastic neutron scattering onsets around $T_{N2} = 18.6$ K. (From reference [13] Creative Commons License.)

and long-wavelength oscillatory magnetic states, and magnetic flux-line lattice or vortex lattice in type-II superconductors. Fig. 9.10(a) shows results of SANS study on a metallic single sample of $Gd_3Ru_4Al_{12}$, obtained using a time-of-flight (TOF) type SANS instrument in zero fields and T = 3.58 K. This material undergoes sinusoidal magnetic ordering in the hexagonal a–b plane at $T_{N2} = 18.6$ K, followed by an onset of three-dimensional helical order at $T_{N1} = 17.2$ K. The SANS results presented in Fig. 9.10(a) are in the form of a two-dimensional Q_x–Q_y map of scattering intensity (a $Q_z = 0$ cut of the full TOF dataset). Three magnetic reflections are observed, which correspond to the multi-domain helical ground state. A fourth reflection (marked by a white circle in Fig. 9.10(a)) is also observed. That originates from a minority crystallographic domain with a different alignment of the c-axis. Fig. 9.10(b) shows the onset of the integrated intensity of elastic neutron scattering around $T_{N2} = 18.6$ K.

The magnetic flux lines in a type-II superconductor are arranged in a parallel periodic formation on a two-dimensional lattice with periodicities of the order of 1000 Å. Neutron scattering from such arrangements of magnetic flux-lines leads to Bragg peaks that can be observed in neutron diffraction at high angular resolution. The symmetry properties of the vortex-lattice can be determined by making scans as a function of crystal orientation. In addition, such experiments can give information on the spatial distribution of the magnetic flux about each line, the bending of the lines due to pinning at defects, and also the lateral flow of flux-lines at high current densities.

9.7 Inelastic Neutron Scattering

In a neutron scattering event, the energy transferred to (or from) the sample under investigation may be taken up (or given off) by a single elementary excitation in the sample, for example, a quantum of a normal vibrational mode. In such a case the variable energy E is often replaced by $\hbar w$, where w is the mode frequency. It is also possible that the scattering event may involve multiple excitations, and there it is more convenient to refer to the total energy E being exchanged. We know that materials with long-range magnetic order exhibit collective excitations known as spin waves. There is a sinusoidal variation in the deviations of the spin components along a particular direction as one moves from atom to atom in the material. The spin waves are quantized, and a single quantum is known as a magnon. This is analogous to phonons, which are the energy quanta of the collective displacement waves for the nuclear positions in a material.

Denoting the scattering angle by θ one can get from Eqn. 9.8:

$$k_0^2 - 2k_0 k_1 cos\theta + k_1^2 = Q^2 \tag{9.49}$$

With the help of Eqn. 9.9, putting $E_0 - E_1 = E$ and eliminating k_0 and k_1 we get:

$$2E_0 - E - 2\sqrt{E_0(E_0 - E)}cos\theta = \frac{\hbar^2}{2m}Q^2 \tag{9.50}$$

Eqn. 9.50 generates a family of curves of E/E_0 versus Q, with θ as a parameter. It is possible to make for any value of E_0 over a range of values of $\vec{Q}$ and E by varying the scattering angle θ. By increasing E, the accessible ranges in both these variables increase. The resolution in $\vec{Q}$ and E, however, tends to get worse with the increase in E_0. Thus, the values of E_0 and θ need to be optimized for each experiment.

Inelastic neutron scattering (INS) spectroscopy covers an enormous energy range, and equivalently a wide range of timescales [14]. Figure 9.11 presents the energy range and timescale available and the associated applications. The whole energy range can be divided into three sectors, but with some overlap between them. In the lowest energy range, ± 10 meV, the applications are mainly quasi-elastic scattering and tunneling spectroscopy. The second range, 0–1000 meV, encompasses the regions of vibrational spectroscopy and magnetic excitations, the latter being our main focus here. The energy range above 1000 meV is in the area of neutron Compton scattering. Apart from spin-wave excitations, there are other types of magnetic excitations and fluctuations that can be measured with inelastic neutron scattering techniques, such as critical fluctuations, crystal field excitations, and moment/valence fluctuations [10].

A very low energy inelastic process is termed quasi-elastic scattering. This usually represents a broadening of the elastic line, and this originates most commonly from

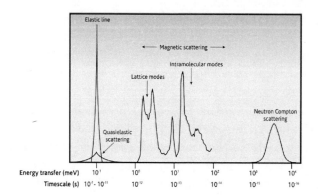

FIGURE 9.11 The energy range and the equivalent timescales and length scales available with inelastic neutron scattering. (From reference [14] with permission from Elsevier.)

the diffusional (i.e., translational or rotational) motion of atoms. It can also originate from the random fluctuations of unpaired electronic spins or magnetic moments of stationary atoms. The energy resolution ΔE of an INS spectrometer and the timescale τ of the motion are related by the Heisenberg uncertainty principle: $\Delta E \tau = \hbar$. This indicates that slower motions require high resolution, and relaxed resolution would be required for rapid motions. The timescale to be probed can be represented by three regimes, and a different technique is used for each of those:

1. $\tau \sim 10^{-11}$s : Δ is 10-100 μeV and direct geometry time-of-flight technique is used.
2. $\tau \sim 10^{-9}$s : Δ is 0.3-20 μeV and backscattering crystal analyzer is used.
3. $\tau \sim 10^{-7}$s : Δ is 0.005-1 μeV and neutron spin echo technique is used.

Direct geometry time-of-flight technique

In a direct geometry quasi-elastic spectrometer, a monochromatic pulsed beam of neutrons, produced by a rotating mechanical chopper system, is incident upon the sample. After the scattering event, the scattered neutrons are timed over a known distance to the detectors which are set at a range of scattering angles. The time-of-flight, T_1, from the sample to detectors is expressed $T_1 = L_1/v_1$, where L_1 is the distance from the sample to the detector. With the incident neutron velocity v_0 being known the energy transfer $\hbar w$ is expressed as [14]:

$$\hbar w = \frac{m}{2}(v_1^2 - v_0^2) = \frac{mL_1^2}{2}\left(\frac{1}{T_1^2} - \frac{1}{T_0^2}\right) \quad (9.51)$$

Here T_0 is the time-of-flight of the elastically scattered neutrons. With the help of Eqn. 9.51 and the known angular position of the detectors the momentum transfer $\vec{Q}$ of the detected neutrons can also be determined.

Backscattering crystal analyzer

The backscattering crystal analyzer is an example of an indirect geometry spectrometer. It is used when a higher resolution is required. The incident neutrons on the sample have a band of energies whose width in energy is greater than the maximum energy transfer to be measured. One neutron energy is selected from the scattered neutrons by Bragg diffraction from a single crystal. Bragg's law states $\lambda = 2d\sin\theta$ where λ is the wavelength of the incident neutron, d is the interplanar spacing in the analyzer single crystal, and θ is the angle of incidence. The resolution is expressed by the differential [14]:

$$\frac{\Delta E}{E} = 2\frac{\Delta \lambda}{\lambda} = 2\left([\Delta\theta \cot\theta]^2 + [\frac{\Delta D}{D}]^2\right)^{0.5} \quad (9.52)$$

In most of the cases, the dominant term is $\cot\theta$ since the uncertainty in the lattice parameter for most single crystals is $\sim 10^{-4}$, which can be ignored. In the limit Bragg angle approaches 90^0, $\cot\theta$ would tend to zero and the energy resolution of the analyzer becomes quite good, being 0.5 μeV for silicon and 10 μeV for graphite. This is the basis of the backscattering spectrometer.

In an inelastic neutron scattering experiment, the magnitudes of $\vec{k}_0$, $\vec{k}_1$, and the scattering angle θ are determined for each recorded scattering event. In measurements involving a single crystal, the direction relative to the crystal lattice, i.e., the direction of $\vec{Q}$ in the crystal coordinate system, is also required to be known. The magnitudes of $\vec{k}_0$, and $\vec{k}_1$ may be estimated by measuring either wavelength or velocity, whereas the directions can be determined from the geometry of the experimental arrangement.

The Energy Range 0–1000 meV and Magnetic Excitations

This energy range encompasses the mid- and near-infrared regions of the electromagnetic spectrum where magnetic excitations are observed. These involve electronic transitions between the energy levels of a metal atom (or ion) and usually involve unfilled d or f orbitals. These transitions are often optically forbidden and can be observed with inelastic neutron scattering because of the interaction of the atomic magnetic moment with the magnetic moment of the neutron. There also exist collective excitations of the atomic magnetic moments in the materials with long-range magnetic order, which result in "spin waves" propagating through the material. Such spin waves can be studied with INS experiments.

The typical energy of the neutrons in this energy is termed thermal neutrons because their typical energies correspond to room temperature. Thermal neutron INS spectroscopy was the first type to be studied with the instrument invented by Bertram Brockhouse in 1952. This instrument is known as the triple-axis spectrometer, and it remains a mainstay of INS instrumentation at reactor sources.

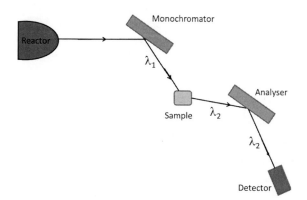

FIGURE 9.12 Schematic digram showing the configuration of an inelastic neutron scattering experiment using a reactor source.

The triple-axis spectrometer is designed to study coherent excitations, which contain information on the relative motions of the nuclei. This method, however, needs the use of large single crystals of the order of several cubic centimetres. The scattering from powders is largely incoherent, which provides information on the motions of the individual nuclei. Different spectrometers are used for such materials.

Spectrometers for coherent spectroscopy

A schematic of the experimental configuration of a triple-axis spectrometer is shown in Fig. 9.12. A monochromatic neutron beam is selected from a constant-flux neutron source by using Bragg reflection from a single-crystal monochromator. This neutron beam is then incident on the sample and gets scattered. Another single-crystal known as the analyzer is positioned in a selected direction in such a way that only a given wavelength can be reflected. A neutron detector is placed after this analyzer, where counting of neutrons would indicate that there has been some process in the sample inducing specific changes of trajectory and wavelength of the incident neutrons.

A given configuration of the triple-axis spectrometer corresponds to a single point in the $(\vec{Q}, E)$ space. The crystals can be rotated and the scattering angle varied to scan the $(\vec{Q}, E)$ space and detect all possible neutron scattering originating vibrational or magnetic processes. A given $(\vec{Q}, E)$ point may be obtained in different ways by varying either k_0 or k_1. The triple-axis spectrometer is enormously flexible and any point in the $(\vec{Q}, E)$ space is accessible in principle. But the point-by-point data acquisition method means that it is slow. Hence, mapping the dispersion of excitation, and sometimes the study of one sample, can take weeks of measurement time.

The implementation of a pulsed source equivalent of a triple-axis spectrometer is also possible. The difference is that instead of a monochromatic beam, a white beam of neutrons is incident on the sample. Similar to the triple-axis spectrometer, Bragg reflection from analyzer crystals are used to select the neutrons scattered by the sample, which will ultimately be detected. However, instead of just one detector, an array of many independent analyzers and detector arms is used for the detection of the scattered neutrons. In a single setting of the instrument and sample, each such arm detects along a parabolic path in the $(\vec{Q}, E)$ space, and this process generates a two-dimensional scan through the $(\vec{Q}, E)$ space for each neutron pulse.

In this energy range, direct geometry spectrometers are mostly found at spallation neutron sources, and these instruments are extensively used to study the magnetic properties of materials. As we know already, magnetic neutron scattering is a relatively low cross-section process, and such experiments require very low background and high sensitivity. Fig. 9.13 shows a schematic diagram for such a direct geometry spectrometer. To block the prompt pulse of very high energy neutrons and gamma rays that are produced when the proton pulse hits the target, a Nimonic chopper for background suppression is used. The core element of the instrument, however, is the Fermi chopper, which consists of a metal drum with a series of slots cut through it. The Fermi chopper is suspended in a vacuum chamber perpendicular to the neutron beam and can be rotated at speeds up to 600 Hz. The incident neutrons are blocked for most of a rotation, except for one particular time when the slots are parallel to the incoming neutrons and they are allowed to go through to the sample. By phasing the opening time of the slots with respect to the neutron pulse, the incident neutron energies in the range from meV to eV can be selected. In a typical time-of-flight inelastic instrument, the detector's coverage is very large, for example, -45 to 135 deg. on the MERLIN spectrometer at the ISIS facility. Hence, in a single experiment, the instrument can map large regions of $(\vec{Q}, \omega)$ space. Vibrational and magnetic excitations have different $\vec{Q}$ dependencies, and hence the large range in Q helps to distinguish between the two kinds of scattering processes. The continuous detector coverage ensures that broad features, which might have been missed in a triple-axis spectrometer, are quite accessible here. In single-crystal samples the scattering takes place along particular directions in space and to take advantage of this He^3 position-sensitive detectors with a vertical resolution of ~ 25 mm are used. The energy resolution involved is between 1% and 2% of the incident energy, and this resolution is constant for all the detector banks with all the detectors being at the same secondary flight path of the neutron.

To determine the incident energy of the neutrons accurately, three low-efficiency detectors are placed in the main beam (Fig. 9.13). This is because of the high neutron flux of incident neutrons. The first detector is placed before the background choppers to monitor the incident flux for normalization. The second and third detectors are positioned just after the Fermi chopper and behind the sample, respectively.

Fig. 9.14 shows an example of the spin-wave dispersion relation with spin-wave energy plotted against wave vector for a single-crystal sample of iron (Fe) at room

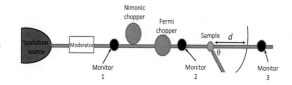

FIGURE 9.13 Schematic of a direct geometry instrument at a pulsed spallation neutron source.

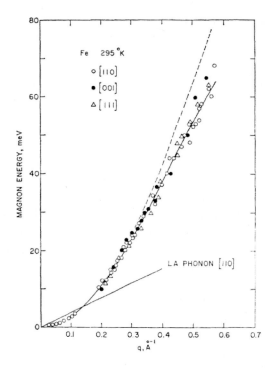

FIGURE 9.14 The spin-wave dispersion relations in iron (Fe) at room temperature obtained with a triple-axis neutron spectrometer. (From reference [17] with permission from APS.)

temperature obtained with a triple-axis spectrometer [17]. The magnon energy appears to be independent of its direction of propagation. The solid line shows the fit for a magnon energy dispersion relation $E = 281Q^2 - 275Q^4$ meV. The dashed line indicates the dispersion relation for a nearest-neighbour Heisenberg spin-coupling model along the [110] direction.

Spectrometers for incoherent spectroscopy

The simplest form of instrument used in a reactor-based neutron source is a filter spectrometer [14]. This is comprised of a monochromator to define the incident neutron energy on the sample and a filter after the sample allowing neutrons only of a given energy to reach the detector. The monochromator can be the same as for the triple-axis spectrometer, but the analyzer is replaced by a cooled beryllium filter. Neutrons with an energy of less than 5 meV only can be transmitted through this filter, whereas the higher energy neutrons are filtered out of the beam with Bragg scattering. These filters are routinely operated at liquid nitrogen temperature to improve the transmission of the neutron beam and the sharpness of the cut-off. With both the incident and final energies of the neutrons are being known, the energy transfer can be easily estimated.

Both indirect and direct geometry instruments are used at a spallation source. The incident white beam contains neutrons with a wide energy range from 2 meV to beyond 2500 meV, and out of that a small fraction of neutrons are inelastically scattered by the sample. The neutrons backscattered through an angle of 135^0 impinge on a graphite crystal, where only one wavelength and its harmonics are Bragg scattered by the crystal and the remaining ones pass through and get absorbed by the shielding. The neutrons at the harmonics of the fundamental wavelength are removed by the beryllium filter so that only neutrons of ~ 4 meV reach the detectors. There are also detector banks positioned for the forward scattering of neutrons scattered at 45 degrees. Thus, the combination of the graphite crystal and beryllium filter acts as a narrow bandpass filter.

The small final neutron energy ensures that for most energies $\vec{k}_0 \gg \vec{k}_1$, and it follows that both the magnitude and direction of $\vec{Q}$ are largely determined by $\vec{k}_0$. Thus, the $\vec{Q}$ is almost parallel to the incoming neutron beam and perpendicular to the sample plane across virtually the entire energy transfer range. This implies that to observe an excitation the vibration must have a component of motion parallel to $\vec{Q}$. This condition will be satisfied for all the vibrations in a randomly oriented powdered polycrystalline sample, and hence all the excitations can be observed in principle.

9.8 Polarized Neutron Reflectometry

Polarized neutron reflectometry is a probe of depth-dependent magnetic structure as well as a characterization of lateral magnetic structures. This technique uses the wave properties of the neutron for the optical study of matter [18]. A neutron beam is reflected from a laterally homogeneous flat material, and the intensity of the reflected beam is recorded at different neutron wavelengths and angles of incidence. The outcome of this process allows an evaluation of the chemical and magnetic depth profile of the material. Polarized-neutron specular reflectometry has been described lucidly in a review article by Ankner and Felcher [18] and we provide below a summary of this technique.

We start with the discussion of the specular reflection process where the incident and reflected neutron beam wave vectors $\vec{k}_i$ and $\vec{k}_f$ enter and exit the surface of the flat material at the same glancing angle α. The momentum of the neutron $|\vec{k}| = 2\pi/\lambda$ can be separated into two components, parallel and perpendicular to the material surface [18]. One can now represent the neutron as a particle with kinetic energy $\hbar k_z^2/2m$, hitting a potential wall of height $U(z)$. Here the potential $U(z)$ describes the laterally homogeneous material and k_z is the perpendicular component of the neutron momentum, which is altered by the potential $U(z)$. The parallel component of the neutron momentum remains unaltered. The neutron bounces back from the material surface if the energy of the neutron is very low. A part of the potential $U(z)$ is related to the scattering length b of the N constituent nuclei per unit volume: $(\hbar^2/2m)N(z)b(z)$. The isotopic scattering length b is constant for thermal and cold neutrons and is conveniently tabulated for all nuclei as well as for the natural isotopic composition of all elements [18].

In the free space above the material surface $U(z) \approx 0$. Considering the neutron being represented by a plane wave incident on the surface from above, the neutron wave function can be represented as [18]:

$$u(z) = e^{ik_z} + re^{-ik_z} \tag{9.53}$$

The Schrödinger wave equation links the wave function inside the material to the potential $U(z)$, and the solution of the Schrödinger equation gives the reflectance r. The experimentally determined quantity in a scattering experiment is the reflectivity $R = |r|^2$.

The wave vector transfer Q_z, which provides a convenient metric for characterizing the specular reflection process, is expressed as [18]:

$$Q_z = k_{zf} - k_{zi} = \frac{4\pi \sin\alpha}{\lambda} \tag{9.54}$$

The momentum Q_z is the quantum mechanical conjugate to position z. This enables the transformation of the depth profile of scattering material $b(z)$ into reflectivity $R(Q_z)$.

The reflectivity in general is unitary for most materials up to a value of $Q_c = \sqrt{16\pi Nb}$ of order 0.01 Å^{-1} [18]. The reflectivity decreases rapidly beyond this limit, with a mean asymptotic Q_z^{-4} dependence.

In the presence of magnetic induction, the interaction potential is expressed as [18]:

$$U(z) = U_n(z) + U_m(z) = (\hbar^2/2m)N(z)b(z) + \vec{B}.\hat{s} \tag{9.55}$$

Here $\hat{s}$ is the neutron spin operator. Now the neutron is a spin-1/2 particle and in an external magnetic field $\vec{H}$, there are two states of quantization with respect to this field. In an experiment, the neutron may be polarized either parallely (+) or antiparallely (-) to $\vec{H}$. If the neutron is polarized in an applied field H_P, the neutron will change its spin state upon encountering an induction $\vec{B}_S$ with a different orientation inside a magnetic material. The final state of the neutron can be analyzed in terms of the polarization with respect to a third field H_a applied customarily parallel to H_P. Four different types of reflectivities can be measured: R^{++}, R^{--}, R^{+-}, R^{-+}.

When all of the magnetic induction is collinear in the neutron path, the spin-dependent Schrödinger equation takes a very simple form. In this situation neutrons remain polarized in the original state ($R^{+-} = R^{-+} = 0$). Neutrons polarized parallely (+) [antiparallely (-)] to H_P encounter a potential $U^{\pm} = (\hbar^2/2m)Nb \pm \mu_n B$, where μ_n is the neutron magnetic moment [18]. The strength of the magnetic scattering in ferromagnetic materials is comparable to that of the nuclear scattering, and an analysis of the reflectivities R^{++}, R^{--} enables a quantitative determination of $B(z)$.

The phenomenology of magnetic reflection is summarized in Fig. 9.15. The orientation and magnitude of the sample magnetization $\vec{M}(z)$ relative to the applied field $\vec{H}$ determine the relative proportions of spin-flip and non-spin-flip scattering. There is no specular spin-flip scattering for a polarized neutron beam incident on a ferromagnetic layer aligned parallely to an external field (Fig. 9.15(a)). A measurement of the intensity of spin-flip scattering relative to non-spin-flip scattering can determine the orientation of an in-plane ferromagnetic layer (Fig. 9.15(b)). There is no difference in refractive index between neutrons polarized parallely and antiparallely to $\vec{H}$ where the magnetization is perpendicular to the surface (see Fig. 9.15(c)). The magnetization state of such kind of sample is indistinguishable in the specularly reflected beam. The presence of domains and their size distribution strongly affects the specularly reflected neutron scattering intensity (see Fig. 9.15(d)). The coherence length of the neutron beam on the surface is about 100 nm, which is larger than the lateral domain size in many samples [18]. Significant scattering of the specular beam is produced by these small domains, and the beam also becomes depolarized. Off-specular methods offer a method of characterizing these domains [18].

Fig. 9.16 shows schematically a polarized neutron reflectometer instrument [18]. A neutron beam of wavelength λ strikes a sample surface at an angle α_i and is reflected from the surface at angle α_f. The instrument acts as a diffractometer with resolution sufficient to separate transmitted and reflected beams at values of Q_z near to which the reflectivity becomes unitary. Specular reflectivity ($\alpha_i = \alpha_f$) is entirely a function of the momentum transfer along the perpendicular-$\hat{z}$ direction. During the experiment a range of Q_z is spanned either by a change of the wavelength while keeping the angle of incidence fixed, or by a change of the angle of incidence at a fixed wavelength. Polarization of the incoming neutrons along an applied magnetic field and analysis of the polarization of the reflected beam are achieved with the help

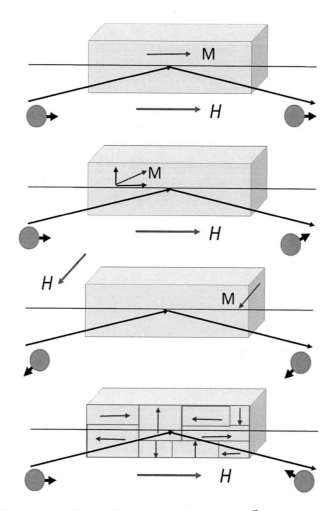

FIGURE 9.15 The phenomenology of magnetic reflection: (a) $\vec{M}(z)$ in the plane of the surface, parallel to $\vec{H}$ produces no spin-flip scattering, but creates different spin-dependent refractive indices for neutrons polarized parallely and antiparallely to $\vec{H}$. (b) $\vec{M}(z)$ oriented at an arbitrary angle in the surface plane produces both spin-flip and non-spin-flip scattering intensity. (c) $\vec{M}(z)$ components normal to the surface have no effect on neutron specular intensity. (d) The presence of domains complicates interpretation of spin-flip and non-spin-flip scattering intensities. (Adapted from reference [18] with permission from Elsevier.)

of appropriate devices such as polarizing mirrors and flat-coil spin flippers. The direction of initial polarization is fixed by convention, and the magnetic sample may change the polarization of the neutron. The reflected neutrons, which are aligned with the polarizer, are detected with an analyzer. Spin-flippers placed before and after the sample cause the reversal of the neutron spin. The reflectivities (R^{++}, R^{--}, R^{+-}, R^{-+}) are then characterized by the sign of the neutron polarization before and after reflection with respect to the reference field [18].

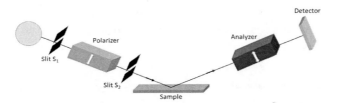

FIGURE 9.16 Schematic representation of a polarized neutron reflectometer. (Adapted from reference [18] with permission from Elsevier.)

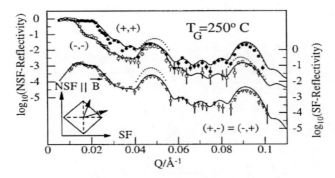

FIGURE 9.17 Spin-polarized neutron specular reflectivity results measured in $H = 30$ Oe from a superlattice with layer thicknesses 52 Å Fe/17 Å Cr prepared at $T_G = 250$ °C. Strong spin-flip scattering which, when modelled with the non-spin-flim scattering intensity, reveals that successive Fe layers align symmetrically with respect to the sample anisotropy axes (see inset) at an unusual angle. (From reference [19] with permission from Elsevier.)

Fig. 9.17 shows the experimentally measured spin-dependent reflectivity of a multilayer of Fe/Cr in which the magnetization of successive Fe layers is non-collinear [19]. The presence of $R^{\pm,\mp}$ specular intensity indicates the presence of magnetic order perpendicular to the applied magnetic field. Fig. 9.17 exhibits a magnetic Bragg reflection at $Q_z = 0.045 \text{Å}^{-1}$, where the peak arises from a series of antiferromagnetically aligned magnetic layers. The non-spin-flip $R^{\pm,\pm}$ and spin-flip $R^{\pm,\mp}$ reflectivities are proportional to the projections of the staggered magnetization parallel and perpendicular, respectively, to the neutron polarization axis.

9.9 Summary

In this chapter we have studied neutron scattering technique as a powerful tool to study the microscopic magnetic structure of magnetic materials. Magnetic neutron diffraction experiments in a solid provide the most direct information on the microscopic arrangement of magnetic moments in a magnetic solid. Magnetic small-angle neutron scattering (SANS) is used for studying magnetic phenomena spanning length scales that are large compared to atomic distances. The SANS technique is particularly useful for exploring magnetic nanocomposites, magnetic domain structures, and long-wavelength oscillatory magnetic states. Inealstic neutron scattering experiments are used to study spin-wave excitations and other types of magnetic excitations and fluctuations like critical fluctuations, crystal field excitations, and magnetic moment/valence fluctuations. Lastly, polarized neutron reflectometry is a technique for probing depth-dependent magnetic structures as well as the characterization of lateral magnetic structures.

Bibliography

[1] Bacon, G. E. (1955). *Neutron Diffraction*. Oxford: Oxford University Press.

[2] Squires, G. L. (1996). *Introduction to the Theory of Thermal Neutron Scattering*. North Chelmsford: Courier Corporation.

[3] Lovesey, S. W. (1984). *Theory of Neutron Scattering from Condensed Matter. Vol. 1. Nuclear Scattering*. Oxford: Clarendon Press.

[4] Windsor, C. G. (1981). *Pulsed Neutron Diffraction*. London: Taylor & Francis Ltd.

[5] Willis, B. T. M., and Carlile, C. J. (2017). *Experimental Neutron Scattering*. Oxford: Oxford University Press.

[6] Boothroyd, A. T. (2020). *Principles of Neutron Scattering from Condensed Matter*. Oxford: Oxford University Press.

[7] Skold, K. and D. L. Price, eds. (1986). *Neutron Scattering*. Elsevier.

[8] Furrer, A., ed. (1996). *Introduction to Neutron Scattering (ECNS '96)*. Villigen, Switzerland: Paul Scherrer Institut. https://inis.iaea.org/collection/NCLCollectionStore/_Public/28/024/28024527.pdf

[9] Prokess, K., and Yokaichiya, F. (2016). Elastic Neutron Diffraction on Magnetic Materials. In K. H. J. Buschow, ed., *Handbook of Magnetic Materials*, Vol. 25. Amsterdam: Elsevier.

[10] Lynn, J. W. (2002). Magnetic Neutron Scattering. In E. N. Kaufmann, ed., *Common Concepts in Materials Characterization; Characterization of Materials*. Hoboken: Wiley, pp. 1–14.

[11] Kennedy, S. J., Murani, A. P., Cockcroft, J. K., et al. (1989). The Magnetic Structure in the Antiferromagnetic Phase of Ce $(Fe_{1-x}Co_x)_2$. *Journal of Physics: Condensed Matter*, 1 (3): 629.

[12] Murani, A. P. (2004). Polarised Neutron Diffraction Measurements on $CeFe_2$ and $LuFe_2$. *Physica B: Condensed Matter*, 345 (1–4): 89–92.

[13] Hirschberger, M., Nakajima, T., Gao, S., et al. (2019). Skyrmion Phase and Competing Magnetic ORDERS on a Breathing Kagomé Lattice. *Nature Communications*, 10 (1): 1–9. Also 10:5831. https://doi.org/10.1038/s41467-019-13675-4.

[14] Parker, S. F. (2016). In J. Lindon, G. E. Tranter and D. Koppenaal, eds., *Encyclopedia of Spectroscopy and Spectrometry*, 3rd edition. Amsterdam: Elsevier, pp. 246–257.

[15] Rietveld, H.M. (1967). Line Profiles of Neutron Powder-diffraction Peaks for Structure Refinement. *Acta Crystallographica*, 22 (1): 151–152.

[16] Vasiliu-Doloc, L., Lynn, J. W., Moudden, A. H., et al. (1998). Structure and Spin Dynamics of $La_{0.85}S_{0.15}MnO_3$. *Physical Review B*, 58 (22): 14913.

[17] Collins, M. F., Minkiewicz, V. J., Nathans, R., et al. (1969). Critical and Spin-wave Scattering of Neutrons from Iron. *Physical Review*, 179 (2): 417.

[18] Ankner, J. F. and Felcher, G. P. (1999). Polarized-Neutron Reflectometry. *Journal of Magnetism and Magnetic Materials*, 200 (1–3): 741–754.

[19] Schreyer, A., Ankner, J. F., Schafer, M., et al. (1996). Correlation between Non-collinear Exchange Coupling and Interface Structure in Fe/Cr (001) Superlattices. *Physica B: Condensed Matter*, 221 (1–4): 366–369.

10
X-ray Scattering

We have seen in the previous chapter that thermal neutrons with their unique combination of wavelength ($\approx Å$), magnetic moment ($\approx \mu_B$), and penetration power in materials can be a very sensitive probe to investigate microscopic magnetic properties. Elastic neutron scattering experiments provide information on the magnitudes and direction of the magnetic moment for complex spin arrangements. Inelastic neutron scattering experiments provide information on spin dynamics such as spin waves or magnons, spin fluctuations, crystal-field excitations, etc.

It had been known for quite some time that X-rays being part of the electromagnetic spectrum can have specific interactions with ordered arrangements of spins in magnetic materials. The absorption cross section and the scattering amplitudes of electromagnetic waves are related through the optical theorem [1], hence magnetic effects are also expected to be observed in X-ray scattering. However, that magnetic interaction is very weak, leading to only very small effects, and as a result for many years, X-rays were not considered as a useful probe of magnetism. In the 1970s–80s in a series of elegant pioneering experiments de Bergevin and Brunel [2, 3] and Gibbs and his collaborators [4, 5] demonstrated that X-ray diffraction effects from magnetic materials could be detected experimentally. These important developments along with the relatively easy access to modern synchrotron radiation sources with the capability of probing small samples with a highly collimated and intense X-ray beam, high degree of photon beam polarization, and energy tunability have made magnetic X-ray diffraction studies a complementary and in some cases competitive technique to neutron scattering studies in the study of magnetic materials [6, 7, 8, 9]. These magnetic X-ray diffraction techniques will be discussed below, and the narrative will closely follow the articles by C. Vettier and L. Paolisini [6, 7].

10.1 Magnetic and Resonant X-ray Diffraction

In a purely nonrelativistic limit, X-rays interact only with the electronic charges in solids, hence no information is obtained on the spin densities. The first calculation of the amplitude of X-rays elastically scattered by a magnetically ordered solid was carried out by Platzman and Tzoar [10] in 1970 within the framework of the relativistic quantum theory. That calculation predicted that magnetic X-ray scattering could be observed. Within two years De Bergevin and Brunel [2] reported the first experimental observation of such effects in a single-crystal sample of antiferromagnetic NiO. They used a conventional sealed X-ray tube to measure two superlattice diffraction peaks, which disappeared above the Neel temperature (~ 530 K) of NiO. The observed intensities of the magnetic Bragg peaks were 4×10^{-8} smaller than the Bragg peaks originating from electronic charge. It took three days' counting time to obtain each magnetic peak. Subsequently, the same group of authors had performed magnetic X-ray diffraction experiments on ferromagnetic and ferrimagnetic compounds [3]. We shall discuss below the process of *magnetic X-ray scattering*, and the narrative will closely follow the article by C. Vettier [6].

In the relativistic limit, the interaction of X-rays with the electrons in a solid depends also on the spin of the electrons, since the charge and spin components cannot be separated in the relativistic theory. A perturbation in the electron dynamics will have an effect, which will depend on its spin state. The amplitude of these effects during the scattering process in a solid can be represented in terms of the change in the wave vector $\vec{k}$ before and wave vector $\vec{k}'$ after the interaction, normalized to the Compton wave vector: $\hbar \vec{Q}/m_e c$, where $\vec{Q}$ is the scattering vector $\vec{Q} = \vec{k} - \vec{k}'$, m_e is electron mass, and c is the speed of light. Such correction, however, vanishes in the forward direction where X-ray absorption or reflection experiments are made.

The magnetic effects observed in the forward direction must be related to resonant effects, which involve characteristic energies of the electrons bound in the atoms. In the event of a photoelectric absorption, an electron is promoted into an empty shell, the symmetry of which depends on the orientation of the local magnetization. This results in new channels for the scattering process, and, in turn, provides new contributions to the X-ray scattering amplitude. Two regimes for the interaction of X-rays with magnetization in solids are distinguished depending on the energy of X-rays [6]: (1) a high energy regime, where the characteristic atomic energies are neglected, leading to the normal magnetic X-ray scattering, and (2) a resonant energy regime in which the X-ray energy is tuned close to the characteristic energies of electrons bound to the atom in the solid so that both the Bragg scattering experiments and absorption measurements become sensitive to electron spin densities.

10.1.1 Classical formalism of magnetic X-ray scattering

In an X-ray diffraction measurement, the directly measured quantity is the differential cross section defined as the probability for a particle of incident energy E_i to be scattered along a given direction, into a unitary solid angle $\Delta\Omega$ with a final energy E_f. A detailed quantum mechanical formalism for the estimation of X-ray scattering cross-section from various sources in a solid has been given by Blume [8]. However, the scattering cross section can also be derived within a purely classical framework [11], and such classical calculation reproduces the quantum mechanical cross section for the spin part quite well, but not the orbital part. We shall first discuss here such simple classical pictures ([3, 6]), and see how an electron with a magnetic moment is accelerated by interacting with the electromagnetic fields associated with the incident X-ray and reradiates electromagnetic components of the scattered wave.

Fig. 10.1 presents a simple schematic picture proposed by De Bergevin and Brunel within a classical model [3]. This picture shows various interaction processes between the incident X-ray and the electrons in a solid. The first process in Fig. 10.1 represents the classical charge or Thompson scattering. Here the X-ray is incident on a free electron, and due to the Coulomb force between the electric field vector and the electronic charge, the electron is accelerated and re-radiates electric dipole radiation. The three other processes in Fig. 10.1 are linked to the electron spin, and those give rise to magnetic X-ray scattering. Out of these, the second process arises from the same Coulomb interaction with the incident electromagnetic wave, where the accelerated spin moment gives rise to magnetic quadrupole radiation. In the third and fourth processes shown in Fig. 10.1, the interaction is between the spin moment and the magnetic field vector.

When the energy of incident X-ray is much larger than the characteristic electron energies in the atom [2, 6] the magnetic part of the scattering amplitude f^{mag} takes the simple form:

$$f^{mag} = -ir_0 \frac{\hbar w}{mc^2} \left(\frac{1}{2} \vec{A}.\vec{L}_j(Q) + \vec{B}.\vec{S}_j(Q) \right) \qquad (10.1)$$

where $\vec{L}_J(Q)$ and $\vec{S}_J(Q)$ represent Fourier transforms (in units of $\hbar$) of the orbital and spin momentum densities, respectively, and $\vec{Q}$ is the momentum transfer and r_0 is the classical electron radius.

The terms $\vec{A}$ and $\vec{B}$ are well-defined complex vectors containing information on the wave vector and polarization states of the incident and scattered X-rays. It is therefore possible to normalize the scattered intensities of X-ray and extract magnetic moment in the units of μ_B. These two vectors are formally not identical, hence the contribution of the spin momentum μ_S and the orbital momentum μ_L to f^{mag} are different and can be distinguished experimentally. Moreover, in contrast to

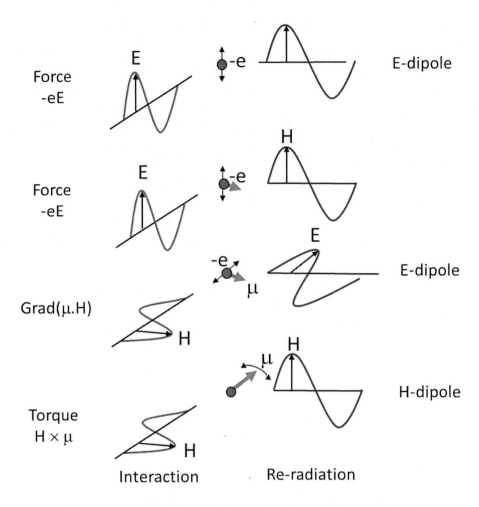

FIGURE 10.1 Schematic representation of the four mechanisms of X-ray scattering. (From reference [3] with permission from IUCr journals.)

the usual Thomson scattering of X-ray, the magnetic X-ray scattering amplitude is not diagonal in the two polarization states and mixes the different polarizations; this is a convenient way to separate the two. The polarization analysis of scattered X-ray intensities also gives information about the direction of the Fourier components of $L(Q)$ and $S(Q)$.

The magnetic amplitude in Eqn. 10.1 can be expressed as a matrix whose basis vectors are the polarization vector perpendicular and parallel to the scattering plane [6]:

$$-i\frac{\hbar w}{mc^2}\begin{pmatrix} A_{\sigma\sigma} & A_{\sigma\pi} \\ A_{\pi\sigma} & A_{\pi\pi} \end{pmatrix}. \tag{10.2}$$

where

$$A_{\sigma\sigma} = \vec{S}(Q).(\hat{k}' \times \hat{k})$$

$$A_{\sigma\pi} = -\frac{Q^2}{2k^2}\left[\left(\frac{\vec{L}(Q)}{2} + \vec{S}(Q)\right).\hat{k}' + \frac{\vec{L}(Q)}{2}.\hat{k}\right] \quad (10.3)$$

$$A_{\pi\sigma} = \frac{Q^2}{2k^2}\left[\left(\frac{\vec{L}(Q)}{2} + \vec{S}(Q)\right).\hat{k} + \frac{\vec{L}(Q)}{2}.\hat{k}'\right]$$

$$A_{\pi\pi} = \left[\frac{Q^2}{k^2}\frac{\vec{L}(Q)}{2} + \vec{S}(Q)\right].(\hat{k}' \times \hat{k})$$

Apart from this purely magnetic contribution, the total scattering cross section, of course, contains the classical Thomson scattering contribution, and an interference term between the magnetic amplitude and the Thomson scattering amplitude. The relative size of magnetic scattering amplitude to the charge scattering amplitude is given by $\frac{\hbar Q}{m_e c}$, which is of the order of a few 10^{-3} in usual diffraction experiments [6]. There is a further reduction in X-ray magnetic scattering in comparison to the normal Thomson charge scattering since only the unpaired electrons take part in that process. Hence, in most cases, the purely magnetic intensity is lower than the Thomson intensity by 7 or 8 orders of magnitude. In the case of ferro or ferrimagnetic materials, there exists in the intensity an interference term between the charge scattering amplitude from the lattice and the magnetic scattering amplitude [6]. Here it is possible to extract the magnetic amplitude by measuring X-ray intensities upon reversal of the magnetization by changing the direction of an externally applied magnetic field. This technique is applicable even with unpolarized X-rays and is equivalent to the polarized neutron scattering method for investigating ferromagnets.

The magnetic X-ray scattering amplitude in Eqn. 10.1 is phase-shifted by $\pi/2$ with respect to the charge scattering. The interference is observed only in circularly polarized X-ray beams in the case of real structure factors (which imply both a centrosymmetric structure and zero absorption). In the case of pure linear polarization, this asymmetry is given by:

$$R = \frac{2Re(Ff^{mag})}{|F|^2} \quad (10.4)$$

where F is the classical crystal structure factor and f^{mag} is the magnetic amplitude.

The scattering amplitude expressed in Eqn. 10.1 also gives rise to magnetic inelastic X-ray scattering, i.e., magnetic Compton scattering [12]. This magnetic Compton scattering provides information on the distribution of unpaired electrons in momentum space. The magnetic Compton signal can be obtained from the total

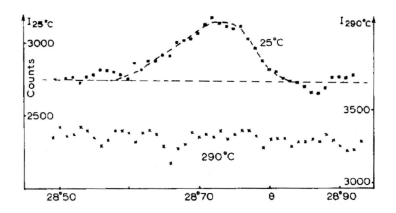

FIGURE 10.2 Magnetic X-ray scattering intensity in NiO diffraction from NiO near the $(\frac{3}{2}, \frac{3}{2}, \frac{3}{2})$ position at T = 25 °C and 290 °C. (From reference [2] with permission from Elsevier.)

X-ray scattering by measuring the asymmetry ratio in Eqn. 10.4 with circularly polarized X-ray beams.

As mentioned above, the first observation (to the best of the author's knowledge) of magnetic X-ray scattering was reported in the early 1970s by De Bergevin and Brunel [2] in a single crystal of antiferromagnetic compound NiO. This compound orders antiferromagnetically around 530 K where the spins parallel to one of the (111) planes are ordered ferromagnetically, and alternate antiferromagnetically from plane to plane. The periodicity is thus doubled in the perpendicular [111] direction. De Bergevin and Brunel [2] could detect the first two superlattice magnetic reflections $(\frac{1}{2}, \frac{1}{2}, \frac{1}{2})$ and $(\frac{3}{2}, \frac{3}{2}, \frac{3}{2})$ along with the first ordinary reflection (111) (see Fig. 10.2). They had used CuK_α radiation from a CGR sealed X-ray tube running at 70 mA and 14 kV with 225 minutes for each count. In Fig. 10.3 we show the results of the magnetic scattering experiment on NiO performed with a synchrotron radiation source (at ESRF, Genoble) with much higher X-ray photon-flux and count rate [13]. In between, the synchrotron radiation was used in the mid-1980s to study the magnetic X-ray scattering from elemental holmium (Ho) [4]. The elemental Ho has a hexagonal close-packed crystal structure with two layers per chemical unit cell. It carries a large magnetic moment of about 10 μ_B/atom and a phase transition to a spiral antiferromagnetic state below $\approx$ 131 K, in which the average moments are ferromagnetically aligned within the basal planes, but rotate from plane to plane [4]. Further down the temperature, around 20 K, there is a first-order phase transition to a magnetic state having a conical spiral structure with a ferromagnetic component along the c-axis. Fig. 10.4 shows magnetic satellite peaks in Ho(004)$^+$ at various temperatures between 40 K and 245 K obtained with synchrotron radiation.

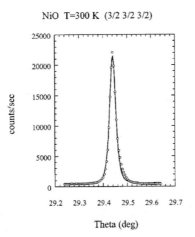

FIGURE 10.3 Rocking curve of the $\frac{3}{2}, \frac{3}{2}, \frac{3}{2}$ magnetic reflection in NiO at T = 300 K. (From reference [13] with permission from APS.)

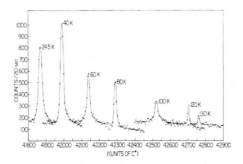

FIGURE 10.4 Temperature dependence of the Ho(004)$^+$ magnetic satellite peaks obtained with synchrotron radiation (lines drawn to guide the eye). (From reference [4] with permission from APS.)

The advantage of using X-rays from a synchrotron radiation source to study magnetic properties of solids are the following [6]:

1. The non-resonant scattering cross section allows experimental determinations of L and S in any type of magnetically ordered compound. The large flux of photons and the high Q-resolution available on synchrotron sources enable one to examine phase transitions and features of magnetic structures in great detail. Even the magnetic form factor can be determined at quite a large Q.
2. The determination of magnetic structures with X-rays in bulk solids appears to be extremely difficult, because of the weak scattering cross section. This is indeed a disadvantage with the non-resonant magnetic X-ray scattering. This problem is somewhat circumvented with the use of resonant scattering. The large enhancements in the scattering intensity can be used to study

systems with weak magnetic moments and small scattering volumes such as micro-crystals or surfaces and interfaces. In addition, the resonant X-ray scattering is element selective. In the synchrotron radiation sources, tunability of the X-ray photon energy makes it possible to turn on and off the resonant signal from one element in a magnetic compound. We shall discuss below the theory of resonant X-ray scattering.

3. The line shape analysis of the resonance provides a new spectroscopic technique to study the excited electronic states. Even the non-magnetic X-ray Thomson scattering can be used to study magnetism through magnetostrictive phenomena. The determination of small lattice changes near magnetic phase transitions becomes possible due to the high Q-resolution offered by synchrotron radiation sources.

10.1.2 Quantum mechanical theory of magnetic and resonant X-ray scattering

In X-ray scattering the fundamental interactions of X-ray photons with bound electrons in magnetic solids are due to the electron charge or caused by the coupling between the electric and magnetic fields of the incident photons with the atomic magnetic moment. The X-rays interact weakly with bulk samples in the photon energy range $\hbar\omega < 100$ keV, and the scattering process can be studied in the lowest Born-Oppenheimer approximation. One can define the X-ray photons by the initial state $|\vec{k}, \hat{\epsilon}\rangle$ characterized by wave vector $\vec{k}$, polarization $\hat{\epsilon}$, and energy $\hbar\omega_k$. On the other hand, the electronic ground state $|g\rangle$ is an eigenstate of the unperturbed electron Hamiltonian H_{el} with energy E_g. The interaction between the X-ray photons and electrons gives rise to a new final eigenstate $|f\rangle$ for the electrons with an energy E_f and a scattered X-ray photon in the state $|\vec{k}', \hat{\epsilon}'\rangle$ with an energy $\hbar\omega_{k'}$. The differential cross section for the coherent elastic X-ray scattering ($\hbar\omega_k = \hbar\omega_{k'}$ and $E_g = E_f$) on single crystals where the individual atomic scattering amplitudes interfere at different lattice sites n can be expressed as [7]:

$$\frac{d\sigma}{d\Omega} = r_0^2 \left| \sum_n e^{\vec{Q}\cdot\vec{R}_n} f_n(\vec{k}, \vec{k}', \hat{\epsilon}, \hat{\epsilon}', \hbar\omega_k) \right|^2 \qquad (10.5)$$

Here $r_0 = e^2/mc^2 \sim 2.82 \times 10^{-5}$ Å is the classical electron radius, and $f_n(\vec{k}, \vec{k}', \hat{\epsilon}, \hat{\epsilon}', \hbar\omega_k)$ is the scattering amplitude of the electrons at site n. $\vec{R}_n$ stands for the position of the nth site in the crystal and $\vec{Q} = \vec{k} - \vec{k}'$ stands for the scattering vector.

Depending on the energy of incident X-ray photon, the scattering amplitude $f_n(\vec{k}, \vec{k}', \hat{\epsilon}, \hat{\epsilon}', \hbar\omega_k) = f_n^0 + f_n^{magn} + f_n' + f_n''$ has two regimes in X-ray scattering experiments:

1. The non-resonant scattering limit: It contains the energy-independent part of the scattering amplitude, which includes the Thomson scattering amplitude f_n^0 and the magnetic scattering amplitude f_n^{magn}.
2. The resonant X-ray scattering regime: It contains the real f' and imaginary f'' parts of the energy-dependent terms or the resonant term of the scattering amplitude. This resonant term $f^{RXS} = f' + if''$ is also termed as the anomalous or dispersive term. This appears when the energy of the incident X-ray photon resides near an absorption edge in a material as a consequence of the photo-absorption effect, which arises due to transitions of core electrons into available electronic states above the Fermi energy.

Blume developed a perturbative method [8] to estimate scattering amplitude $f_n(\vec{k}, \vec{k}', \hat{\epsilon}, \hat{\epsilon}', \hbar\omega_k)$ starting from an unperturbed system described by the electron and non-interacting photon Hamiltonians. With further assumptions that the unperturbed radiation is monochromatic, and the vector potential which characterizes the radiation field is a linearly polarized plane wave, the scattering amplitude can be expressed in terms of non-resonant and resonant scattering amplitudes [7, 9]:

$$f_n(\vec{k}, \vec{k}', \hat{\epsilon}, \hat{\epsilon}', \hbar\omega_k) = - \langle g| \sum_j e^{i\vec{Q}\cdot\vec{r}_j} |g\rangle \, \hat{\epsilon}^{*\prime}.\hat{\epsilon}$$

$$- i\frac{\hbar\omega_k}{mc^2}\hat{\epsilon}^{*\prime}. \langle g| \sum_j e^{i\vec{Q}\cdot\vec{r}_j} \left(\frac{\vec{Q}\times\vec{p}_j}{\hbar Q^2}.\vec{A} - \vec{s}_j.\vec{B}\right) |g\rangle .\hat{\epsilon}$$

$$+ \frac{1}{m}\sum_c \frac{E_g - E_c}{\hbar\omega_k}\left(\frac{\hat{\epsilon}^{*\prime}. \langle g| \tilde{O}^\dagger(\vec{k}') |c\rangle \langle c| \tilde{O}(\vec{k}) |g\rangle .\hat{\epsilon}}{E_g - E_c + \hbar\omega_k - i\Gamma_c/2}\right.$$

$$\left. - \frac{\hat{\epsilon}. \langle g| \tilde{O}(\vec{k}) |c\rangle \langle c| \tilde{O}(\vec{k}') |g\rangle .\hat{\epsilon}^{*\prime}}{E_g - E_c - \hbar\omega_k}\right) \quad (10.6)$$

The first term in Eqn. 10.6 represents the Thomson contribution f^0 associated with the isotropic charge density distribution. The second term represents the non-resonant X-ray magnetic scattering amplitude f^{magn}, which is associated with the distribution of orbital and spin densities. Here m, $\vec{p}_j$, and $\vec{s}_j$ denote the electron mass, the momentum, and the spin operators, respectively. The vectors $\vec{A}$ and $\vec{B}$ contain the coupling between the X-ray photon polarization and the scattering wave vectors for the orbital and the spin part of the non-resonant X-ray magnetic scattering amplitude, respectively. The resonant contribution $f^{RXS} = f' + if''$ is represented by the last two terms. In the resonant process, the initial X-ray photon promotes an electron from core levels into a partially occupied or empty valence shell, which subsequently decays into the same initial state resulting in an elastically re-emitted photon. Here Γ_C is the width of the excited level $|c\rangle$ with energy E_c and $2\pi\hbar/\Gamma_C$ its lifetime. The current operator $\tilde{O}(\vec{k}) = \sum_j e^{i\vec{Q}\cdot\vec{r}_j}(\vec{p}_j - i\hbar \times \vec{s}_j)$ probes the intermediate states $|c\rangle$ when the incident energy is tuned across a characteristic energy $\hbar\omega = E_g - E_c$ [7].

The sums in Eqn. 10.6 are extended to those electrons, which interact with the radiation field: (i) all the core and valence electrons Z for the Thomson term ($f^0 \propto Zr_0$), (ii) only those electrons occupying the partially filled shells in the case of the magnetic term ($f^{magn} \propto Z^{magn}r_0$), and (iii) the resonant electrons ($f^{RXS} \propto Z^{res}r_0$) for resonant terms. It may be noted that for simplicity it is assumed that only one state is populated at $T=0$ and hence the thermal population factor is omitted in Eqn. 10.6.

Non-resonant X-ray scattering

In the Thomson term $f_0 = Zr_0$ all the core and valence electrons interact with the radiation field. The amplitude of the Thomson scattering depends on the Fourier transform of the electron charge density multiplied by the scalar product of incident and scattered photon polarization $\hat{\epsilon}^{\star\prime}\cdot\hat{\epsilon}$. Since the Thomson scattering is isotropic, it does not cause any change in the photon polarization state. In comparison to the Thomson scattering, the amplitude of non-resonant magnetic scattering is very weak. This is because of the relativistic character of the electron–photon interaction, which introduces a scale factor $\hbar Q/2\pi mc = 2(\lambda_c/\lambda)\sin\theta$ between magnetic scattering and Thomson scattering amplitudes [7]. Here $\lambda_c = 2\hbar/mc = 0.002426$ nm is the Compton wavelength of the electron. Thus, the pure magnetic scattering amplitude is reduced significantly by a factor of $\hbar\omega_k/mc^2$ in comparison to the classical charge scattering. The magnitude of purely magnetic scattering is even more reduced due to the rapidly decreasing magnetic form factor, the smaller number N_m of electrons contributing to magnetic properties, and any fluctuations that reduce the magnetic order parameter $<M>$. The ratio between the cross sections of magnetic and charge scattering can be expressed as [7]:

$$\frac{\sigma_{mag}}{\sigma_{charge}} \simeq \left(\frac{\hbar\omega_k}{mc^2}\right)^2 \left(\frac{Z^{magn}}{Z}\right)^2 <M>^2 \left(\frac{f_m}{f}\right)^2 \quad (10.7)$$

Here $mc^2 = 511$ keV is the rest mass of an electron, Z^{magn} represents the number of electrons per atom contributing to the magnetic properties, Z the number of electrons per atom, and f_m and f are the magnetic and charge form factors, respectively. Magnetic order parameter $<M>$ is equal to one only at low temperature when the system is magnetically saturated [7]. In a sample having Mn^{2+}-ion with the assumption $\frac{f_m}{f} \approx 1$, the ratio $\frac{\sigma_{mag}}{\sigma_{charge}} \approx 10^{-6}$.

The amplitude of non-resonant X-ray magnetic scattering f^{magn} can be expressed in terms of the Fourier transforms of orbital and spin magnetization densities $L(\vec{Q}) =$

$\langle g | \sum_j \vec{I}_j e^{i\vec{Q}.\vec{r}_j} | g \rangle$ and $S(\vec{Q}) = \langle g | \sum_j \vec{s}_j e^{i\vec{Q}.\vec{r}_j} | g \rangle$, respectively. The f^{magn} amplitude then assumes the operative form [7]:

$$f^{magn}(\vec{Q}) = -i \frac{\hbar \omega_k}{mc^2} (L(\vec{Q}).\vec{P}_L + S(\vec{Q}).\vec{P}_S) \qquad (10.8)$$

Here $\vec{P}_L$ and $\vec{P}_S$ are vectors describing the polarization and wave vector dependence of the magnetic scattering process, and can be expressed as [7]:

$$\vec{P}_L = -2\sin^2\theta \left(\vec{Q} \times [(\hat{\epsilon}'^* \times \hat{\epsilon})] \times \vec{Q} \right)$$
$$\vec{P}_S = \hat{\epsilon}'^* \times \hat{\epsilon} + (\hat{k}' \times \hat{\epsilon}'^*)(\hat{k}.\hat{\epsilon}) - (\hat{k} \times \hat{\epsilon})(\hat{k}.\hat{\epsilon}'^*) - (\hat{k}' \times \hat{\epsilon}'^*) \times (\hat{k} \times \hat{\epsilon}) \qquad (10.9)$$

The above expressions can be simplified with the choice of an appropriate reference coordinate system for the polarization vectors with respect to the scattering geometry. It is evident in X-ray magnetic scattering that by taking advantage of the geometrical prefactors for spin and orbital magnetic moments, it is possible to separate the spin $\vec{S}$ and orbital $\vec{L}$ moment contributions in the total magnetization density. The prefactors can be adjusted by changing either the scattering geometry or the X-ray polarization channels. Polarization analysis thus enables a clear-cut distinction between charge and magnetic scattering.

The non-resonant terms f^0 and f^{magn} dominate in Eqn. 10.6 in the energy range far from the characteristic energy, i.e., $\hbar\omega \gg E_c - E_g$. This condition is valid when the incident energy of the X-ray photon is tuned above the absorption edges. Most of the non-resonant magnetic scattering experiments, however, are performed below the absorption edges because of experimental reasons. As a result of this, the experimental outputs can be affected by the presence of the resonant contribution, and thus it is necessary to evaluate carefully the resonant contamination on a case-to-case basis.

10.1.3 Resonant X-ray scattering

One is in the resonant X-ray scattering regime when the photon energy is tuned to such characteristic energy where the resonant denominator $\frac{1}{E_g - E_c + \hbar\omega_k - i\Gamma_c/2}$ in Eqn. 10.6 plays an important role. In this energy regime, there is an enhancement of the magnetic intensities across the absorption edge. The resonant X-ray scattering process causes the transfer of an electron from core levels into a partially occupied or empty valence shell. The subsequent decay into the same initial electron state results in an elastically re-emitted X-ray photon, where Γ_c is the width of the excited level $|c\rangle$ with energy E_c and $2\pi\hbar/\Gamma_c$ its lifetime. The theory of resonant X-ray scattering (RXS) is based on the general theory of photon–electron interaction in the non-relativistic limit where non-resonant X-ray magnetic scattering terms are neglected. This theory applies to both absorption and diffraction regimes, since the signal intensities depend on the matrix elements of the interacting operator between the ground state and the excited state wave functions [7].

As shown in Eqn. 10.6 the amplitude of RXS $f^{RXS} = f' + if''$ depends on the intermediate states $|c\rangle$ of the atomic system, and it depends specifically on the incident photon energy $\hbar\omega$. In the analysis of the RXS diffraction data, it is possible to use zero-angle values for $f' + if''$, which are derived from absorption data or theoretical calculations in the dipole approximation. This is because the RXS process arises mostly from inner-shell electrons, which are less sensitive to the atomic size effects.

The amplitude of resonant scattering can be written with some approximations after starting from second-order terms in Eqn. 10.6 containing the resonant contribution as [7]:

$$f^{RXS} \approx -\frac{1}{m} \sum_c \frac{E_g - E_c}{\hbar\omega_k} \cdot \frac{\langle g| \sum_j e^{-i\vec{k}'\cdot\vec{r}_j} \hat{\epsilon}'^{\star}\cdot\vec{p}_j |c\rangle \langle c| \sum_j e^{i\vec{k}\cdot\vec{r}_j} \hat{\epsilon}\cdot\vec{p}_j |g\rangle}{E_g - E_c + \hbar\omega_k - i\Gamma_c/2} \qquad (10.10)$$

The magnetic X-ray scattering cross section expressed in Eqn. 10.1 vanishes at zero momentum transfer and does not contribute to absorption effects. Hence, the normal magnetic X-ray cross section cannot take into account the magnetization sensitive dichroic effects that take place near X-ray absorption edges. These correspond to resonant scattering processes, which were predicted by Blume through a detailed quantum mechanical formalism [8]. This was subsequently validated by actual experimental observation of a resonant magnetization-sensitive scattering in ferromagnetic Ni [14] and antiferromagnetic Ho [5].

The same mechanism of promotion of an inner shell electron (in the concerned solid) into available electronic states by the incident X-ray is at the origin of magnetic dichroism and the enhancement of the magnetic scattering amplitude in a solid. Pauli's exclusion principle allows the electronic transitions only to unoccupied electronic states, and this introduces a sensitivity to exchange interaction in the magnetization of the final states. This, in turn, makes this additional contribution to the cross section arising from electric transitions a sensitive probe of the local magnetic moments. In comparison with the non-resonant magnetic X-ray scattering, which arises from the magnetic couplings, the resonant magnetic X-ray scattering and the dichroism can be analyzed in terms of electric multipole transitions between an atomic core electron level and either an unfilled atomic shell or a narrow electron band.

The resonant magnetic scattering amplitude can be written in a simple and symmetric dependence on the magnetization direction and the polarization of the photon beams [6, 15]. This will have all the symmetric terms that can arise with combinations of magnetization direction, the polarization vector, and the wave vectors of the incident and scattered X-ray beam. The coefficients will be dictated by the nature of the interaction, namely the promotion of an electron into an available electronic state above the Fermi level. Restricting only to the electric dipole transitions, the resonant part of the scattering amplitude can be expressed as [6, 15]:

$$f_{E1}^{mag-res} = \frac{3\lambda}{8\pi}\{e_f^*.e_i[F_{11}+F_{1-1}] - i(e_f^* \times e_i).z_j[F_{11}+F_{1-1}] \\ + (e_f^*.z_j)(e_i.z_i)[2F_{10} - F_{11} - F_{1-1}]\} \quad (10.11)$$

where $\vec{e}_i$ and $\vec{e}_f$ is the polarization vectors, which were contained implicitly in the vector $\vec{A}$ and $\vec{B}$ in Eqn. 10.1 for the non-resonant magnetic scattering. The quantity z_j represents the magnetization direction of the local moment on-site j. The resonant amplitude given by Eqn. 10.11 is governed by the F_{lm}. These coefficients contain matrix elements coupling the ground state to the excited magnetic levels and also a resonant denominator, and are expressed as [6, 15]:

$$F_{lm} = \sum_{a,b} \frac{p_a p_a(b) \Gamma_x(aMb;EL)}{(E_a - E_b - \hbar w) - \frac{i\Gamma}{2}} \quad (10.12)$$

where a and b are the initial and excited states of energy E_a and E_b, respectively, p_a gives the statistical probability for the initial state a, and $p_{a(b)}$ gives the probability that the excited state b is vacant for transitions from a. Γ is the total width for b, and $\Gamma_x(aMb;EL)$ gives the partial width for electric 2^L-pole resonance (EL) decay in a magnetic ion decay from b to a when summed over M. The coefficients F_{lm} can be calculated by modelling the density of states and wave functions.

The resonant amplitude $f_{E1}^{mag-res}$ involving dipolar transitions expressed in Eqn. 10.11 contains three different terms:

1. The first term corresponds to the classical anomalous scattering.
2. The second term, which is linear in z_j, leads to magnetic scattering amplitude at the same place in reciprocal space as the usual magnetic scattering. In addition it describes the X-ray magnetic circular dichroism in absorption experiments ($k_i = k_f, e_i = e_f$).
3. The third term is quadratic in z_j, and it gives second-harmonic magnetic Bragg peaks. It does not vanish with real e_i and e_f in the absorption case and leads to linear dichroism.

While the dipolar transitions are usually the strongest, the quadrupolar terms need to be considered also. This quadrupolar scattering amplitude contains terms up to the fourth order in z_j, which can lead to diffraction harmonics up to the fourth order [6, 15]. The existence of fourth-order magnetic Bragg peaks near resonance in a scattering experiment is a signature of quadrupole transitions. Easy separation of the multipolar contributions to the resonance is thus allowed in X-ray resonance scattering experiments.

Based on this resonant X-ray scattering theory, the prediction was made on the occurrence of large resonant magnetic scattering in the vicinity of the $L_2 - L_3$ and $M_4 - M_5$ absorption edges in the rare earth and actinides and at the K and L edges in the transition metals [15]. Such resonances were shown to result from electric multipole transitions, with sensitivity to magnetization arising from exchange effects. The calculated magnetic scattering was shown to be comparable to the charge scattering for some transitions. Soon after this theoretical development, very large resonance effects were experimentally observed in the antiferromagnet UAs [16]. This compound with NaCl crystal structure undergoes a magnetic ordering at 120 K as a type-1 antiferromagnet with a magnetic moment of approximately 2 μ_B per uranium atom. Further down, in temperature at $T = 63$ K, UAs exhibits a first-order phase transition into a type-1A 2-q structure, where the Fourier components of the magnetization are $(\frac{1}{2},0,0)$ and $(0,\frac{1}{2},0)$ with the net moment lying along the [110] direction. There are three possible magnetic domains associated with this structure and each domain gives rise to two sets of magnetic Bragg peaks. The domain labelled (xy) results in the $(0,\pm\frac{1}{2},2)$ and $(\pm\frac{1}{2},0,2)$ satellite peaks about the $(0,0,2)$ charge peak, while the (yz) domain gives rise to the $(0,\pm\frac{1}{2},2)$ and $(0,0,2\pm\frac{1}{2})$ satellite peaks. The transition from the type-1 to type-1A structure in UAs corresponds to a doubling of the magnetic unit cell [16]. Total integrated intensity vs incident X-ray energy at the M_V absorption edge of uranium along with the temperature dependence of the reflections from the type-1 $(0,0,1)$ and type-1A $(0,\frac{1}{2},2)$ structures are shown in Fig. 10.5. The intensity of the $(0,0,1)$ magnetic reflection vanishes at the transition temperature, while the intensity of the $(0,\frac{1}{2},2)$ magnetic reflection has a sharp onset. At the peak of the resonance curve, the magnetic scattering intensity is $\sim 0.01\%$ of the intensity at the charge peak or 10^5 times larger than the expected non-resonant magnetic scattering intensity.

X-ray magnetic scattering is now a routine technique for the investigations of magnetism and magnetic materials, especially with the availability of synchrotron radiation sources. This technique nicely complements the magnetic neutron scattering technique regarding Q-resolution, single magnetic domain studies, separation of orbital and spin contributions to the magnetic moment, and magnetic ordering in materials with high neutron absorption. In addition, the chemical selectivity of X-ray resonant scattering allows the separation of the contribution of different elements to the magnetism in alloys and intermetallic compounds.

10.1.4 X-ray magnetic circular dichroism and X-ray magnetic linear dichroism

X-ray magnetic circular dichroism (XMCD) spectroscopy is a powerful technique that measures the difference in the absorption of left- and right-circularly polarized

X-rays by a magnetic sample [17]. This technique has the advantage of elemental specificity, which comes with all core electron spectroscopies. XMCD can provide quantitative information about the distribution of spin and orbital angular momenta with the help of simple sum rules. There are also other strengths of XMCD, which include the capacity to determine spin orientations from the sign of the XMCD signal, and the ability to separate magnetic and non-magnetic components in heterogeneous samples.

The key components for an XMCD experimental setup are [17]:

1. A source of circularly polarized X-rays: The possibility of X-ray circular polarizers had been demonstrated with Cu-K_α radiation [18], but the resultant beams were not bright enough for any practical applications. This problem is of course solved with the availability of synchrotron X-ray sources, especially the third generation synchrotron sources with insertion devices. The elliptical undulator in the third generation light sources is one of the most successful devices for the production of circularly polarized synchrotron radiation. Here, as an electron passes through the insertion device the magnetic field vector rotates, causing the particle to spiral about a central axis. The peak energy of the X-ray at the undulator output can be varied by changing the vertical separation between the magnet assemblies of the undulator, while the polarization is varied by changing the relative positions of the adjacent rows of magnets.

2. A monochromator and necessary optics: In synchrotron sources, this arrangement is called a beamline, and the major synchrotron sources in the world have dedicated beamlines for XMCD experiments. Grazing incidence mirrors and gratings are generally used in the soft X-ray region. The source polarization is almost completely preserved as it passes through the optics in such geometries [17]. A typical soft X-ray beamline provides $\sim 10^{12}$ photons/sec between 50 and 2000 eV with $\Delta E/E \sim 20,000$ at the lower energy. Crystal monochromators are generally suitable in the hard X-ray region, but the effect of the crystal optics on beam polarization can be considerable. The perfect circular polarization from a hard X-ray elliptical undulator could be degraded by the crystal optics to the extent that in some cases pure horizontal polarization is produced. The solution here is to start with linear polarization and convert it to circular polarization with a retarder, commonly a quarter-wave plate [17].

3. A facility for providing an external magnetic field to the sample under study: Electromagnets are used where a relatively small field (say up to ± 2.5 T) is required. Electromagnets are easy to automate and versatile systems can be made with sample manipulators allowing 180 polar and 360 azimuthal rotation of the sample, which permits any geometry of magnetic field, sample

surface, and incident photon direction [17]. For experiments in a higher magnetic field, superconducting magnets are required.

4. A system for detecting the X-ray absorption: XMCD is essentially a measurement of relative absorption coefficients, and all of the detection methods used for conventional X-ray absorption spectroscopy can in principle be used. In practice, the three most important modes are transmission, fluorescence, and electron yield [17].

Transmission is the simplest measurement, where the X-ray beam intensity before reaching the sample (I_0) is usually measured using the electron yield or photocurrent from a partially transmitting metal grid. The intensity of the X-ray beam after transmission through the sample (I) can be recorded using a second grid, a solid metal plate, or an Si photodiode. Electron-based detection methods are preferred for concentrated inorganic samples. The electron yield can be measured directly with a channeltron electron multiplier or indirectly as the photocurrent of electrons flowing to the sample from the ground. Electron methods, however, may suffer from experimental artifacts if switching of the magnetic field is involved in the measurement of the XMCD effect. The trajectories of emitted photoelectrons will depend on the applied magnetic field, and the apparent absorption cross section will vary if changing the magnetic field affects the fraction of photoelectrons that are accepted by the collector. Thus, electron-based XMCD measurements are best done with the experimental setup where the polarization of the X-ray beam can be changed. It may, however, be noted that even polarization-switched experiments suffer from potential experimental errors. The effective source point for left- and right-circularly polarized X-ray beams may be slightly different, which can transform into an energy difference between the two beams at a given monochromator position. The slight mismatch can then result in a derivative-shaped contribution to the XMCD spectrum that is stronger for sharper features [17]. This effect is independent of the applied field, hence one can check that there is indeed no XMCD effect in the absence of sample magnetization.

The problem of magnetic field switching can be circumvented with the experiments done in fluorescence mode since the magnetic field will not affect the X-ray photons once they are emitted. Most fluorescence detectors, however, convert X-rays into electrons, hence the detector resolution or gain may get influenced by a strong magnetic field. Hence, the potential detector sensitivity needs to be examined before varying the magnetic field in XMCD measurements.

The concepts of XMCD spectroscopy, pioneered by Schütz et al. [19], are represented in Fig. 10.6 with the example of a *d*-transition metal. In the subsequent

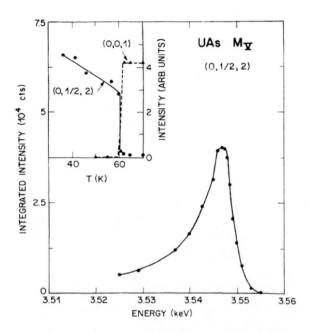

FIGURE 10.5 Total integrated intensity vs incident X-ray energy at the M_V absorption edge of uranium. Inset shows the temperature dependence of the reflections from the type-1 (0,0, 1) and type-I A (0, $\frac{1}{2}$, 2) structures. (From reference [16] with permission from APS.)

narrative, we will closely follow the review article by Stöhr et al. [20]. Here it is assumed that the d shell has a spin moment, which is due to the imbalance of spin-up and spin-down electron states below the Fermi level E_F or, equivalently, by the imbalance of spin-up and spin-down hole states above E_F. Here one needs to use a spin-dependent X-ray absorption process to measure the difference in the number of d holes with spin up and spin down. This is achieved through the use of right circularly polarized (RCP) and left circularly polarized (LCP) X-ray photons, which transfer their angular momentum, $+\hbar$ and $-\hbar$, respectively, to the excited photoelectron. The photoelectron carries the transferred angular momentum as a spin or an angular momentum or both [20]. If the photoelectron originates from a spin–orbit split level, for example, $p_{\frac{3}{2}}$ level(L_3 edge), the angular momentum of the photon can be transferred in part to the spin through the spin–orbit coupling. The RCP X-ray photons transfer the opposite momentum to the electron from LCP X-ray photons, and hence photoelectrons with opposite spins are created in two cases. The $p_{3/2}(L_3)$ and $p_{1/2}(L_2)$ levels have opposite spin–orbit coupling ($l+s$ and $l-s$, respectively), hence the spin polarization will be opposite at the two edges. In the absorption process, spin-up and spin-down are defined relative to the photon helicity or photon spin [20]. The spin flips are forbidden in electric dipole transitions governing X-ray absorption. As a result spin-up (spin-down) photoelectrons from the

p core-shell can only be excited into spin-up (spin-down) d hole states. The spin-split valence shell will act as a detector for the spin of the excited photoelectron, and the transition intensity will be simply proportional to the number of empty d states of a given spin.

The direction of magnetization in the sample under study will be the reference for the quantization axis of the valence shell detector. The size of the dichroism effect scales as $\cos\theta$, where θ is the angle between the photon spin and the magnetization direction [20]. As shown on the right side of Fig. 10.6, the maximum dichroism effect (typically 20%) is observed when the spin direction of the X-ray photon and the magnetization directions are parallel and antiparallel. If the X-ray photon spin and the magnetization directions are perpendicular, the resonance intensities at the L_3 and L_2 edges lie between those for parallel and antiparallel alignments. The size, direction, and anisotropies of the atomic magnetic moments in a solid material can be correlated quantitatively by simple sum rules to the differences in the intensities at the L_3 and L_2 edges for parallel and antiparallel X-ray photon spin and magnetization orientation in the sample. Hence, all these quantities can be determined by XMCD spectroscopy to a great degree of accuracy.

It is well known that linearly polarized X-rays can probe the orientation of molecular orbitals [20]. The electric field vector E of the linearly polarized X-rays acts as a detector for the number of valence holes in different directions of the atomic volume or the Wigner-Seitz cell in a solid. Another effect, which arises from the presence of magnetic anisotropy in the sample, forms the basis of X-ray magnetic linear dichroism (XMLD) spectroscopy. We will study this phenomenon of XMLD taking the example of NiO (see Fig. 10.7). The charge distribution around the atoms in NiO solid is nearly spherical because of the cubic symmetry of the lattice. No linear dichroism effect is expected in the absence of magnetic interactions. However, NiO is an antiferromagnet below 530 K, where the Ni spins are oriented in the (1,1,1) plane along the $[\bar{2}, 1, 1]$, $[1, \bar{2}, 1]$ or $[1, 1, \bar{2}]$ axes in the fcc lattice. Although there is no net magnetic moment because of an equal number of spins pointing to opposite directions, there is a preferential magnetic axis in NiO. This magnetic anisotropy breaks the cubic symmetry of the charge through the spin–orbit coupling, and, as a consequence, the charge exhibits a small anisotropy in the unit cell. This charge anisotropy, in turn, leads to an asymmetry of the X-ray absorption signal. The maximum XMLD effect is obtained for the geometries: electric field(E) parallel versus E perpendicular to the magnetic anisotropy axis [20]. The XMLD effect has a $\cos^2\theta$ dependence, where θ is the angle between E and the magnetic anisotropy axis. The XMLD effect is in general small in $3d$ metals due to the small size of the spin–orbit interaction and the large bandwidth, which results in a small charge anisotropy when the d-electron states are summed over the Brillouin zone. A sizeable XMLD effect of the order 10% may, however, be observed in the presence of multiplet

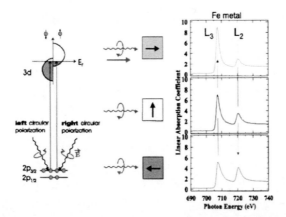

FIGURE 10.6 Principles of X-ray magnetic circular dichroism spectroscopy, illustrated for the case of L edge absorption in d band transition metal Fe. (From reference [20] with permission from World Scientific Publishing.)

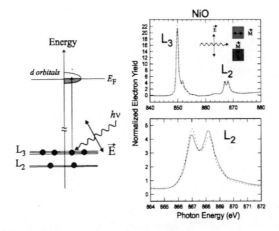

FIGURE 10.7 Principles of X-ray magnetic linear dichroism spectroscopy, illustrated for NiO. (From reference [20] with permission from World Scientific Publishing.)

splitting in transition metal oxide compounds like NiO. Multiple fine structures in the L_3 and L_2 resonances have been observed in the X-ray absorption spectrum of NiO, and this is shown in Fig. 10.7. The XMLD effect is particularly visible at the Ni L_2 edge, where a significant difference is observed for the geometries E parallel versus E perpendicular to the magnetic anisotropy axis.

10.2 Summary

In this chapter we have studied the magnetic X-ray scattering technique, which in comparison to neutron scattering is a relatively new tool to study the microscopic magnetic structure of magnetic materials. The resonant X-ray technique nicely complements the magnetic neutron scattering technique regarding Q-resolution, single magnetic domain studies, separation of orbital and spin contributions to the magnetic moment, and magnetic ordering in materials with high neutron absorption. Furthermore, the chemical selectivity of X-ray resonant scattering allows separation of the contribution from different elements to the magnetism in alloys and intermetallic compounds. X-ray magnetic circular dichroism (XMCD) spectroscopy techniques provide quantitative information about the distribution of spin and orbital angular momenta in a magnetic material with the help of simple sum rules. The other strengths of XMCD include the capacity to determine spin orientations, and the ability to separate magnetic and non-magnetic components in heterogeneous samples.

Bibliography

[1] Jackson, J. D. (2007). *Classical Electrodynamics*. Hoboken: John Wiley.
[2] De Bergevin, F., and Brunel, M. (1972). Observation of Magnetic Superlattice Peaks by X-ray Diffraction on an Antiferromagnetic NiO crystal. *Physics Letters A*, 39 (2): 141–142.
[3] De Bergevin, F., and Brunel, M. (1981). Diffraction of X-rays by Magnetic Materials. I. General Formulae and Measurements on Ferro- and Ferri-magnetic Compounds. *Acta Crystallographica Section A: Crystal Physics, Diffraction, Theoretical and General Crystallography*, 37 (3): 314–324.
[4] Gibbs, D., Moncton, D. E., d'Amico, K. L., et al. (1985). Magnetic X-ray Scattering Studies of Holmium using Synchrotron Radiation. *Physical Review Letters*, 55 (2): 234.
[5] Gibbs, D., Harshman, D. R., Isaacs, E. D., Vettier, C., et al. (1988). Polarization and Resonance Properties of Magnetic X-ray Scattering in Holmium. *Physical Review Letters*, 61 (10): 1241.
[6] Vettier, C. (1984). Magnetic X-ray Scattering. *Acta Physics Polonica*, 86: 521.
[7] Paolasini, L. (2014). Resonant and Magnetic X-ray Diffraction by Polarized Synchrotron Radiation. *École thématique de la Société Française de la Neutronique*, 13: 03002.
[8] Blume, M. (1985). Magnetic Scattering of X-rays. *Journal of Applied Physics*, 57 (8): 3615–3618.
[9] Blume, M. (1994). In G. Materlik, C. J. Sparks and K. Fischer, eds. *Resonant Anomalous X-ray Scattering: Theory and Applications*. Amsterdam: Elsevier.
[10] Platzman, P. M., and Tzoar, N. (1970). Magnetic Scattering of X-rays from Electrons in Molecules and Solids. *Physical Review B*, 2 (9): 3556.
[11] Durbin, S. M. (1998). Classical Theory of Magnetic X-ray Scattering. *Physical Review B*, 57 (13): 7595.

[12] Schülke, W. (1991). In G. Brown and D. E. Moncton, eds., Chapter 15, *Handbook on Synchrotron Radiation*, Vol. 3. Amsterdam: Elsevier Science.

[13] Fernandez, V., Vettier, C., De Bergevin, F., Giles, C., and Neubeck, W. (1998). Observation of Orbital Moment in NiO. *Physical Review B*, 57 (13): 7870.

[14] Namikawa, K., Ando, M., Nakajima, T., and Kawata, H. (1985). X-ray Resonance Magnetic Scattering. *Journal of the Physical Society of Japan*, 54 (11): 4099–4102.

[15] Hannon, J. P., Trammell, G. T., Blume, M., and Gibbs, D. (1988). X-Ray Resonance Exchange Scattering. *Physical Review Letters*, 61 (10): 1245.

[16] Isaacs, E. D., McWhan, D. B., Peters, C., et al. (1989). X-ray Resonance Exchange Scattering in UAs. *Physical Review Letters*, 62 (14): 1671.

[17] Funk, T., Deb, A., George, S. J., et al. (2005). X-ray Magnetic Circular Dichroism—a High Energy Probe of Magnetic Properties. *Coordination Chemistry Reviews*, 249 (1–2): 3–30.

[18] Hart, M. (1978). X-ray Polarization Phenomena. *Philosophical Magazine B*, 38 (1): 41–56.

[19] Schütz, G., Wienke, R., Wilhelm, W., et al. (1989). Spin-dependent Absorption at the K-and L2, 3-edges in Ferromagnetic $Fe_{80}Pt_{20}$ alloy. *Physica B: Condensed Matter*, 158 (1–3): 284–286.

[20] Stöhr, J., Padmore, H. A., Anders, S., et al. (1998). Principles of X-ray Magnetic Dichroism Spectromicroscopy. *Surface Review and Letters*, 5 (06): 1297–1308.

11
Microscopic Magnetic Imaging Techniques

Magnetic imaging techniques enable one to have a direct view of magnetic properties on a microscopic scale. One of the most well-known magnetic microstructures is the magnetic domain. The other example of magnetic microstructures is the nucleation and growth of a magnetic phase across a first-order magnetic phase transition. Such structures can be observed in real space, and their distribution as a function of material and geometric properties can be investigated in a straightforward manner. In this chapter, we will discuss three different classes of magnetic imaging techniques, namely (i) electron-optical methods, (ii) imaging with scanning probes, and (iii) imaging with X-rays from synchrotron radiation sources. There are numerous scientific papers and review articles on these subjects. Instead of going into detail about the individual techniques, this chapter will provide a general overview of the working principles of various magnetic imaging techniques. There are not many specialist books, monographs, or review articles covering all these magnetic imaging techniques under the same cover, but the present author has found the book *Modern Techniques for Characterizing Magnetic Materials* [1] and the article "Magnetic Imaging" [2] to be quite useful while writing this chapter.

11.1 Electron-Optical Methods

Electron-optical methods and electron microscopy encompass a large body of techniques for magnetic imaging. The advanced electron microscopy techniques today can provide images with very impressive resolutions of the order of 1 nm, and show high contrast and sensitivity to detect small changes in magnetization in a material. The particle-like classical picture of the Lorentz force acting on an electron in a magnetic field form the basis of magnetic images of materials observed in various modes of electron microscopy. Electrons are charged particles, and hence electromagnetic fields are utilized as lenses for electrons. A magnetic lens consists of copper wire coils with an iron bore. The magnetic field generated by this assembly acts as a convex lens, which can bring the off-axis electron beam back to focus. Change in the trajectory of an electron in the magnetic field of a magnetic sample results in magnetic contrast and, in turn, provides information on the local magnetization in the material. However, the correct interpretations of the results in many cases involve a wave-like quantum mechanical picture of electrons. The detectable magnetic signal in a sample of magnetic material arises due to the phase change or phase shift of the electron wave going through the vector potential generated by the local magnetic field in the sample.

In an electron microscope, certain physical properties of the material under investigation are manifested as an image, which contains the information in the form of contrast or intensity modulation. A source signal or electron beam travels through the electron microscope, which gets modified by the microscope components and the sample under study, and finally gets detected in a photographic film or CCD camera. The electron microscope family can be divided mainly into two classes: scanning electron microscope and transmission electron microscope.

11.1.1 Scanning electron microscopy

The scanning electron microscope (SEM) provides information from the surface of a sample. The schematic of an SEM is shown in Fig. 11.1. The source of the electron is a filament producing electrons by thermionic emission or a field-emission gun. Electromagnetic lenses control the trajectory of electrons. The final image is obtained on the fluorescent screen or a CCD camera. The microscope operates in a high vacuum since the electrons need to propagate freely in the microscope. After hitting the sample platform the electron beam either passes through the sample unaffected or interacts with it. There can be in general three different types of interactions of electrons with an atom in the sample (Fig. 11.2).

1. The electrostatic interaction with the positively charged nucleus of an atom scatters an electron of the primary beam at an angle of more than 90^0. This is known as backscattering, and the scattered electrons have practically the same energy as the ones of the primary beam.
2. The electrostatic interaction with the positively charged nucleus of an atom scatters an electron of the primary beam at an angle of less than 90^0. The electrons do not lose energy and they are referred to as elastically scattered electrons.

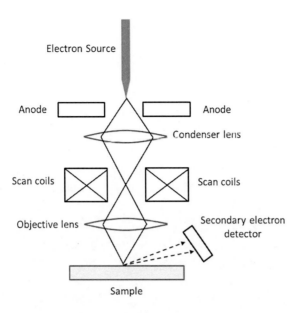

FIGURE 11.1 Schematic representation of a scanning electron microscope.

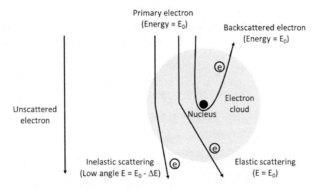

FIGURE 11.2 Three different types of interactions of electrons with an atom in a solid sample.

3. Interacting with the "electron cloud" of the atom, the electrons in the primary beam can also lose energy. Such electrons are known as inelastically scattered electrons. This interaction can lead to various processes. Loosely bound electrons, for example, in the conduction band of the material, can easily be ejected by the incoming primary electron beam, thus giving rise to "secondary electrons". On the other hand, the inner-shell ionization of an atom creates a hole, which is filled by an electron of an outer shell. The surplus energy either generates characteristic X-rays or is transferred to another electron, which is emitted as an Auger electron. Besides these, the deceleration of the electrons in the electrostatic Coulomb field of the nucleus

can generate Bremsstrahlung or continuum X-rays. The primary beam of the electrons can also generate heat and oscillations of loosely bound electrons, i.e., plasmons.

In an SEM a finely focused electron beam is scanned over the sample, and the backscattered and secondary electrons generated from the sample surface are recorded in an electron detector. Two types of magnetic imaging are possible with a conventional SEM, which depends on the Lorentz force acting on either the secondary electrons or the backscattered electrons. The former is known as type-I contrast and the latter as type-II contrast [1, 3]. When the low energy (0–50 eV) secondary electrons ejected from the sample are deflected by the magnetic stray field (due to sample magnetization) existing over the sample surface, this event generates the secondary electron contrast. This type-I magnetic contrast increases with decreasing electron energy, and it is very sensitive to the magnetization at the probe position on the sample surface with respect to the detector [1]. On the other hand, type-II contrast is produced when high energy incident electrons (with typical energy >30 kV) get deflected by the magnetic flux density in the sample. As the electrons travel through the sample, depending on the local magnetization of the sample, the deflection can be towards or away from the sample surface. This gives rise to the backscattering yields. This method is suitable for the study of soft magnetic materials with low anisotropy since it does not require the presence of a stray internal field from the sample. The image contrast represents the magnetic domains. The domain contrast is low and it depends strongly on the angle of incidence of the electron beam. The contrast varies typically from tenths of a per cent for conventional SEM energies of 30 kV to 1 per cent in a 200 kV SEM. The spatial resolution is limited to about half a micron. A probing depth of up to 15 microns may, however, be achieved with high incident electron energy, which enables the study of domain structure in the bulk of the sample.

Fig. 11.3 presents the schematic of an experimental setup for magnetic imaging in SEM. The Everhart–Thornley detector or ET-detector (ETD) is used for conventional type-I and type-II imaging. However, the asymmetric geometry of the ETD can only visualize limited orientations of magnetic domains [3]. A symmetric annular in-lens detector (ILD) can be used to overcome this problem (Fig. 11.3). The ILD enables an easy way for observing a surface domain structure over a wide region, even for complex magnetic domains such as the mazy patterns produced in uniaxial Co-Pt and Fe-Pt alloys [3]. Fig. 11.4 shows an SEM image of a bulk Fe-40 at% Pt sample acquired with an ILD. The image clearly shows mazy magnetic domains all over the regions. Along with the maze pattern, this experiment reveals a spotty contrast (indicated by single arrows in Fig. 11.4) that represents small, branching magnetic domains magnetized in the opposite direction, i.e., a type of 180° domain [3]. Fig. 11.4 also presents the surface topography and magnetic domain structure

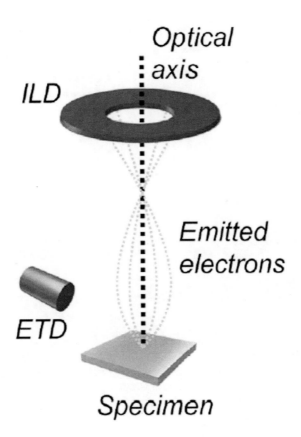

FIGURE 11.3 Schematic of an experimental setup for magnetic imaging in SEM. The non-annular Everhart–Thornley detector (ETD) is located outside the column of SEM, which is tilted away from the optical axis. The annular in-lens detector (ILD) is placed inside the SEM column and centred on the optical axis. (From reference [3] under Creative Common License.)

imaged on the same bulk Fe-40 at% Pt sample by a conventional magnetic force microscope (to be discussed in Section 11.2), which highlights the utility of the magnetic imaging with SEM using an ILD.

Spin-polarized scanning electron microscopy

Spin-polarized scanning electron microscopy (Spin-SEM) or scanning electron microscopy with polarization analysis (SEMPA) can provide information on surface magnetization with high spatial resolution. In this technique, the spin-polarized secondary electrons excited from the sample surface are detected in a spin-polarized analyzer. The working principle of the Spin-SEM or SEMPA is presented schematically in Fig. 11.5. Spin-polarized secondary electrons at the surface of a magnetic sample are excited by a finely focused beam of medium-energy electrons.

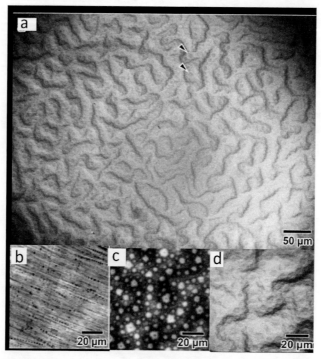

FIGURE 11.4 SEM and MFM images of the Fe-40 at% Pt bulk sample: (a) ILD image of bulk Fe-40 at% Pt sample showing mazy magnetic domains and spotty reverse domains (marked by single arrowes), (b) surface topography, (c) MFM images and (d) ILD image, which clearly shows the domain walls and spotty reverse domains. (From reference [3] under Creative Common License.)

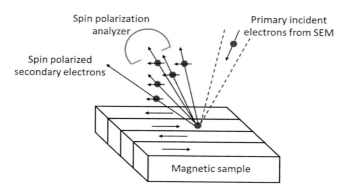

FIGURE 11.5 Schematic presentation of the working principle of a SEMPA.

This phenomenon is utilized in Spin-SEM or SEMPA to get a magnetic image of the sample surface with high spatial resolution, especially the domain patterns.

A Spin-SEM setup requires a high-brightness electron gun producing a narrow beam of primary electrons and an efficient spin-polarization detector [1, 4]. Thus, most of the equipment in laboratories uses a commercially available SEM to which a spin analyzer is attached. The sample surface is scanned by a focused beam of high energy unpolarized electrons, which are scattered by the electrons near the sample surface predominantly by an inelastic scattering mechanism. Ultimately a primary electron loses all its energy, and the process generates a cascade of excited low energy electrons. Out of these excited secondary electrons, a considerable number travels back to the surface undergoing additional elastic and inelastic events and may ultimately exit the sample if the energy left is sufficient enough to overcome the vacuum level. Since the number of such emitted secondary electrons depend on the curvature of the sample surface as well as the local work function, an image of the sample topography and its chemistry can be obtained by recording these electrons for each sample position of the incoming beam of the primary electron. In a ferromagnetic sample the spin-polarization of the emitted secondary electron can be defined as $P = (N\uparrow - N\downarrow)/(N\uparrow + N\downarrow)$, where $N\uparrow$ ($N\downarrow$) is the number of spins per unit volume aligned parallely (antiparallely) to the net magnetization of the magnetic sample. With the measurement of spin-polarization P along a certain direction in the space, a map can be obtained of the magnetization component of the sample in that direction.

The spin-polarization of the secondary electrons depends strongly on the energy of the secondary electrons [1]. At secondary electron energies typically above 10 eV, the polarization P is close to that calculated from the spin imbalance of electron bands near the Fermi level. These values of P are 28%, 19%, and 5% for the 3d-transition metals Fe, Co, and Ni, respectively, and these reasonably agree with the measured value [1]. The spin polarization P increases at lower energies and peaks near the vacuum level with a value reaching $P = 50\%$ in the case of Fe. Preferential inelastic scattering of spin-$\downarrow$ electrons at the low energy range gives rise to a higher escape probability of the spin-$\uparrow$ electrons, and this, in turn, leads to an enhancement in P. At the lowest energies the intensity of the secondary electrons is also the highest. An efficient instrument for magnetic imaging collects the abundance of electrons from a typical energy window 0–10 eV exploiting their high spin-polarization [1]. This interesting combination of polarization enhancement and high intensity makes it possible to image magnetic patterns in ultrathin films of only a single atomic layer thickness.

A Spin-SEM is insensitive to the spurious signals arising out of the fluctuations in the incoming beam current and from the variation in the secondary electron emission at different topographies of the sample. This is because the spin-polarization P is a normalized quantity, and hence the changes in the number of emitted electrons get cancelled out. The total number of electrons emitted $N = N\uparrow + N\downarrow$ presents a direct map of the secondary electron yield. As a result, the topographic map of the sample comes for free when the spin-polarization is determined. In an experiment two individual images are obtained in the spin-analyzer of the SEM that is sensitive

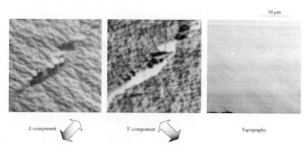

FIGURE 11.6 Two in-plane magnetization-component (X, Y) and one topography SEMPA images of Co (0001) at room temperature. (From reference [5] under Creative Common License.)

to the magnetization direction in the sample, that is one for $N \uparrow$ and one for $N \downarrow$. The topographic feature such as defects and scratches are observed in both images. The superimposed image is a contrast that reverses between the two images. The maps of the sum and the normalized differences are calculated from these two images. No magnetic contrast remains superimposed on the topographic image, whereas the defects and scratches which are present in the topographic image are visible in the magnetic image. The Spin-SEM thus provides topographic and magnetic information simultaneously, while separating them completely. As an example, two in-plane magnetization-component (X, Y) Spin-SEM images and a topography image of Co (0001) at room temperature are shown in Fig. 11.6. An easy direction of magnetization of this sample lies along the [0001] crystal axis, which is in the direction perpendicular to the sample plane. However, due to the magnetostatic energy, the magnetization inclines in the in-plane direction, and closure domains form at the sample surface [5].

Spin-polarized low energy electron microscopy

A Spin-polarized low energy electron microscopy (SPLEEM) consists of a low energy electron microscope (LEEM) with a spin-polarized electron source. The working principle of a SPLEEM instrument is presented in Fig. 11.7. Low energy spin-polarized electrons photo emitted from a GaAs photocathode are projected on the sample surface through an illumination column. The electrons after passing through the illumination column are decelerated in the objective lens, and they finally hit the sample surface with normal incidence. A magnetic beam splitter separates the backscattered electron beam from the incoming electron beam. The backscattered electron beam after passing through the imaging column is focused on a phosphorous screen to obtain a magnified image of the surface. Slow electrons can probe matters based on their strong spin–spin interaction. In addition, the slow electrons have a short inelastic spin-dependent mean free path in solids and the penetration depth in the solids is small. As a result, SPLEEM is extremely surface

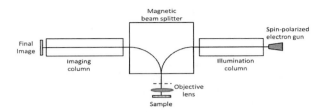

FIGURE 11.7 Schematic representation of the working principle of a SPLEEM instrument.

sensitive, and the properties of the top few atomic layers of the sample dominate image contrasts [6]. Thus, spin properties of the solid can be probed only in the reflection experiments. This enables probing spin in a low energy electron microscope equipped with a source of spin-polarized electrons. Normal incidence and reflection of electrons are utilized in this type of microscope.

Magnetic contrast in SPLEEM arises from two sources [1]:

1. The spin–spin interaction or exchange interaction that takes place between the spin-polarized incident electron and the spin-polarized electrons in the material under study. The scattering potential of the atoms depends on the relative orientation of the incident slow electron and of the unpaired electrons of the atom. Spin-dependent scattering potentials cause spin-dependent backscattering intensities for electrons with spin parallel or spin antiparallel to the unpaired electrons of the ferromagnetic materials under study.
2. There is a difference in the inelastic mean free paths of electrons with spin parallel and antiparallel to the unpaired electrons in the material. This difference in the mean free paths contributes to the magnetic contrast.

SPLEEM images are two-dimensional, whereas magnetization is a vector with three components. Mapping all components of the magnetization vector is possible by measuring image contrast in several SPLEEM images of the same surface area. The spin manipulator can be aligned in three different settings, with spin polarization of the electron illumination aligned along either of two orthogonal directions parallel to the sample surface or in the direction perpendicular to the sample surface. The SPLEEM images are acquired with these three orthogonal settings of the spin manipulator. A comparison of magnetic contrast in these three images can then determine the sample magnetization vector with good spatial resolution and good angular resolution.

SPLEEM is useful for the study of magnetization in the topmost layers of single-crystal surfaces, epitaxial thin films, and thin films with strong preferred crystal orientation. This is because in single crystals the backscattered intensity is focused into a few diffracted beams, in particular into the specular beam, which is generally used in imaging. This also happens in the epitaxial thin films

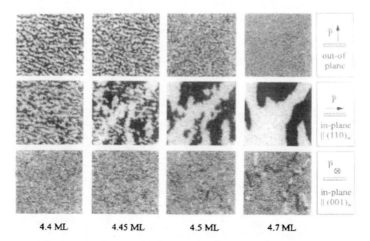

FIGURE 11.8 SPLEEM images at different stages of the spin-reorientation transition in a Co film on a 10 monolayer (ML) thick Au(111) layer on W(110). (From reference [8] with permission from Oxford University Press.)

and for the specularly scattered beam in polycrystalline thin films with strong texture. SPLEEM instruments can record images showing topographic and magnetic contrast simultaneously. This feature enables the determination of the correlation of atomic step structure and magnetic domain formation during film growth. In low-dimensional ferromagnetic systems, the interplay between magnetocrystalline, magnetoelastic, and shape anisotropies influences the direction of spontaneous magnetization [7]. This balance between different anisotropy energies can vary in magnetic films and layered structures with changes in temperature, sample thickness, or surface composition, which often can lead to spin-reorientation transitions (SRT) of the magnetization. SPLEEM is a powerful tool for studying such SRT in low-dimensional structures. Fig. 11.8 presents the evolution of SRT in Co/Au(111) films grown on W(110). There is an SRT from out-of-plane magnetization at small film thickness to in-plane magnetization at large film thickness [8].

11.1.2 Transmission electron microscopy

Transmission electron microscopy (TEM) is very suitable to image the local magnetic domain structure with very high spatial resolution. TEM has been in use since the early 1950s in microscopic investigation of magnetic materials. A large number of interactions take place when a beam of fast electrons hits a thin solid specimen, which can provide detailed information on the compositional, electronic, structural, and magnetic properties. The typical image resolution for structural information is 0.2–1.0 nm, 1–3 nm for compositional information, and 2–20 nm for imaging magnetic domain structure.

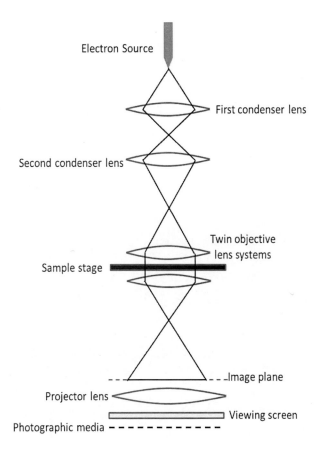

FIGURE 11.9 Schematic representation of the working principle of a TEM.

Magnetic structure imaging by TEM

Fig. 11.9 shows a schematic presentation of a conventional TEM. It has an electron gun, an electron lens system, a sample stage, and various detectors. The sample stage is attached with a goniometer to manipulate sample orientation. There are three segments in the lens system: the condenser lenses, the objective lenses, and the projector lens. Beam formation and illumination are controlled by condenser lenses. The objective lenses are responsible for imaging, whereas the projector lens is responsible for magnification alterations. A scanning transmission electron microscope (STEM) is a particular class of TEMs that scans a focused electron beam across the sample.

In a TEM the sample under investigation is usually exposed to the high magnetic field (typically 0.5–1.0 T) of the objective lens. This may be sufficient to severely distort or even completely eradicate most of the micro-magnetic structures of interest. To overcome this problem of the high magnetic field in the sample region several strategies have been devised [15]. Such strategies include:

1. Switching of the standard objective lens.
2. Changing the position of the sample so that it is no longer immersed in the field of the objective lens.
3. Retaining the sample in its standard position but changing the pole pieces once again to provide a non-immersion environment.
4. Introduction of super mini-lenses in addition to the standard objective lens, which is once again switched off.

In recent times an objective lens system has been designed that realizes a magnetic-field-free environment at the sample position. In combination with a higher-order aberration corrector, atom-resolved imaging with sub Å spatial resolution can be achieved in such systems with a residual magnetic field of less than 0.2 mT at the sample position [9].

TEM can be used to investigate mainly thin samples with thickness usually not exceeding 100 nm. Thus, it is useful to study magnetic thin films and multilayers. However, the bulk substrates need to be removed or severely thinned before observations can be made. The problem arises with bulk magnetic materials with the necessity of thinning the sample, which is often time-consuming. Moreover, the domain structure in a thinned sample is quite unlikely to be the same as that in the bulk material, although it is often possible to extract some information on the interplay between magnetic domain walls and structural defects.

Lorentz transmission electron microscopy

Lorentz transmission electron microscopy (LTEM) is the general name given to the modes of TEM used to image magnetic domains, some of which provide primarily qualitative information and some others more quantitative information [10]. Several different imaging and diffraction modes are available, including Foucault and Fresnel modes, differential phase contrast, and a range of holographic techniques as well as low angle electron diffraction. In addition, an imaging mode to obtain quantitative maps of the magnetic induction distribution is also possible using a scanning transmission electron microscope (STEM). Samples can be studied in their as-grown state, in magnetic remanent states, in the presence of applied fields or currents, and also as a function of temperature. The images are used to derive basic micromagnetic information, the nature of domain walls, points at which magnetic domain nucleation occurs, the importance of domain wall pinning sites, and many other related phenomena [10]. There are many specialized literature available on LTEM, but for a concise tutorial of the subject matter to a general audience the present author finds these articles by De Graef [14], Petford-Long and De Graef [10], and Chapman and Scheinfein [15] pretty useful. Much of the following discussion is borrowed from the article by Chapman and Scheinfein [15], including some figures for illustrating the principle.

The functional principle of LTEM is based on the classical Lorentz interaction. The primary electron beam experiences a deflection due to Lorentz force while passing through the region of magnetic induction in the sample. The direction of the deflected beam depends on the magnetic induction direction within the micro-magnetic structure being imaged and is perpendicular to it. The Lorentz deflection angle β_L is given by [15]:

$$\beta_L = \frac{e\lambda t(\vec{B} \times \vec{n})}{h} \quad (11.1)$$

Here $\vec{B}$ is the induction averaged along an electron trajectory, $\vec{n}$ is a unit vector parallel to the incident beam, t is the specimen thickness, λ is the electron wavelength, and h is Planck's constant. The Lorentz deflection angle is thus proportional to the product of the averaged magnetic induction and the thickness of the sample. Substitution of typical values into Eqn. 11.1 reveals that β_L is typically a few tens of microradians, and it can be < 1 μrad in the case of some magnetic multilayers or films with low values of saturation magnetization. The problem is even more when the magnetization vector lies perpendicular to the plane of the sample. In this situation, β_L will be zero unless the sample is tilted with respect to the primary electron beam to introduce an induction component perpendicular to the direction of travel of electrons. However, because of the small magnitude of β_L there is no danger of confusing magnetic scattering with the more familiar Bragg scattering where angles are typically in the range 1–10 mrad [15].

The discussion above so far takes into account only particle properties of electrons and is essentially classical in nature. A quantum mechanical description of the interaction between the electron beam and sample (which takes into account the wave properties of the electron) is, however, necessary for a full quantitative description of the spatial variation of magnetic induction. In this quantum mechanical approach, the magnetic sample is be considered as a phase modulator of the incident electron wave, and the phase gradient $\nabla \phi$ of the specimen transmittance can be expressed as [15]:

$$\nabla \phi = \frac{2\pi e t(\vec{B} \times \vec{n})}{h} \quad (11.2)$$

The magnetic contribution to the phase shift is relatively large in comparison to the electrostatic contribution for higher accelerating voltages. However, from Eqn. 11.1 one can see that the Lorentz deflection angle β_L is proportional to electron wavelength and hence inversely proportional to incident electron beam energy. A good compromise here is the accelerating voltage range 200–400 kV [10].

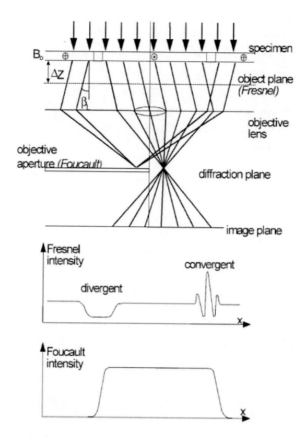

FIGURE 11.10 Schematic representation of Fresnel and Foucault mode imaging. (From reference [15] with license from Elsevier.)

Substitution of typical numerical values in Eqn. 11.2 shows that magnetic thin film samples may normally be regarded as strong, albeit slowly varying, phase objects [15]. The phase change, for example, involved in crossing a domain wall usually exceeds π rad.

Fresnel and Foucault modes of imaging

The first and most commonly used techniques in LTEM for revealing magnetic domain structures using a conventional TEM machine are the Fresnel and Foucault imaging modes. Fig. 11.10 shows the schematic representation of these two modes. A sample with three domains separated by two 180° domain walls is assumed here for the purpose of illustration. In Fresnel mode, the imaging lens is simply defocused so that the object plane is no longer coincident with the sample. The magnetic domain walls are then imaged as alternate bright and dark lines in a contrast-free background. The bright lines appear when the domain walls are such

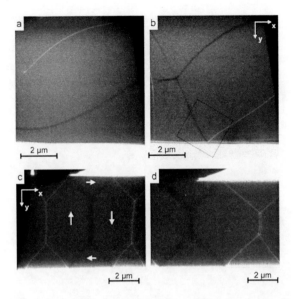

FIGURE 11.11 Fresnel defocus images of FeSiBNbCu alloy ribon samples: (a) and (b) rapidly annealed samples, (c) and (d) stress annealed samples. (From reference [11] under Creative Commons Licence.)

that the magnetization on either side of the wall deflects the electrons towards the wall. The convergent wall images consist of sets of fringes running parallel to the wall if a coherent electron source is used. Detailed analysis and simulation of the fringe patterns can give information about the domain wall structure [10]. Fig. 11.11 shows Fresnel defocus images recorded with an LTEM from rapid annealed and stress annealed FeSiBNbCu alloy ribbons after saturation of the samples using an out-of-plane magnetic field. A random magnetic domain pattern is visible in the rapidly annealed sample without any external stress [11]. The rectangle in Fig. 11.11(b) marks a region that has been studied [11] using off-axis electron holography (Fig. 11.17).

In Foucault mode imaging, a contrast forming aperture is needed in the plane of the diffraction pattern (Fig. 11.10). This is used to obstruct one of the two components into which the central diffraction spot is split due to the deflections suffered as the primary electron beam passes through the sample. This aperture under normal conditions will be the objective aperture, but if the reverse mode is used, the back-focal plane of the objective lens is co-planar with the selected area aperture [10]. The contrast then arises from the magnetization within the magnetic domains. The areas where the magnetization orientation in domains is such that electrons are deflected through the aperture will appear as bright areas. On the other hand, the dark areas will correspond to those domains where the orientation of magnetization is oppositely directed. If the relative directions of the aperture and

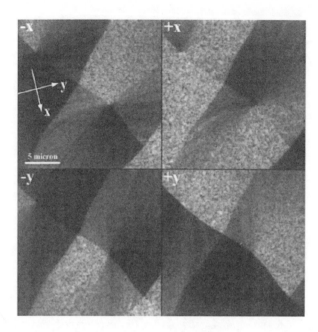

FIGURE 11.12 Foucault mode LTEM images of antiparallel domains in a 50 nm thick Ni-20%Fe Permalloy thin film for four aperture shifts for the same sample region. (From reference [14] with license from Elsevier.)

image are known, the direction of magnetization within the various domains of the sample can be determined qualitatively.

Fig. 11.12 presents the Foucault mode images in a 50 nm thick Ni-20%Fe Permalloy film for four aperture shifts for the same sample region. With the knowledge that the direction of the Lorentz deflection is normal to the corresponding in-plane magnetization direction, these four images can be used to obtain a qualitative picture of the magnetization configuration in the sample [14].

Fresnel and Foucault mode imaging is relatively simple to implement and they provide a clear picture of the overall domain geometry as well as a useful indication of the directions of magnetization at least in the bigger domains. These attributes make them the preferred techniques for in situ experiments. In the Foucault mode, it is difficult to position the contrast-forming aperture reproducibly. In addition in both the imaging modes the relation between image contrast and the spatial variation of magnetic induction is usually not linear. This makes the extraction of quantitative data unreliable, more so in the regions where there is a rapid variation of magnetic induction. It may also be noted that contrast arising from the physical microstructure is often present in the images along with the magnetic contrast. It is thus necessary to reduce the contribution from lattice diffraction within the image to a minimum in order to obtain the most appropriate profile of the domain wall. This is possible with the help of a post-experimental image processing technique

[12]. A suitable choice of defocus in the Fresnel mode can reduce the influence of the high spatial frequency components in the image if the physical microstructure is on a significantly smaller scale than its magnetic counterpart [10].

Low angle electron diffraction

The low angle diffraction (LAD) technique involves direct observation of the structure in the split central diffraction spot. This technique can be used to obtain semiquantitative information about the magnetic domain structure in the sample and also to follow magnetization processes. The basic requirement for LAD necessitates that the magnification of the intermediate and projector lenses of the TEM is sufficient enough to make the small Lorentz deflections visible. For this purpose, camera constants defined as the ratio of the displacement of the beam in the observation plane to the deflection angle itself are required to be typically in the range 50–1000 m [15]. If the detailed structure of the LAD pattern is to be observed accurately, a high spatial coherence is a must in the illumination system. This requires that the angle subtended by the illuminating radiation at the sample is considerably smaller than the Lorentz angle of interest, which is quite easy to fulfil in a TEM equipped with a field emission gun (FEG). It may be noted here that LAD provides global information from the whole of the illuminated specimen area rather than local information.

Differential phase contrast microscopy

The differential phase contrast (DPC) microscopy is an in-focus imaging technique, which uses the scanning transmission electron microscope (STEM). This technique can be thought of as a local area LAD technique using a focused probe. Fig. 11.13 shows a schematic of how DPC contrast is generated. The local Lorentz deflection at the position of the electron probe is determined with the help of a quadrant detector sited in the far-field. Direct information on the two components of the Lorentz deflection β_L is obtained out of the difference signals from opposite segments of the detector. The difference signals are measured as the probe is scanned in a regular raster across the specimen, and this leads to semiquantitative images with a resolution approximately equal to the diameter of the electron probe. A probe size of < 1 nm can be achieved in an FEG-STEM with the sample being placed in field-free space.

In DPC microscopy two difference-signal images are collected simultaneously. A map of the component of magnetic induction perpendicular to the electron beam can be constructed from these images. A third image can also be formed using the total signal falling on the detector. This is a standard incoherent bright-field image and contains structural information. Thus, in DPC microscopy, it is possible

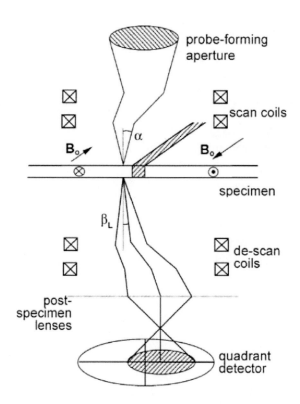

FIGURE 11.13 Schematic representation of the DPC imaging technique using an STEM instrument (From reference [15] with license from Elsevier.)

to get simultaneous information and possible correlation of magnetic and physical microstructure. This simple approach, however, assumes that the different images collected on the quadrant detectors arise entirely as a result of the magnetic induction–electron beam interaction. However, in reality, contrast arising from the physical microstructure is always present. This effect can be nullified to some extent by using an annular quadrant detector in place of the solid quadrant detector in the so-called modified DPC imaging [10, 15]. We present an example of DPC imaging in Fig. 11.14 where DPC images of planar FeRh thin films are presented.

DPC images contain information about magnetization directions within the domains, and they are not well suited to measurements of absolute values of phase shift [10]. In this sense, the DPC images are semiquantitative in nature. There is also an increase in instrumental complexity and operational difficulty compared with the Foucault and Fresnel imaging modes. Furthermore, a relatively long time is needed to record each image. Thus, recording real-time series of images during a magnetization reversal cycle is not possible unless the magnetization reversal process is not significantly affected by time-dependent effects.

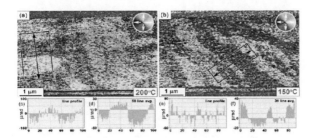

FIGURE 11.14 DPC imaging of the planar FeRh thin films: (a) 200 °C (Sample 1), and (b) 150 °C. The direction of magnetization is shown in the colour wheels (see inset). (From reference [13] Creative Commons License.)

Electron holography

Electron holography employs the interference of two coherent electron waves [15]. The image contrast is a direct consequence of the quantum nature of the electron wave. An electron wave is split, and one part of the wave interacts with the sample whereas the second part passes through free space. The two-electron waves are recombined coherently, and electron phase shifts due to the enclosed electromagnetic fields are observed. These phase shifts are also accompanied by electron deflections as in the case of the imaging modes discussed above, but the holographic interference effect does not exist in a classical world where the value of Planck's constant is effectively zero.

A schematic diagram illustrating the principle of electron holography using an STEM is shown in Fig. 11.15. A real partially coherent electron source (represented by a solid arrow in Fig. 11.15) is split into two virtual electron sources through the use of an electron biprism. The electron biprism (represented by a hollow circle in Fig. 11.15) is a drawn 1 μm diameter quartz fibre with a thin coating of evaporated metal [15]. The virtual images of the biprism form the two coherent virtual sources (represented by hollow arrows in Fig. 11.15), and they are focused by the electron optics into the two fine electron probes coherently illuminating the sample under investigation. The separation of these two virtual sources can be varied by changing the voltage applied to the biprism and by changing the excitation of electron lenses. A shadow image of a large area of the sample can be observed by keeping the electron beam stationary and operating the objective lens at a relatively large defocus. The wave packets interact with the electromagnetic fields of the sample and recombine to form an interference pattern or hologram [15]. This appears as a fringe modulated twin-image in the detector plane and is recorded on a slow-scan CCD camera. A diffractogram with two characteristic sidebands can be obtained from a fast Fourier transform of the hologram whose separation depends upon the fringe spacing. The amplitude and phase of the electron wave emerging from the sample can be obtained by isolating one of the sidebands and taking its inverse Fourier transform.

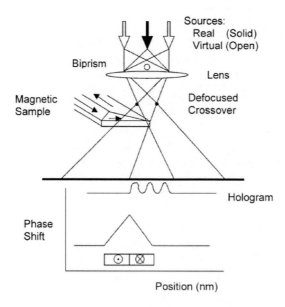

FIGURE 11.15 Schematic representation of the absolute mode of STEM holography. (From reference [15] with license from Elsevier.)

Information on the thickness and bulk inelastic mean-free path of electrons in the solid is contained in the amplitude image. The relative phase difference $\Delta\phi$ due to magnetic induction present in the sample region can be expressed as [15]:

$$\Delta\phi = -\frac{2\pi e}{h} \int \vec{B}.\vec{dS} \qquad (11.3)$$

Here dS is an element of the area between the two enclosing electron paths. This phase shift is estimated quite accurately since it is extracted from electron interference patterns.

There are two distinct modes of STEM holography, namely the absolute and differential modes. In the absolute mode (Fig. 11.15), one of the two-electron probes travels through the sample and the other passes through a vacuum. The phase shift generated by the electromagnetic fields present in the sample can be estimated absolutely with the assumption of zero-phase contribution from the vacuum. The differential mode is shown schematically in Fig. 11.16. Here both of the split electron probes pass through the sample. The separation of the beam paths is adjustable and it can be made as small as 10 nm [15]. There is no requirement for a vacuum reference phase in the differential mode. The area defining the enclosed magnetic flux here is approximately constant for every point in the hologram. In this mode the phase shift in a uniformly magnetized region in a specimen of constant thickness is constant. This is in contrast to the absolute mode. The differential mode has the advantage over the absolute mode for the investigation of the magnetic domain wall profiles,

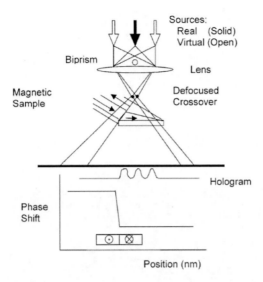

FIGURE 11.16 Schematic representation of the differential mode of STEM holography. (From reference [15] with license from Elsevier.)

which are determined directly to within a resolution given by the probe diameter. Electron holography can also be performed on a fixed-beam TEM. In the standard off-axis mode the biprism is located after the sample and is used to mix those parts of the electron beam that passed through the sample and those passing through the adjacent region of free space.

Holographic techniques require a high level of instrumental sophistication including a field emission gun. Also, the performance of such techniques to obtain magnetic interferograms decreases rapidly with increasing inelastic scattering [15]. This limits the maximum thickness of the sample that can be investigated with electron holography. Within such constraints, however, electron holography remains a very versatile technique, which offers a complete description of the detailed distribution of magnetic induction. Fig. 11.17(a) presents an off-axis electron hologram of a rapidly annealed FeSiNbCuB alloy ribbon sample. Fresnel defocus images of the same sample is presented in Fig. 11.11. Figure 11.17(b) presents the resulting magnetic induction map determined from this off-axis electron hologram, where local changes in magnetization direction were inferred from changes in the direction of the phase contours across domain walls [11].

11.2 Imaging with Scanning Probes

In scanning probe imaging techniques, a probe, which sometimes is in the form of a sharp tip, is laterally moved above the sample surface. The size of an interaction

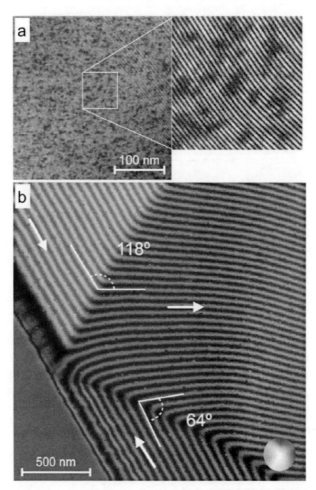

FIGURE 11.17 (a) Off-axis electron hologram of a rapidly annealed FeSiNbCuB alloy ribbon sample. The enlarged region shows the bending of the electron holographic interference fringes, which results from local variations in projected magnetic flux density in the specimen. (b) Magnetic induction map inferred from the region of the off-axis electron hologram from the region indicated in (a). (From reference [11] Creative Commons License.)

between the probe and the sample is measured. A feedback loop is usually used to adjust the probe height in such a way as to keep the strength of this interaction between the probe and the sample constant.

11.2.1 Magnetic force microscopy

Atomic force microscopy (AFM) is a very well-known technique for the imaging of surface morphology of materials at the nanoscale. The sample is imaged by scanning it using a sharp tip placed at the end of a miniature cantilever. The deflection of the cantilever is recorded through an optical detection system. In AFM either the repulsive binding force upon close approach between tip and sample or the van der Waals force at larger distances is used to regulate the tip–sample distance. In the close approach, the sample surface morphology is obtained by recording the cantilever deflection during the scanning. This is known as contact mode, and here the interaction between the sample surface and the tip is continuous during the scanning. The other possible mode is intermittent contact or tapping mode, where sample surface morphology is obtained by monitoring the amplitude of oscillation of the cantilever set into vibration. The interaction between tip and sample is limited here to a fraction of the period of the cantilever oscillation.

In magnetic force microscopy (MFM) the tip has a magnetic coating, and the force experienced by this magnetic tip from the stray field emanating from the sample surface is measured. The easy implementation of MFM into the existing AFMs, the high resolution, and the insensitivity to the surroundings make MFM one of the most widely used techniques for magnetic imaging. However, MFM is an indirect imaging technique, since it is the stray field above the surface of the magnetic domain pattern, and not the magnetic domain itself, that is imaged.

The working principle of an MFM is shown schematically in Fig. 11.18. A piezo-scanner on which the sample is mounted enables spatial movement of the sample in x, y, z directions and mapping of the sample surface. The cantilever holds the tip at one end. At the beginning of the experiment, the morphology of the sample is mapped in standard AFM tapping mode. Then a second scan of the surface is taken at a fixed tip–sample distance Δz (marked as the lift height in Fig. 11.18) following the topography line obtained from the first scan. During this oscillation is induced in the cantilever at or near its first free resonance frequency f_0 [16]. In the presence of magnetized samples, the magnetic tip experiences a force. If the z-axis is assumed to be normal to the sample surface, then the component of the force normal to the surface is F_z. This force causes a variation in the amplitude A and a

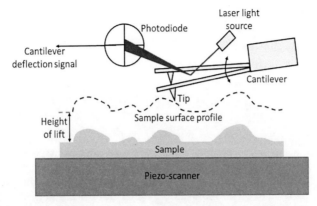

FIGURE 11.18 Schematic presentation of the working principle and experimental setup and of magnetic force microscopy (MFM).

shift in the phase ϕ of the oscillation of the cantilever and also in its first resonance frequency f_0 [16]:

$$\Delta A = \frac{A_0 Q_c}{2k_c^2}\left(\frac{\partial F_z}{\partial z}\right)^2 \qquad (11.4)$$

$$\Delta \phi = \frac{Q_c}{k_c}\left(\frac{\partial F_z}{\partial z}\right) \qquad (11.5)$$

$$\Delta f_0 = -\frac{f_0}{2k_c}\left(\frac{\partial F_z}{\partial z}\right) \qquad (11.6)$$

Here k_c and Q_c are the cantilever spring constant and the quality factor of the cantilever first resonance in air, respectively, and A_0 is the amplitude at the resonance frequency f_0 in air and without external forces. In usual MFM experiments, the magnetic images of a sample are obtained in lift mode by recording the values of $\Delta\phi$ and/or Δf_0.

The left-hand side of Fig. 11.19(a) shows schematically a magnetic thin film in which two oppositely magnetized domains are separated by a Néel wall. A stray field as indicated by arrows will be generated outside the sample due to magnetic poles at both ends of the domain wall. When a magnetic tip with magnetization direction as sketched in Fig. 11.19(a) is moved across the domain wall, the tip will experience an attractive force on the left side of the domain wall, and a repulsive force on the right side. An image obtained from the cantilever oscillation frequency is shown qualitatively on the right-hand side of the figure, where a brighter contrast denotes a higher oscillation frequency corresponding to a repulsive force [2].

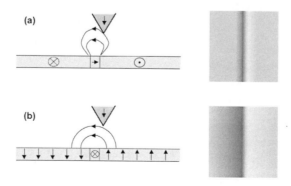

FIGURE 11.19 Schematic representation of the image contrast in magnetic force microscopy: (a) Thin-film sample with in-plane magnetization. The stray field originating from a Néel wall between two domains with opposite magnetization direction leads to a black/white contrast in the MFM image as shown at the right. (b) Thin-film sample with out-of-plane magnetization. The stray field originating from the edges of the domains leads to a contrast as shown at the right. (From reference [2] with permission from Springer Nature.)

Fig. 11.19(b) shows the case of a magnetic thin film sample with out-of-plane magnetization. In a thin film sample of the infinite extent with a single domain with perpendicular magnetization, the magnetic flux density is completely closed inside the film by the demagnetizing field. Such a sample will not have a stray field outside the sample. A part of the flux, however, is also closed outside the sample as shown in Fig. 11.19(b). This part of the flux causes the reduction of the demagnetizing energy, which gives rise to stripe domains in films with perpendicular anisotropy. This means that close to the domain walls there is a magnetic force acting on the tip in MFM experiments. The perpendicular component of this force changes the sign at the centre of the domain wall, which gives rise to an image as shown at the right-hand side of Fig. 11.19(b). The domains will be represented with a constant alternating contrast in the final image if the width of the stripe domains is comparable to the tip–sample distance, i.e., approximately 50 nm or more. It should, however, be realized that magnetic images by MFM are not directly corresponding to the magnetization distribution in the sample. The image contains only indirect information about the magnetic domains, and the actual domain pattern needs to be reconstructed from the force images. This is quite straightforward in the case of narrow out-of-plane domains, but can be quite involved in the case of in-plane domains with different magnetization angles [2]. An additional complication arises from the long-range nature of dipolar interactions. Another source of complication arises from the interaction of the stray field of the magnetic tip, which may affect the sample magnetization in soft magnetic samples.

Fig. 11.20(a) shows a topographic image of the polycrystalline $La_{0.7}Sr_{0.3}MnO_3$ film with a granular structure. Figs. 11.20(b) and (c) show the corresponding phase-detection MFM image and frequency detection MFM images, respectively, of the same sample. The phase-detection MFM image and its corresponding signal profile are well characterized as those of the perpendicular magnetic field gradient with the areas at the non-magnetic grain boundary phase showing only the bright or black contrast [17]. The frequency detection image shows some white–black dipole characteristics at the non-magnetic grain boundary phase. These results indicate that a part of the LSMO magnetic grains is magnetically antiparallel to each other [17].

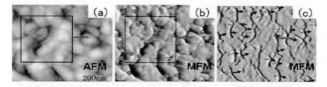

FIGURE 11.20 (a) Topographic image, (b) phase detection image, and (c) frequency detection image for an $La_{0.7}Sr_{0.3}MnO_3$ granular film with in-plane magnetization. (From reference [17] with permission from Royal Society of Chemistry.)

11.2.2 Spin-polarized scanning tunneling microscopy

In a scanning tunneling microscopy (STM) the size of the tunnel current between the tip and a conducting sample represents the tip–sample interaction. In the spin-polarized scanning tunneling microscopy (SP-STM) tunnel current between the tip and a magnetic sample depends in general on the relative orientation of the magnetization in the tip and the sample. This spin-sensitive tunnel current or tunneling magnetoresistance is used for magnetic imaging.

The principle of the working of SP-STM is based on the conservation of electron spin during the tunneling process between the magnetic tip and a magnetic sample. The relative orientation of the local magnetization of the magnetic tip and the sample determines the tunneling current flowing between a tip and a sample. In addition spin polarization of the electronic states of tip and the sample influences the tunneling current. Fig. 11.21(a) provides a schematic explanation of this tunneling magnetoresistance in SP-STM. We know that due to the quantum mechanical exchange interaction between electrons, the density of states in a magnetic material

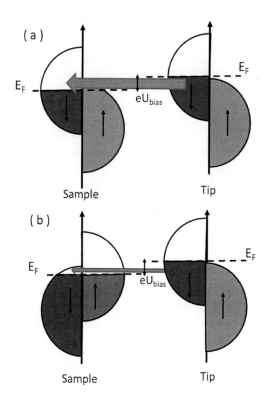

FIGURE 11.21 Schematc representation of the working principle of spin-polarized scanning tunneling microscopy. The tunneling of electron takes place between electronic states of the tip and the sample of the same-spin direction.

splits up into minority and majority states. The diagram on the left-hand side of Fig. 11.21(a) shows the spin-resolved or spin-split density of states of the magnetic sample under investigation. The horizontal line represents the Fermi level of the sample, and the majority (minority) states are shown on right (left) of the vertical energy axis. The cartoon in the right of Fig. 11.21(a) represents the spin-split density of states of an SP-STM tip magnetized in such a way that its magnetization direction is parallel to the magnetization of the sample. A bias voltage U_{bias} is applied between the sample and tip, and this represents an energy difference between the Fermi levels of the sample and the tip. Spin-flip scattering is not allowed and the electron spin is preserved during the tunneling process. As a result, electron tunneling takes place between majority (minority) states of the tip and majority (minority) states of the sample. It may be noted that for a positive bias voltage as shown in Fig. 11.21(a), electrons tunnel from occupied states of the tip into unoccupied states of the sample. The cartoon on the left-hand side of Fig. 11.21(b) shows the density of states of the magnetic sample where majority and minority states are exchanged. Here tip and sample magnetizations are antiparallel. In this situation, minority electrons from the tip can tunnel into the majority states of the sample, and vice versa. The tunnel current will generally be different in these two cases with tip magnetization parallel or antiparallel to the sample magnetization. This phenomenon is more commonly known as tunnel magnetoresistance and is utilized for magnetic imaging in an SP-STM experiment.

The tunnel current also depends exponentially on the distance between the tip and the sample apart from its dependence on the relative orientation of sample and tip magnetization. This former phenomenon is the normal contrast mechanism used in STM to obtain topographic information of a sample. In an SP-STM it is needed to separate topographic contribution from the magnetic contribution in the tunnel current. This exercise needs to be done at each position during an image scan.

There are several different types of operation modes, all of which have advantages and disadvantages, depending on the particular type of application [18]. Here we discuss where the magnetic signal is identified after modulating a parameter using a lock-in amplifier to separate the relatively weak magnetic signal from the much larger topographic signal. The modulated parameter is bias voltage U_{bias} in one method, whereas the direction of tip magnetization is modulated in the other.

The first method involves the measurement of the local differential tunneling conductance $dI/dU(U, x, y)$ as a function of bias voltage U and spatial coordinates x and y [18]. This method is also termed as spin-polarized scanning tunneling spectroscopy. In the experiment, a small modulation voltage U_{mod} is added to the applied bias voltage U_{bias}. The resultant tunneling current modulation is the frequency and phase selectively detected using a lock-in amplifier. The energy resolution of the spectroscopic measurement is determined by the amplitude of the

FIGURE 11.22 Tip covered by a magnetic thin film, used in spin-polarized scanning tunneling spectroscopy by modulating the bias voltage between tip and sample. (a) shows geometry with sensitivity for in-plane magnetization, while (b) presents geometry with sensitivity for out-of-plane magnetization.

bias voltage modulation, provided the thermal energy broadening is minimized by the operation of the STM at low temperatures. Tips coated with a thin film of a ferromagnet or antiferromagnet are used as spin-polarized tips. Fig. 11.22 shows such a tip covered with a thin magnetic film. The magnetization at the tip apex can be either perpendicular to the tip axis or along the tip axis (Fig. 11.22). The configuration presented in Fig. 11.22(a) and 11.22(b) probes samples with in-plane (out of plane) magnetization. The energy derivative of the tunnel current can then be estimated from the output lock-in amplifier signal. The energy derivative of the tunnel current will in general depend on the relative orientation of tip and sample magnetization as in the case of the tunnel current itself. The magnitude of the tunnel current can be utilized to control the distance between the tip and the sample, and the lock-in amplifier signal to obtain information on the sample magnetization. The problem in this approach is that the magnetic signal varies in size as a function of bias voltage. On the other hand, the advantage is that any existing STM can be used once a tip is covered with a magnetic film. To make the technique relatively insensitive to the application of external magnetic fields, antiferromagnetic tips can be useful.

The second method of magnetic imaging with SP-STM is schematically presented in Figs. 11.23(a) and (b). In this approach, the tip is made of magnetic material and the tip magnetization is periodically reversed by the current passing through a coil. The tip material needs to be magnetically soft and with very low magnetostriction. A normal tip presented in Fig. 11.23(a) is used to detect the out-of-plane component of sample magnetization. A special tiny washer-like ring design of the tip (Fig. 11.23(b)) is used for detection of the in-plane component of sample magnetization. Here the electrons tunnel between the sample and the bottom-most atoms of the ring. The sample magnetization is modulated and the resulting tunnel magnetoresistance is detected with the lock-in amplifier. The signal in this method can be used for magnetic imaging at any bias voltage.

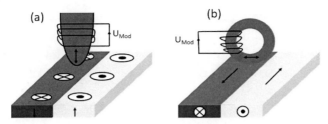

FIGURE 11.23 Schematic representation of an SP-STM experimental setup using the method of modulated tip magnetization. (a) An out-of-plane-sensitive magnetic tip and (b) an in-plane-sensitive ring shaped magnetic probe.

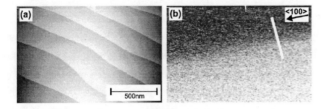

FIGURE 11.24 SP-STM image of (a) the topography and (b) the magnetic signal of an Fe whisker. (From reference [19] with permission from AIP Publishing.)

Imaging with SP-STM regularly resolves less than 1 nm. However, SP-STM is very surface sensitive and has to be performed under ultrahigh vacuum conditions. The high surface sensitivity also implies that the buried layers cannot be accessed by SP-STM. Figure 11.24 shows the topography and the magnetic contrast of an Fe whisker, imaged with SPSTM [19]. Perfectly grown, defect free Fe whiskers have simple domain patterns consisting of 180° domain walls along <100> direction and 90° domain walls along <110> direction. The topographic image shows monotonic steps and terrace widths between 200 and 400 nm (Fig. 11.24(a)). Fig. 11.24(b) presents the image of a magnetic signal, which clearly reveals two domains separated by a 180° domain wall running along the $<100>$ direction. The spin signal contains no features of the topography and no magnetic structure of the 180° domain wall can be seen in the topography.

11.2.3 Scanning Hall probe and scanning SQUID microscopy

Scanning Hall probe microscopy (SHPM) is an interesting non-invasive technique to have quantitative measurements of surface magnetic field profiles with a relatively high spatial resolution (less than a few μm) under variable temperature and magnetic field operation. SHPM incorporates a miniature Hall-bar sensor as the probe, which is placed near the sample under investigation and is scanned over the sample surface with the help of a dedicated scanner. The spatial images of the magnetic profile on the sample surface are obtained by recording the output signal of the Hall sensor.

The functional principle of SHPM is based on the Hall effect, which is the appearance of a transverse voltage (known as the Hall voltage V_H) when a current-carrying conductor is placed in a perpendicular magnetic field. The Hall voltage V_H is thus proportional to the magnetic field and the current, and it is inversely proportional to the carrier density n of the conductor. A Hall probe can be used to measure the magnetic field if the density of charge carriers n and the geometrical factors are known. The probe in SHPM detects the stray field coming out of the sample, which is proportional to the magnetization of the sample. Thus, SHPM works as a magnetometer, based on the Hall effect.

There are various types of Hall sensors available for the operation of SHPM, and some materials show better functionality than others depending on the specific measurement. One of the best choices while working at low temperatures, GaAs/AlGaAs heterostructures is a two-dimensional electron gas below the surface. This is due to the special combination of small density and high mobility of the charge carriers, which leads to a high signal-to-noise ratio. InSb, however, is the material of choice for room temperature applications because of its highest mobility among the semiconductors around room temperature. Even at low temperatures, InSb is the material suitable for measurements in magnetic fields above 5 kOe since GaAs/AlGaAs-based probes often show de Hass–van Alfen oscillations in the field regime above 1 kOe. Nowadays in many of the designs, the spatial resolution of SHPM (about 1 micron or less) is dictated by the size of conductive channels on Hall probe.

The spatial resolution in SHPM is in principle limited by the active region of the Hall probe and its separation from the sample surface, and not by the scanning technique. SHPM can use various kinds of scanning techniques. In one of the techniques, the Hall probe is brought into proximity with the sample surface using scanning tunneling microscopy (STM) positioning techniques [20]. Fig. 11.25 shows a schematic representation of such an SHPM. In this SHPM the Hall probe is mounted on the piezotube of a custom-manufactured low-temperature STM with a stick-slip coarse approach mechanism and is tilted $\approx 1^0$–2^0 with respect to the sample [20]. The active Hall sensor is patterned about 13 μm away from the probe corner, which is coated with a thin layer of gold to serve as a tunneling tip. The sample is approached to establish tunneling first, and then the Hall probe is scanned across the sample surface to measure the magnetic field and the surface topography simultaneously. Alternatively, the sample is retracted about 0.5 μ and the Hall probe can be scanned much more rapidly to get the magnetic field profile with a slightly lower spatial resolution. In another scanning scheme, the SHPM consists of a mechanical positioning sub-micron-XY stage and a flexible direct contact to

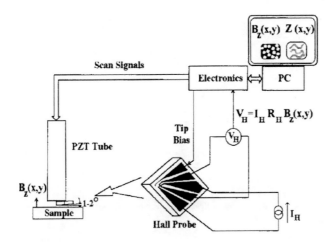

FIGURE 11.25 Schematic diagram of a scanning Hall probe microscope using scanning tunneling microscopy (STM) positioning techniques. (From reference [20] with permission from AIP Publishing.)

the sample without a feedback control system for the z-axis [21]. The three-axis actuators driven by stepper motors at room temperature are controlled by a personal computer. A raster-scan scheme was used to achieve optimal repeatability [21]. This SHPM has a large scan range up to millimetres, high spatial resolution (≤ 4 μm), and high field sensitivity in a wide range of temperature (4.2 K–300 K) and magnetic field (10^{-7} T–1 T). Fig. 11.26 presents scanning Hall-probe images mapping an isothermal magnetic field induced antiferromagnetic to ferromagnetic transition in intermetallic compound Ru-doped $CeFe_2$ using a similar scanning Hall-probe system with 5 μm square InSb Hall sensors [22].

The scanning superconducting quantum interference device (SQUID) microscope (SSM) is a powerful tool for imaging magnetic fields above sample surfaces with high sensitivity and bandwidth. The SQUID sensor measures local magnetic flux through a micron-sized pickup loop. The ultimate field sensitivity of the SQUID microscope is given by the SQUID flux noise divided by the effective pickup area. For typical values of flux noise of $2\times 10^{-6} \phi_0/\text{Hz}^{1/2}$, a pickup area of $(7$ $\mu m)^2$ corresponds to an effective field noise of $\approx 1 \times 10^{-10}$ T/Hz$^{1/2}$. Here superconducting flux quantum $\phi_0 = h/2e = 2.07 \times 10^{-15}$ Tm^2. On the other hand, a pickup area of $(1 \text{ mm})^2$ corresponds to an effective field noise $\approx 4\times 10^{-15}$ T/Hz$^{1/2}$.

Niobium(Nb)-based SQUID sensors offer a sensitivity of one order of magnitude better than high-temperature superconductor (HTSC) SQUIDs with equivalent spatial resolution. Nb-based SQUID is more commonly used in SSM for its reliable junction quality and good SQUID performance. However, the Nb SQUID sensor used for SSM needs to be kept at a stable temperature well below its superconducting

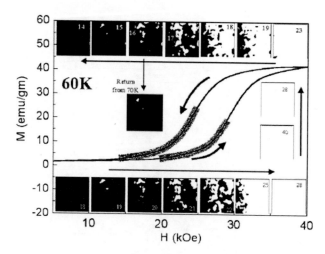

FIGURE 11.26 Isothermal M versus H plot and representative Hall-probe images at 60 K in Ru-doped CeFe$_2$. Starting counterclockwise from the bottom left-hand corner, the images represent the antiferromagnetic state in the ascending field cycle, the antiferromagnetic–ferromagnetic transition regime in the ascending field cycle, the ferromagnetic state, the ferromagnetic–antiferromagnetic transition regime in the descending field cycle, and the final antiferromagnetic state at the end of the cycle. The frame below 16 kOe in the top row shows the effect of temperature cycling on the supercooled ferromagnetic state. (From reference [22] with permission from APS.)

transition temperature 9 K. This somewhat restricts the operation of the Nb-based SSM with the necessary requirement of cryogenic temperature below 9 K. A very similar scanning scheme as described earlier in the context of SHPM can be employed for SSM as well. This scanning scheme consists of a mechanical positioning sub-micron-XY stage and a flexible direct contact to the sample without a feedback control system for the z-axis. Using the same scheme for scanning a scanning probe microscope for magnetic imaging can be built, which can interchangeably be used both as an SHPM and an SSM [21]. Furthermore, a scanning SQUID probe microscope has been developed with STM and AFM to measure fine magnetic field distribution on the conductor and insulator samples, respectively [23].

11.3 Magnetic Imaging Using Synchrotron Radiation Sources

We have earlier discussed in Section 10.1.4 on X-ray magnetic circular dichroism (XMCD) and X-ray magnetic linear dichroism (XMLD). The magnetic dichroism in X-ray absorption can be utilized as a contrast mechanism for magnetic imaging in magnetic materials. The elemental sensitivity of this magnetic contrast mechanism is an added advantage in this magnetic imaging technique. We have studied earlier that XMCD originates from the spin-polarization of resonant transitions between occupied core states and spin-split unoccupied states just above the Fermi level excited by circularly polarized X-ray photons.

Polarized synchrotron radiation is usually used for X-ray absorption spectroscopy. The X-rays from a synchrotron radiation source are linearly polarized if the beamline only accepts radiation through a horizontal slit positioned in the plane of the electron storage ring [24, 25]. On the other hand, if the horizontal aperture is placed below the orbit plane of the ring, the electrons in the ring appear to rotate counterclockwise, and the transmitted radiation is right-circularly-polarized (RCP). Left-circularly-polarized (LCP) X-ray photons can be obtained for an aperture placed above the plane of the storage ring.

In integrated XMCD measurements the X-ray absorption is detected either by measuring the amount of emitted electrons at the sample surface or by measuring the transmitted intensity of X-rays after transfer through the sample. The former method is usually known as total electron yield detection of X-ray absorption. The other method is a more direct way of measuring the X-ray absorption, but it would require appropriate thin samples. These two approaches are implemented in imaging experiments by combining resonant excitation by circularly polarized X-rays and some existing microscopy techniques. One can use a photoelectron emission microscope to get a magnified image of the sample from the emitted electrons at the surface after absorption of X-ray photons. An image of the sample surface is obtained with X-ray in transmission microscopy from the locally different transmission of circularly polarized X-rays, which are tuned to elemental absorption edges. We shall now discuss below three experimental methods for magnetic imaging.

11.3.1 Scanning X-ray microscopy

The working principle of scanning X-ray microscopy is shown in Fig. 11.27. A monochromatic X-ray beam is focused to the smallest possible spot size. The X-ray intensity transmitted through the sample or the fluorescent X-ray or electron intensity from the sample is monitored at various focused beam positions on the sample. The energy resolution is determined by the monochromator in the beamline at the synchrotron radiation source and the spatial resolution is determined by the size of the X-ray spot [24]. The reflected and focused beam from grazing incidence mirrors or the diffracted and focused beam from a multilayer mirror or a zone plate can be used to get small X-ray spots. In Fig. 11.27 a zone plate is used. The zone plate is composed of tiny lithographically fabricated concentric ring structures, which acts like an X-ray lens. In the zone plates, the width of the alternating open and closed rings varies as a function of the radius, which gives rise to a diffraction of the X-rays. The focal length of a zone plate is determined by the wavelength of the X-rays, whereas the attainable resolution is determined by the width of the

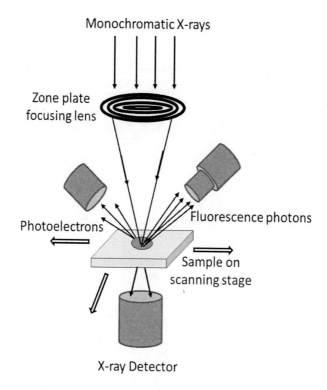

FIGURE 11.27 Schematic diagram representing experimental setup and working principle of scanning X-ray microscopy.

outermost ring of the zone plate. A resolution of 20 nm and below can be achieved in zone plates with the help of state-of-the-art lithography. In comparison to electron microscopy, X-ray transmission or fluorescent microscopes are well suited for studies in the presence of a magnetic field. Moreover, these techniques are bulk sensitive.

11.3.2 Transmission X-ray microscopy

The working principle of transmission X-ray microscopy is shown in Fig. 11.28. A condenser zone plate in conjunction with an order-sorting aperture (typically with a 10–20 μm diameter) creates a monochromatic X-ray spot on the sample [25]. A magnified image of the sample is generated by a micro-zone plate onto a phosphor screen or X-ray-sensitive CCD camera. The width of the outermost zones of the microzone plate determines the spatial resolution. This method is also well suited for studies in the presence of magnetic fields. Fig. 11.29 presents images obtained with magnetic transmission X-ray microscopy showing a variation of the domain patterns with aspect ratio in rectangular 50-nm-thick permalloy $Fe_{80}Ni_{20}$ samples in different external magnetic fields applied along the long edge of the samples [26].

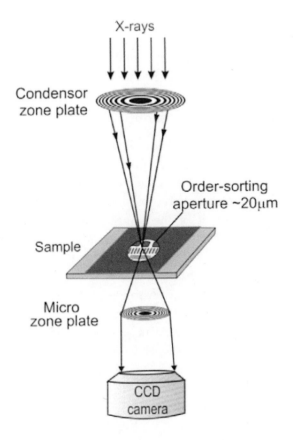

FIGURE 11.28 Schematic diagram representing experimental setup and working principle of transmission X-ray microscopy. (From reference [25] Creative Commons licence.)

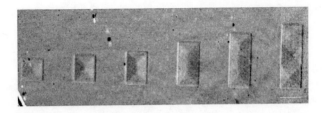

FIGURE 11.29 Variation of the domain patterns with aspect ratio in rectangular 50-nm-thick permalloy $Fe_{80}Ni_{20}$ samples in varying external magnetic fields. (From reference [26] with permission from Elsevier.)

Magnetic X-ray transmission microscopy or M-XTM (sometime also termed as magnetic transmission X-ray microscopy (M-TXM)) is the transmission counterpart to X-ray photoelectron microscopy (to be discussed below) for the laterally resolved detection of XMCD [2]. The resolution is usually higher in M-XTM, and it is not sensitive to external magnetic fields. It is thus ideally suited for the imaging of magnetization reversal processes. In comparison to the X-ray photoelectron microscope, which is more sensitive for signals of ultrathin films or interfaces, information from more deeply buried layers can easily be obtained by M-XTM.

11.3.3 X-ray photoelectron microscopy

In Section 11.1.1 above we have spin-polarized low energy electron microscopy (SPLEEM) technique, which uses spin-polarization detection of low energy secondary electrons emitted from the surface of a magnetic sample for magnetic imaging. In a very similar way, using wavelength-tunable X-rays from a synchrotron radiation source magnetic contrast of the sample surface is obtained through the XMCD or XMLD effects on the emitted secondary electron from the sample surface. The potential of photoelectron emission spectroscopy (PEEM) for the spectro-microscopic detection of X-ray absorption is recognized more with the availability of X-rays with a tunable wavelength from synchrotron radiation sources [1].

The working principle of X-ray photoelectron microscopy is shown in Fig. 11.30. The sample is exposed to a monochromatic X-ray beam that is only moderately focused to match the maximum field of view of a photoelectron microscope. Electrons emitted from the sample are projected with magnification onto a phosphor screen and the image can be viewed in real time at video rates. The spatial resolution is determined by the electron optics in the photoelectron microscope, which is limited by three quantities: spherical aberration, chromatic aberration, and diffraction [24, 25]. In the sample, X-ray excitation generates photoelectrons with characteristic energy distribution. The electron intensity is dominated by the secondary electron tail in the 0–20 eV kinetic energy range. Here zero kinetic energy corresponds to the vacuum level of the sample. The intensity and resolution of X-ray photoelectron microscopy (X-PEEM) are determined by the secondary electron signal. The X-PEEM signal is usually large, and it closely follows the X-ray absorption spectrum of the sample under investigation. However, the energy spread of the inelastic tail (about 5 eV for most materials) spoils the resolution through chromatic aberrations [24]. The effective width of the energy spread is somewhat reduced by placing a suitable aperture in the back focal plane of the PEEM (Fig. 11.30).

XMCD gives rise to a different X-ray absorption by magnetic domains in the sample with magnetization parallel and antiparallel to the helicity of the incoming synchrotron X-rays. Two images are obtained for the opposite helicity of the circularly polarized X-rays. Those are subtracted to eliminate all topographic contrast and obtain a magnetic domain image of the sample. To eliminate the effects of different illumination across the image, the subtracted result is usually

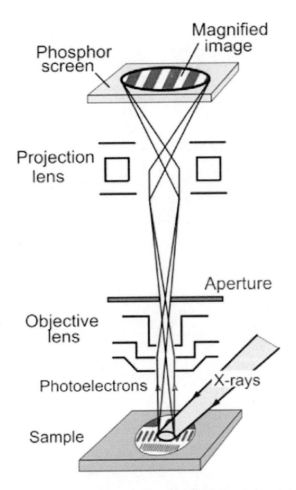

FIGURE 11.30 Schematic diagram representing experimental setup and working principle of X-ray photoelectron microscopy. (From reference [25] Creative Commons licence.)

normalized with respect to the sum of the two images. We already learned that XMCD as an integral method is widely used in connection with a set of sum rules. Thus, it is possible to get quantitative magnetic information like the effective spin moment or the orbital moment per atom from a pair of X-ray absorption spectra for opposite helicity. A quantitative analysis of this micro-spectroscopic data set is also possible if the energy of X-ray photon energy is scanned in small steps and a set of PEEM images is gathered for several energies around the absorption edges for both helicities. This exercise will give rise to microscopic images of the spin and orbital moments [1]. Summarizing, one can say that magnetic imaging by XMCD-PEEM provides element-selective magnetic domain images.

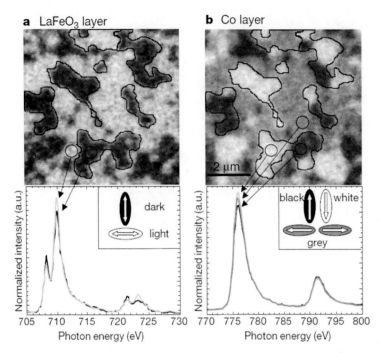

FIGURE 11.31 Images and local XMLD and XMCD spectra from the antiferromagnetic and ferromagnetic layers for 1.2-nm Co on $LaFeO_3/SrTiO_3$(001): (a) Fe L-edge XMLD image; (b) Co L-edge XMCD image. (From reference [27] with permission from Springer Nature.)

Apart from XMCD, it is also possible to use linearly polarized X-rays for magnetic imaging. The absorption of linearly polarized X-rays is influenced by the orientation of the electric field vector with respect to a uniaxial deviation from cubic symmetry of the electronic states involved in the transitions. The magnetic order in a sample can cause such a deviation. The absorption spectra for parallel and perpendicular alignment of polarization and magnetization axes show a difference at the absorption edges. This linear magnetic dichroism can be appreciable in insulating materials. However, XMLD is much smaller in metals, typically more than a factor of 10 smaller in comparison to the XMCD. XMLD can still be used for magnetic imaging with a careful design of the experiment and incorporating an appropriate exposure time.

Fig. 11.31 shows X-PEEM (both X-MCD and X-MLD) images obtained on thin Co film grown on the top of a 40 nm $LaFeO_3$ film grown on $SrTiO_3$(001) substrate [27]. In Fig. 11.31(a) the contrast in the images arises from antiferromagnetic domains in $LaFeO_3$, whereas in Fig. 11.31(b) it is from the ferromagnetic domains in Co. The in-plane orientations of the antiferromagnetic axis and ferromagnetic spins are indicated below the images. Fig. 11.31 also shows XMLD and XMCD spectra

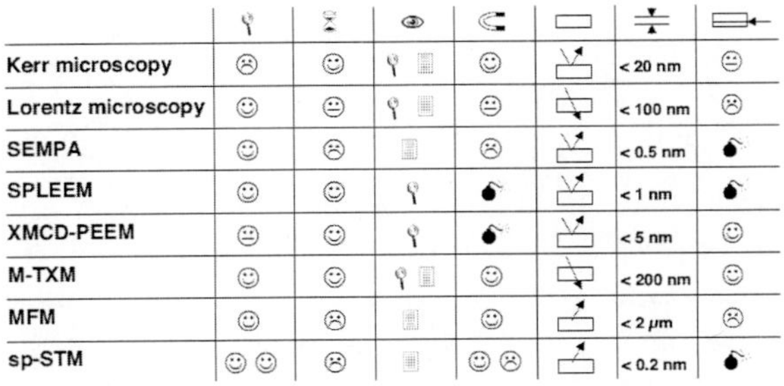

	🔍	⏱	👁				
Kerr microscopy	☹	☺	? ▦	☺	⬉	< 20 nm	☺
Lorentz microscopy	☺	☺	? ▦	☺	⬇	< 100 nm	☹
SEMPA	☺	☹	▦	☹	⬉	< 0.5 nm	●
SPLEEM	☺	☺	?	●	⬉	< 1 nm	●
XMCD-PEEM	☺	☺	?	●	⬉	< 5 nm	☺
M-TXM	☺	☺	? ▦	☺	⬇	< 200 nm	☺
MFM	☺	☹	▦	☺	⬆	< 2 μm	☹
sp-STM	☺☺	☹	▦	☺ ☹	⬆	< 0.2 nm	●

FIGURE 11.32 Summary of the features of various magnetic imaging techniques. From left to right the columns with pictograms refer to resolution, image acquisition speed, type of imaging (parallel imaging or scanning), sensitivity to applied magnetic fields, type of depth information (surface based or transmission), information depth, and possibility of obtaining depth selective information. (From reference [2] with permission from Springer Nature.)

for LaFeO$_3$ and Co, respectively, depicting the origin of the intensity contrast in the respective images.

11.4 Summary

This chapter is focused on various techniques for actual imaging of magnetic micro-structures. Three different classes of magnetic imaging techniques have been discussed, namely (i) electron optical methods, (ii) imaging with scanning probes, and (iii) imaging with X-rays from synchrotron radiation sources. Electron-optical methods can provide images with a very impressive resolution of the order of 1 nm, and show high contrast and sensitivity to detect small changes in magnetization in a material. In the category of scanning probes, MFM is possibly the most widely used technique for magnetic imaging. This is because of high resolution, ease of use under environmental conditions, and the commercial availability of instruments. An additional advantage of MFM is the possibility of obtaining topographic information within the same instrument. SP-STM is presently the magnetic imaging technique with very high resolution of less than 1 nm. However, it is very surface sensitive, and it needs to be operated under ultrahigh vacuum conditions. The core principle of synchrotron radiation-based techniques involves magnetic dichroisms in X-ray absorption as contrast mechanism for the magnetic imaging. The particular strength of such techniques is the elemental sensitivity of this magnetic contrast mechanism.

We will now summarize the features of the various techniques for magnetic imaging discussed in this chapter with the help of pictograms in Fig. 11.32 borrowed from the reference [2]. In addition to the various techniques discussed in this chapter, the Kerr microscopy discussed in Chapter 8 is also included for comparison. The resolution is referred in the first column headed by a magnifying glass. The ranges span between that of Kerr microscopy (500 nm) and SP-STM (< 1 nm), the latter being characterized by a double smiley. The resolution of XMCD-PEEM is somewhat worse compared to SPLEEM despite the identical electron imaging technique. This is possibly due to chromatic aberration.

A summary of the speed of image acquisition is presented in the second column, headed by an hourglass. The speed depends on image size, resolution, and the actual system under study. In general the scanning probe techniques and SEMPA as the scanning technique with intensity loss in the spin detection process is considered as relatively slow. The third column headed with an eye symbol lists the type of image acquisition. The page with lines pictogram refers to scanning imaging, while the magnifying glass pictograms indicate parallel imaging.

The fourth column headed by a horseshoe magnet symbol gives an overview of the sensitivity to external magnetic fields. The techniques SPLEEM and XMCD-PEEM using low energy secondary electrons for the imaging are marked by bomb symbols, which indicate a high sensitivity to external magnetic fields. In SEMPA also slow electrons are detected. However, the image is determined by the high energy primary electrons, and moderate magnetic fields are tolerated in certain dedicated setups. Two different symbols are shown in the field for SP-STM, indicating two different experimental methods. Sensitivity to external magnetic fields is higher when the magnetization of a soft magnetic tip is modulated than in the method where the sample bias voltage is modulated.

The fifth column headed by a rectangular box presents the way the magnetic information is obtained. The different pictograms indicate on-surface acquisition and transmission measurements. For the scanning probe techniques MFM and SP-STM, modified on-surface symbols are used. This indicates that unlike in the other techniques the surface is not probed by an incident electron or photon beam. The approximate corresponding length scales, information depth in the case of on-surface measurements, and maximum sample thickness in the case of transmission measurements are listed in the sixth column, titled with a distance pictogram [2].

The last column summarizes the possibility of obtaining depth-resolved information, for example, in multilayered magnetic samples. This is readily achievable in the XMCD-based techniques, XMCD-PEEM and M-TXM. Due to their element-selectivity, these techniques can address different magnetic layers at different depths separately. A careful adjustment of the phase, a depth sensitivity can also be obtained in Kerr microscopy. Lorentz microscopy as a transmission technique can provide information from all depths of a sample. In MFM, due to the long-range dipolar interaction, it is possible to obtain information about magnetic domain walls in more deeply buried layers. In the more surface-sensitive techniques like SEMPA, SPLEEM, and SP-STM, no depth-resolved information can be obtained from the images.

Bibliography

[1] Zhu, Y., ed. (2005). *Modern Techniques for Characterizing Magnetic Materials*. Dordrecht: Kluwer Academic Publishers, pp. 455.

[2] Kuch, W. (2006). Magnetic Imaging. In *Magnetism: A Synchrotron Radiation Approach*. Heidelberg: Springer, pp. 275–320.

[3] Akamine, H., Okumura, S., Farjami, S., et al. (2016). Imaging of Surface Spin Textures on Bulk Crystals by Scanning Electron Microscopy. *Scientific Reports*, 6 (1): 1–8.

[4] Koike, K., (2013). Spin-polarized Scanning Electron Microscopy. *Microscopy*, 62 (1): 177–191.

[5] Kohashi, T. (2018). Magnetization Analysis by Spin-Polarized Scanning Electron Microscopy. *Scanning*, 2018. https://doi.org/10.1155/2018/2420747.

[6] Rougemaille, N., and Schmid, A. K. (2010). Magnetic Imaging with Spin-polarized Low-energy Electron Microscopy. *The European Physical Journal-Applied Physics*, 50 (2).

[7] Blundell, S. (2001). Magnetism in Condensed Matter. *Condensed Matter Physics*. (Oxford Series Publications, 2001), 29.

[8] Duden, T., and Bauer, E. (1998). Spin-polarized Low Energy Electron Microscopy of Ferromagnetic Layers. *Microscopy*, 47 (5): 379–385.

[9] Shibata, N., Kohno, Y., Nakamura, A., et al. (2019). Atomic Resolution Electron Microscopy in a Magnetic Field Free Environment. *Nature Communications*, 10 (1): 1–5.

[10] Petford-Long, A. K., and de Graff, M. (2012). In E. N. Kaufmann, ed., *Characterization of Materials*, John Wiley and Sons, Inc., pp. 1787.

[11] Kovács, A., Pradeep, K. G., Herzer, G., et al. (2016). Magnetic Microstructure in a Stress-annealed $Fe_{73.5}Si_{15.5}B_7Nb_3Cu_1$ Soft Magnetic Alloy Observed Using Off-axis Electron Holography and Lorentz Microscopy. *AIP Advances*, 6 (5): 056501.

[12] Wong, B. Y. and Laughlin, D. E. (1996). Direct Observation of Domain Walls in NiFe Films Using High-resolution Lorentz Microscopy. *Journal of Applied Physics*, 79 (8): 6455–6457.

[13] Almeida, T. P., Temple, R., Massey, J., et al. (2017). Quantitative TEM Imaging of the Magnetostructural and Phase Transitions in FeRh Thin Film Systems. *Scientific Reports*, 7 (1): 1–11.

[14] De Graef, M., 2001. 2. Lorentz Microscopy: Theoretical Basis and Image Simulations. *Experimental Methods in the Physical Sciences*, 36: 27–67.

[15] Chapman, J. N., and Scheinfein, M. R. (1999). Transmission Electron Microscopies of Magnetic Microstructures. *Journal of Magnetism and Magnetic materials*, 200 (1–3): 729–740.

[16] Passeri, D., Dong, C., Reggente, M., et al. (2014). Magnetic Force Microscopy: Quantitative Issues in Biomaterials. *Biomatter*, 4 (1): 29507. DOI: 10.4161/biom.29507.

[17] Li, Z., Wei, F., Yoshimura, S., Li, G., et al. (2013). Quantitative Analysis of the Magnetic Domain Structure in Polycrystalline $La_{0.7}Sr_{0.3}MnO_3$ Thin Films by Magnetic Force Microscopy. *Physical Chemistry Chemical Physics*, 15 (2): 628–633.

[18] Wiesendanger, R. (2009). Spin Mapping at the Nanoscale and Atomic Scale. *Reviews of Modern Physics*, 81 (4): 1495.

[19] Schlickum, U., Wulfhekel, W., and Kirschner, J. (2003). Spin-polarized Scanning Tunneling Microscope for Imaging the In-plane Magnetization. *Applied Physics Letters*, 83 (10): 2016–2018.

[20] Oral, A. H. M. E. T., Bending, S. J., and Henini, M. (1996). Real-time Scanning Hall Probe Microscopy. *Applied Physics Letters*, 69 (9): 1324–1326.

[21] Tang, C. C., Lin, H. T., Wu, S. L., et al. (2014). An Interchangeable Scanning Hall Probe/Scanning Squid Microscope. *Review of Scientific Instruments*, 85 (8): 083707.

[22] Roy, S. B., Perkins, G. K., Chattopadhyay, M. K., et al. (1994). First Order Magnetic Transition in Doped Alloys: Phase Coexistence and Metastability. *Physical Review Letters*, 92: 147203.

[23] Miyato, Y., Hisayama, K., Matsui, Y., et al. (2013). Scanning SQUID Probe Microscope with STM and AFM. In *2013 IEEE 14th International Superconductive Electronics Conference (ISEC)*, 1–4. DOI: 10.1109/ISEC.2013.6604263.

[24] Stöhr, H. A., Padmore, S., Anders, T., et al. (1998). Principles of X-Ray Magnetic Dichroism Spectromicroscopy. *J Surface Review and Letters*, 5: 1297.

[25] Streubel, R., Fischer, P., Kronast, F., et al. (2016). Magnetism in Curved Geometries. *Journal of Physics D: Applied Physics*, 49 (36): 363001.

[26] Fischer, P. (2003). Magnetic Soft X-ray Transmission Microscopy. *Current Opinion in Solid State and Materials Science*, 7 (2): 173–179.

[27] Nolting, F., Scholl, A., Stöhr, J., et al. (2000). Direct Observation of the Alignment of Ferromagnetic Spins by Antiferromagnetic Spins. *Nature*, 405 (6788): 767–769.

12

Nano-Scale Magnetometry with Nitrogen Vacancy Centre

We have studied in earlier chapters that spin-based techniques like neutron scattering, muon spin resonance spectroscopy, and nuclear magnetic resonance can give detailed information on the magnetic structure of a material down to the atomic scale. These techniques, however, cannot provide a real-space image of the magnetic structure and are not sensitive to samples having nanometre-scale volumes. On the other hand, techniques like magnetic force microscopy, scanning hall bars, and superconducting quantum interference devices (SQUIDs) enable real-space imaging of the magnetic fields in nanometre-scale samples. But they have a constraint of finite size, and also they act as perturbative probes working in a rather narrow temperature range. A relatively new technique of magnetometry based on the electron spin associated with the nitrogen-vacancy (NV) defect in diamond combines the powerful aspects of both these classes of experiments. A very impressive combination of capabilities has been demonstrated with NV magnetometry, which sets it apart from other magnetic sensing techniques. That includes room-temperature single-electron and nuclear spin sensitivity, spatial resolution on the nanometre scale, operation under a broad range of temperatures from ~ 1 K to above room temperature, and magnetic fields ranging from zero to a few tesla, and most importantly it involves a non-perturbative operation [1]. Here we present a concise introduction to NV magnetometry. There are a few excellent review articles [2, 3] and tutorial article [4] on NV magnetometry, and readers are referred to those for a more detailed exposure to the subject.

12.1 Physics of the Nitrogen-Vacancy (NV) Centre in Diamond

Figure 12.1 presents an NV centre in the crystal lattice of a diamond. It is a point defect consisting of a substitutional nitrogen atom and a missing carbon atom in the neighborhood. The NV centres can have negative (NV$^-$), positive (NV$^+$), and neutral (NV^0) charge states of which NV$^-$ is used for magnetometry [2]. An NV$^-$ centre has six electrons, five of which come from the dangling bonds of the three neighbouring carbon atoms and the nitrogen atom. The negative charge state arises from one extra electron captured from an electron donor. The NV axis is defined by the line connecting the nitrogen atom and the vacancy. There can be four NV alignments depending on the four possible positions of the nitrogen atom with respect to the vacancy.

Figure 12.2 presents an energy-level diagram for NV electronic spin-triplet states ($^3A_2, ^3E$) and spin-singlet states (^1E, $^1A_1$). There are three possible sub-levels with magnetic quantum number $m = 0, \pm 1$ for each spin-triplet state where the quantization axis is set by the NV axis [2]. The optical transition between spin-triplet states $^3A_2$ and ^3E can be excited with light with wavelength $\lambda \approx 450$ nm to 637 nm. A diode-pumped solid-state laser with 532 nm wavelength can be used to drive this transition. The optical transition between the triplet states has a 637 nm wavelength. On the other hand, the transition between ^1E and $^1A_1$ has a wavelength 1042 nm, which corresponds to infrared light. However, lattice vibrations in the diamond

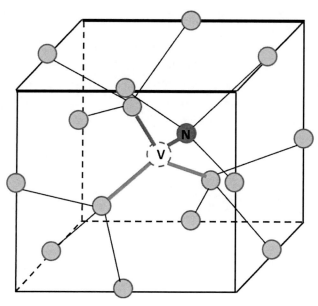

FIGURE 12.1 Nitrogen-vacancy centre with a vacancy and nitrogen atom in diamond crystal lattice.

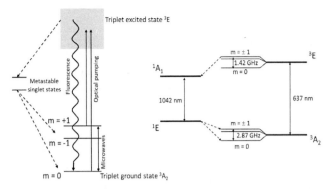

FIGURE 12.2 Electronic energy levels of nitrogen-vacancy centre and the associated optical transitions.

lattice introduce phonon sidebands to both these transitions, which broaden the NV absorption and fluorescence spectra by hundreds of nanometre [2]. The fluorescence arising due to the decay from the triplet state ^3E to the triplet state $^3A_2$ is in the range of 637 nm to approximately 800 nm wavelength. These optical transitions are mostly spin-conserving. However, there is also an inter-system crossing between the spin-singlet and spin-triplet states. There is non-radiative decay from ^3E to $^1A_1$, and the intersystem crossing rate is higher for the $m = \pm 1$ states than for the $m = 0$ [2]. There is also a non-radiative intersystem crossing from ^1E to $^3A_2$.

The electron spin of the NV defect in diamond lattice acts as a point-like magnetic field sensor, which can be optically probed and readout through its spin-dependent photoluminescence. Let us assume that the NV centre is illuminated with resonant light, which drives the transition from the spin-triplet ground state to the spin-triplet excited state. Since the optical transition is spin-conserving, the NV will be excited to the spin-triplet excited state ^3E, $m = 0$ state if it is initially in the $^3A_2$, $m = 0$ ground state. Subsequently, the NV centre decays back to the spin-triplet ground state with $m = 0$ sub-level while emitting fluorescence. This transition is cyclic if the NV centre is illuminated continuously and there will be a high fluorescence intensity. If, on the other hand, NV is initially in one of the $m = \pm 1$ sub-levels of spin-triplet ground-state $^3A_2$, it will be excited to the ^3E, $m = \pm 1$ states, and there will be a finite possibility of undergoing inter-system crossing to the singlet states. The NV centre first decays to the ^1E state from the $^1A_1$ state, which has a 200 ns lifetime at room temperature [2]. The NV centre subsequently undergoes inter-system crossing to the spin-triplet $^3A_2$ state. After several excitation cycles, the probability is more that the NV centre ends up in the $m = 0$ cycling transition, and one can then say that the NV centre has been optically pumped to the $m = 0$ state. If the NV centre is initially in one of the $m = \pm 1$ sub-levels of the spin-triplet ground state, then it emits less fluorescence from the ^3E excited state to $^3A_2$ ground state transition, as it decays largely nonradiatively through the singlet states. Summarizing, we can say that by measuring fluorescence, it is possible to read out the NV spin state from its fluorescence intensity.

The spin-triplet ground state $^3A_2$ of NV centre plays an important role. Due to electron spin–spin interaction this state has a zero-field-splitting parameter $D \sim 2.87$ GHz between the $m = 0$ and the $m = \pm 1$ sub-levels (Fig. 12.2). The NV centre couples to the external magnetic fields through the Zeeman effect. This is described by the Hamiltonian expressed in units of hertz (Hz) as [2]:

$$H = \gamma \vec{B}.\vec{S} \qquad (12.1)$$

Here, $\vec{S}$ stands for the dimensionless spin-projection operator for the NV electronic spin and $\gamma \approx 28$ GHz/T is the electron gyromagnetic ratio. If the NV axis is chosen as the z direction, and a magnetic field $B = B_z \hat{z}$ is aligned along this z direction, then the energies of the m-sub-levels are expressed as [2]:

$$E(m) = Dm^2 + \gamma B_z m \qquad (12.2)$$

Fig. 12.3 presents the energy-level diagram for an NV centre as a function of the magnetic field. It is important to note here that the energies of the $m = \pm 1$

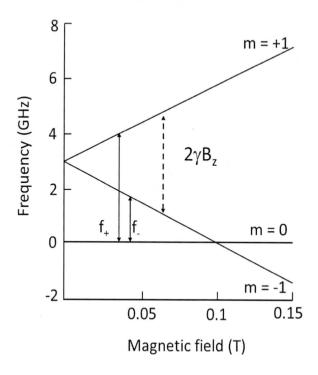

FIGURE 12.3 Energy-level diagram for a nitrogen-vacancy centre as a function of applied magnetic field. The magnetic field is aligned with the axis of the nitrogen-vacancy centre. The energies of magnetic resonance transitions are expressed as f_+ and f_-.

sub-levels depend linearly on the magnetic field, and the optical detection of this energy shift forms the basis of *nitrogen-vacancy* magnetometry. The energies of the magnetic resonance transitions between $m=0$ level and $m=\pm 1$ levels are expressed as:

$$\Delta E = D \pm \gamma B_Z \tag{12.3}$$

These energies are shown in Fig. 12.3 in frequency units as $f_\pm$. The vector magnetic field can be determined from these experimentally measurable resonance frequencies.

There are hyperfine structure associated with an NV centre due to the nuclear spin of the nitrogen atom, which has two stable isotopes, ^{14}N (99.6% natural abundance) with nuclear spin 1 and ^{15}N (0.4% natural abundance) with nuclear spin 1/2. This nuclear spin state introduces additional terms to the Hamiltonian expressed in units of hertz as [2]:

$$H_I = PI_Z^2 + \gamma_N \vec{B}.\vec{I} \tag{12.4}$$

$$H_{hf} = A_\| S_Z I_Z + A_\perp (S_X I_X + S_Y I_Y) \tag{12.5}$$

Here $\vec{I}$ is the dimensionless spin-projection operator for the nuclear spin. When considering an ^{14}N nucleus, $P \approx -4.95$ MHz is the quadrupole splitting, $\gamma_N \approx 3.077$ MHz/T is the nuclear gyromagnetic ratio, and $A_\| \approx -2.16$ MHz and $A_\perp \approx -2.7$ MHz are the parallel and perpendicular hyperfine coupling parameters, respectively [2]. The nuclear Zeeman term $\gamma_N \vec{B}.\vec{I}$ is a small term and is often be neglected. Each of the magnetic resonances f_+ and f_- splits into three transitions with conservation of m_I due to this hyperfine coupling with the ^{14}N nuclear spin. Magnetic resonance spectra show this hyperfine splitting, provided the resonance line width is narrower than the splitting.

12.2 A Brief Introduction to the Principle of NV Magnetometry

In the area of condensed matter and materials science, NV magnetometry can be used to study static magnetic textures such as domain walls in ferromagnets and skyrmions, magnetic excitations such as spin waves in ferromagnets, static current distributions such as superconducting vortices, and electrical noise currents in metals [1]. Let us first study the elementary considerations for the use of NV centres in imaging magnetic fields generated by various kinds of samples. The NV electron spin resonance splitting is sensitive to the first order to the projection of the magnetic field $\vec{B}$ on the NV spin quantization axis $B_\|$, and this is the quantity that is usually measured in the NV magnetometry experiments. The full vector field $\vec{B}$ can be reconstructed by measuring any of its components in a plane at a distance

d from the sample. However, this component should not be parallel to the plane of measurement.

Summarizing the discussion in the section above, we can say that the negatively charged NV centre has an $S=1$ electron spin that can be excited by incoherent optical excitation and readout through spin-dependent photoluminescence. Sensitivity to external magnetic fields is provided by the Zeeman effect. There can be three complementary sensing protocols, which cover a dynamic frequency range spanning from direct current (DC) up to ~ 100 GHz [1]. The first one is a measurement of electron spin resonance (ESR) frequencies $\omega_\pm$, which is usually done by sweeping the frequency of a microwave drive field and monitoring the spin-dependent photoluminescence. The DC magnetic field $\vec{B}$ can then be obatined using Eqn. 12.3 for a magnetic field oriented along the NV axis. The second protocol deals with the transverse spin relaxation rate, which is characterized by periodically flipping the phase of a spin superposition using microwave π-pulses. This is sensitive to magnetic fields at the spin-flip frequency. In the third technique known as NV relaxometry, the longitudinal spin relaxation rates are measured by preparing a spin eigenstate and monitoring the spin populations as a function of time. The longitudinal spin relaxation rates are sensitive to power spectral density of the magnetic field $g(\omega_\pm) = \int_{-\infty}^{+\infty} \overline{B_i(\tau).B_i(0)} e^{-i\omega\tau} d\tau$ [1]. Here the horizontal bar denotes a thermal average over the degrees of freedom of the material. This technique makes the NV centre a field-tunable spectrometer, which enables measurements of frequencies all the way up to ~ 100 GHz.

With the use of single NVs, the typical sensitivities of magnetic field sensing range from tens of μTHz$_z^{-1/2}$ for DC fields to tens of nTHz$^{-1/2}$ for alternating current (AC) fields. These numbers, however, strongly depend on experimental parameters such as the NV centre coherence time, the NV spin manipulation technique, the photon collection efficiency, and the use of specialized spin readout protocols [1].

12.3 Diamond Materials and Microscopy

NV centres can exist stably only a few nanometres below the diamond surface. This is important for NV magnetometry experiments on solid materials as the NV-sample distance d plays a key role in determining the ability to sense and resolve the magnetic properties. Natural diamonds contain NV centres and other defects, but synthetic diamonds are a better choice because of the controlled and reproducible manufacturing method. Different types of NV experiments require different samples [2]:

1. NV ensemble experiments with many NV centres being investigated in a single sample. These NV centres can either be located in a thin sheet in

the diamond or they are distributed throughout the entire volume of the diamond material.
2. Single-NV experiments, where a diamond sample with few defects is used. A particular NV is selected for the experiments.
3. NV nanodiamond experiments, using nanodiamonds containing one or many NV centres. Such nanodiamonds can be attached to the cantilevers of atomic force microscope or trapped in optical dipole traps.

Diamond samples to suit the needs of such experiments can be synthesized in several ways. These manufacturing techniques include high-pressure high-temperature (HPHT) growth, chemical vapour deposition (CVD) growing diamond layer-by-layer in a gaseous environment, and explosives detonation producing nanodiamonds with high nitrogen and NV densities [2]. Furthermore, with HPHT and CVD growth it is possible to obtain diamonds of various qualities, such as single crystal, polycrystals, optical-grade, and electronic-grade diamonds.

The NV centres are often probed optically with confocal microscopy setups. Fig. 12.4 shows a schematic of such a setup where a pump laser beam reflected off a dichroic mirror is focused through a lens or microscope objective and illuminates the diamond sample containing NV centres. The laser used is often a 532 nm solid-state laser, and the dichoric mirror is a mirror that selectively reflects short optical wavelengths but transmits long-wavelength lights. The fluorescence from the NV centres is collected through the same lens and passed through the dichroic mirror into a photodetector [2]. A pinhole is used to filter out fluorescence from the out of focus regions in the sample. NV confocal microscopy can be used in several situations [2]:

1. The pump laser beam can be focused onto the diamond and a diffraction-limited volume of NV centres can be investigated by measuring the collected fluorescence intensity with a single-pixel photodiode. The volume under investigation can be scanned with the help of scanning galvanometer mirrors or a piezo-driven objective or sample mount to obtain a wide-field image one pixel at a time.

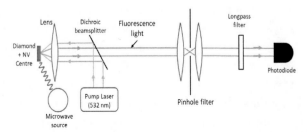

FIGURE 12.4 Schematic diagram of confocal microscopy setup for performing experiments with nitrogen-vacancy centres in diamond.

2. In another way to obtain a wide-field image, a wide area on the diamond is illuminated and the fluorescence is imaged with a camera. With this technique, a diffraction-limited spatial resolution ($< 1\mu$) is possible for imaging individual NV centres and thus mapping the magnetic field over a spatial region.
3. It is also possible to use diamonds and NV centres in atomic force microscopy (AFM) and use optical readout to investigate the NV centres, which can yield a few-nm spatial resolution. The fluorescence light from a diamond AFM probe containing an NV centre is collected in a microscope objective, and then by moving the probe it is possible to investigate different spatial regions without significantly influencing the light collection. The light collection can also be enhanced with the help of a diamond nanopillar probe by directing fluorescence light to the objective.

12.4 Optically Detected Magnetic Resonance

The techniques for NV magnetometry often require a bias magnetic field of a certain amplitude and direction that can be measured with the continuous wave (CW) optically detected magnetic resonance (ODMR) technique. Let us consider a situation where a sample containing a single NV centre is continuously illuminated with 532 nm green light within the confocal volume in a typical confocal microscopy setup shown in Fig. 12.4. The green light optically pumps the NV centre into the $m = 0$ sub-level and a high level of fluorescence from the NV centre is detected. The NV centre is also subjected simultaneously to microwave (MW) radiation, the frequency of which is varied within a certain range containing the magnetic resonances. The fluorescence level decreases when the MW frequency is on resonance with one of the transitions between $m = 0$ and ± 1 levels. This is because the MW field disturbs the NV optical pumping by transferring NV centres to the $m = \pm 1$ sub-levels, which fluoresce less brightly. Fig. 12.5 presents a typical spectrum showing the fluorescence intensity as a function of MW frequency, and that is known as the ODMR spectrum. In this schematic, it is assumed that the magnetic field is pointing along the NV axis, and the magnetic field amplitude can be determined from the resonance frequencies. The fractional difference in fluorescence intensity between MW on- and off-resonance is called the fluorescence contrast C [2], and a $C \approx 0.2$ contrast is typical for single NV centres. The linewidth of the magnetic resonance $\Delta\nu$ is another important parameter, which is related to the inhomogeneously broadened transverse spin relaxation time $T_2^* = 1/(\pi\Delta\nu)$. The typical value of T_2^* is approximately 100 ns for NV centres in HPHT diamond material rich in nitrogen and a few micro (μ) seconds for NV centres in diamonds with low nitrogen concentration [2]. It is beneficial to have a large contrast and

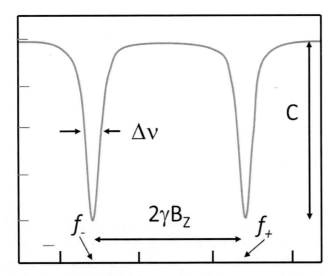

FIGURE 12.5 A typical optically detected magnetic resonance spectrum for a single nitrogen-vacancy centre. It is assumed that the magnetic field is along the nitrogen-vacancy axis. (From reference [2] with permission from Springer Nature.)

a narrow linewidth for measuring the transition frequencies, and, in turn, getting information on the magnetic field accurately.

When one is dealing with an ensemble of NV centres in the confocal volume, it is possible to have in total eight magnetic resonances due to the four possible alignments of the NV centre. Some of these resonances are degenerate in certain directions of the magnetic field. One may select one NV alignment by applying a bias magnetic field along the NV axis so that changes in the magnetic field projection along this axis affect the resonance frequencies approximately linearly [2]. The eight ODMR frequencies have a more complicated dependence on the magnetic field $\vec{B}$ if all four NV alignments are used, and this option will generate vector information about the magnetic field.

12.5 NV Magnetometers

12.5.1 Samples for NV Magnetometry

The key to high-resolution NV magnetometry is getting NV centres close to the sample. The most straightforward way to achieve a small distance between the NV centre and a sample is to deposit the sample directly on a diamond containing NV centres implanted to shallow depths. The magnetic fields emanating from the sample are then measured at the sites of these NV centres. However, there are some special materials that require epitaxial growth on a lattice-matched substrate. Such materials cannot be deposited directly on the diamond. In this case, NV-containing diamond particles, such as nanocrystals or microfabricated nanostructures, can be deposited on the material [1].

It is possible to do scanning-probe magnetic imaging in several ways [1]:

1. Attaching an NV-containing nanodiamond to an atomic force microscope tip.
2. Microfabricating an all-diamond tip hosting a shallow NV centre.
3. Synthesizing the sample itself on a sharp tip.

The precise information of the distance d between the NV centre and the sample is very important for the correct interpretation and quantitative estimation of the measured field data. There are various ways to determine d with a precision of a few nanometres. One way is to take account of the strong dependence on d of the NMR signal from protons located at the diamond surface [1]. In other methods, one estimates d by measuring the field from a calibration sample.

12.5.2 DC magnetometer

It is possible with an NV DC magnetometer to measure magnetic fields with frequency components from DC up to the bandwidth of the magnetometer (BW). This bandwidth can be determined with the application of oscillating magnetic fields of increasing frequencies while recording the magnetometer response. The signal in an NV magnetometer decreases with increasing frequency, and the bandwidth is defined as the frequency where the signal has decreased by a factor of two [2].

In the case where the magnetic field is measured with CW ODMR, the microwave (MW) frequency is scanned across the magnetic resonance within a certain time τ_{scan}. A value for the magnetic field can be obtained during each such time interval, and the bandwidth (BW) of the magnetometer will then be BW $\sim 1/\tau_{scan}$ [2]. The bandwidth may be rather small (BW $\leq$ 100 Hz), since many microwave generators can only scan the frequency slowly ($\tau_{scan} \geq$ 10 ms). The measurement also becomes vulnerable to low-frequency technical noise or drifts in the experimental apparatus because of such a slow scan. For a sensitive measurement, a modulation technique can be quite helpful, one possibility of which involves the application of an oscillating magnetic field using an external coil along with lock-in detection. The other possibility is using microwave generators with built-in frequency modulation. Such a case is presented in Fig. 12.6 where the microwave frequency is modulated with the frequency f_{mod} around a central value f_c. The maximum excursion from the centre frequency is termed as the frequency deviation f_{dev}. The centre frequency f_c is usually chosen around one of the magnetic resonance frequencies, say $m = 0 \rightarrow$ -1 transition. A lock-in amplifier referenced to the modulation frequency is then used to demodulate the detected fluorescence signal. The demodulated signal S_{LI} has a dispersive lineshape with a zero-crossing at $f_c = f_{res}$ and is linear $S_{LI} \approx \alpha \left(f_c - f_{res}\right)$,

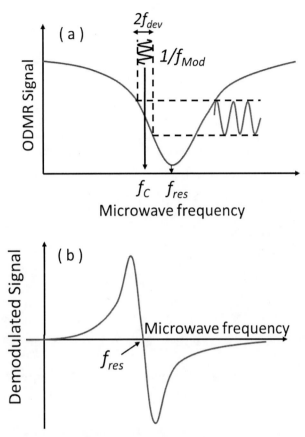

FIGURE 12.6 (a) Schematic presentation of an optically detected magnetic resonance signal. The microwave frequency is modulated at the frequency f_{mod} around the central value f_c with a frequency deviation of f_{dev}. (b) A dispersive line shape of the demodulated signal while sweeping the central frequency. (Adapted from reference [2] with permission from Springer Nature.)

when $f_c - f_{res}$ is well within the linewidth of the magnetic resonance $\Delta\nu$. Here α is a proportionality constant. A slowly varying magnetic field can be represented by the relation $B(t) = B_0 + \Delta B(t)$, where B_0 is defined by the set value of the centre frequency f_C by the formula $f_c = D - \gamma B_0/2\pi$. The linear relationship between the magnetic field deviation and demodulation signal can be expressed as [2]:

$$\delta B(t) = -2\pi S_{LI}(t)/(\alpha\gamma) \tag{12.6}$$

Thus, the demodulated signal can be used to measure slowly varying magnetic fields. Optical pump power and the microwave power set the bandwidth of NV magnetometer, and for high powers, a bandwidth as large as a few MHz has been demonstrated [2].

A complementary technique to the CW magnetometer is a pulsed NV magnetometer for DC magnetic fields. Within this technique, one way is to initialize the NV centres to the $m = 0$ sub-level, then switch off the pump laser before applying a frequency-dependent microwave π-pulse to probe the $m = 0 \to +1$ or $m = 0 \to -1$ resonance frequency. Subsequently, the final-state fluorescence is readout with a second laser pulse. The microwave π-pulse duration τ_π and the inhomogeneity-broadened transverse spin relaxation time (T_2^*) contribute to the linewidth. A longer π-pulse can lead to a narrower Fourier width, but that would excite fewer NV centres. A choice of $\tau_\pi = T_2^*$ gets the best sensitivity. In another way one can measure the ODMR frequency with Ramsey interferometry [2]. In this technique magnetic field is extracted from ODMR frequency measured by using two short $\pi/2$-pulses separated by an interaction time ($\approx T_2^*$) to accumulate magnetic-field-dependent phase.

12.5.3 AC Magnetometer

In contrast to the DC magnetometer, an AC magnetometer is sensitive to synchronized magnetic fields or a synchronized magnetic signal within a narrow bandwidth around a specific frequency [2]. NV AC magnetometry has been significantly influenced by NMR techniques where pulse sequencing or dynamical decoupling of RF or microwave radiation is used to remove magnetic inhomogeneity and extend coherence lifetime. The T_2^* coherence time can be extended by decoupling sequences to a considerably longer T_2 coherence time. Decoupling sequences can be used to sense coherent AC fields and incoherent AC fields or magnetic noise at kHz-MHz frequencies.

The role of a dynamical decoupling or DD sequence in NV AC magnetometry is illustrated in Fig. 12.7. The NV centres are initialized to an equal superposition of two Zeeman sub-levels with the first $\pi/2$-pulse. The AC magnetic field B_{AC} with frequency f_{AC} and period T_{AC} with projection along the NV axis then induces faster or slower Larmor precession, depending on its instantaneous sign [2]. The phase accumulation the NV centres acquire from B_{AC} can be maximized with the synchronization of the experiment and the AC magnetic field, and suitable choice of the pulse spacing $\tau = T_{AC}/2$. Asynchronous AC magnetic noise detection is also possible with decoupling sequences. If there is strong magnetic noise at f_{AC}, the choice of the pulse spacing $\tau = T_{AC}/2$ would tend to destroy the NV coherence, while a choice of $\tau = T_{AC}$ would cause the NV coherence to be immuned to magnetic noise with frequency f_{AC} [2].

There are other detection schemes like correlation spectroscopy that can be used to sense incoherent nuclear and paramagnetic AC magnetic fields. This technique uses two DD sequences separated by a time $\tilde{\tau}$, each of which will accumulate phase from an AC magnetic field, and, in turn, enables the study of the phase correlations in a nuclear spin bath with frequency resolution approximately T_2^{-1}. In another scheme known as double electron–electron resonance (DEER), it is possible to sense

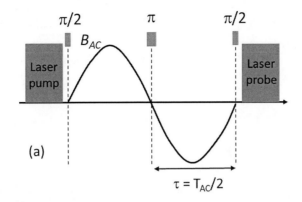

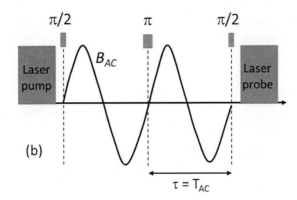

FIGURE 12.7 Schematic representation of a DD sequence for AC magnetometry. (a) The π-pulse allows one to retain the absolute value of the phase accumulation from B_{AC}. This, in turn, allows one to get the most decoherence while sensing noise. (b) Here, $\tau = T_{AC}$, and the phase accumulation gets cancelled irrespective of the relative phase and amplitude. This leads to the situation where the NV centres become immuned to magnetic noise at f_{AC}. (Adapted from reference [2] with permission from Springer Nature.)

electronic spins using simultaneous NV and electronic π-pulses. Similarly, one can detect NV decoherence when driving NV Rabi oscillations at a frequency that matches the desired AC frequency or driving both the NV centres and a target spin at the same Rabi frequency [2]. The presence of an AC field is detected from the resultant NV decoherence.

12.5.4 Sensitivity of NV magnetometers

NV magnetometers are currently less sensitive than other technologies such as SQUID magnetometers, but their main strength lies in the high spatial resolution.

Nevertheless, it is worth having a short discussion here on the sensitivity of NV magnetometers.

The magnetic field sensitivity (in units of $T/\sqrt{Hz}$) is the smallest change in magnetic field detectable with a measurement bandwidth of 1 Hz. The sensitivity of an NV magnetometer has certain quantum limits, and the most fundamental of those is due to the spin projection noise associated with the finite number of NV centres probed. This spin projection noise-limited sensitivity is expressed as [2]:

$$\delta B_{PN}\sqrt{\tau_m} \approx \frac{1}{\gamma\sqrt{N_{NV}T_2}} \qquad (12.7)$$

Here δB_{PN} represents the uncertainty in the magnetic field due to the spin projection noise, γ stands for the electron gyromagnetic ratio, N_{NV} is the number of NV centres in the sample, and T_2 is the coherence time.

The magnetic field sensitivity is also limited by the photon shot noise of the light used to readout the NV spin state. This photon-shot-noise-limited sensitivity is expressed as [2]:

$$\delta B_{SN}\sqrt{\tau_m} \approx \frac{\Delta\nu}{\gamma C\sqrt{R}} \qquad (12.8)$$

Here δB_{SN} represents the uncertainty in the magnetic field due to the photon shot noise, R is the rate of detected photons, $\Delta\nu$ is the full width half maximum (FWHM) of the magnetic resonance, and C stands for the fluorescence contrast. The FWHM of the magnetic resonance and the coherence time are related through the equation $T_2 = 1/\pi\Delta\nu$. On the other hand, the rate of detected photons R is related to the detected power P through the equation $R = P/(hc/\lambda)$, where h is Planck's constant, c the speed of light, and λ is the wavelength of the detected light. The overall magnetometer signal S increases linearly with the rate of detected photons, i.e., $S \propto R$, whereas the uncertainty ΔS due to the photon shot noise increases only as of the square-root, i.e., $\delta S \propto \sqrt{R}$. As a result the sensitivity of the magnetometer improves as $\Delta S/S \propto \sqrt{R}/R = 1/\sqrt{R}$.

Depending on the experiment and diamond sample, the parameters used to calculate photon shot noise and spin projection noise can vary [2]. In comparison to DC magnetometers, the coherence time is generally longer for AC magnetometers. The coherence times T_2 and T_2^* are also a function of the particular DD sequence and depend strongly on the diamond material used. The rate of detected photons depends on the number of NV centres in the probed sample volume, the input pump power, and the light detection efficiency, which is typically quite low (a few per cent) for a confocal setup [2]. On the other hand, the number of NV centres used depends on the NV density and the size of the sample volume probed. The NV density depends on the diamond material, the irradiation dose, and the annealing procedure. The fluorescence contrast C can be up to 20% for a single NV center.

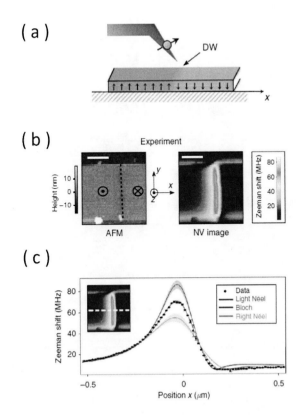

FIGURE 12.8 (a) Scanning the stray field above a DW in a 1.5-μm-wide magnetic wire made of a Ta (5 nm)/Co$_{40}$Fe$_{40}$B$_{20}$ (1 nm)/MgO (2 nm) trilayer stack, with the help of a single NV defect hosted in a diamond nanocrystal grafted on the tip of an atomic force microscope; (b) left-hand panel shows the AFM image and and the right-hand panel shows the corresponding Zeeman shift map recorded on a 1.5-μm wide stripe comprising a single DW; (c) a linecut of the experimental data across the DW together with the predicted curves for right-handed Néel DW (solid line at the bottom), left-handed Néel DW (solid line at the top) and Bloch DW (solid line in the middle). Black dots represent experimental data points. The Bloch DW hypothesis is the one that matches best with the experimental data. (From reference [5] with permission from Springer Nature.)

The contrast C for ensembles of NV centres, however, is usually much smaller due to background fluorescence.

Magnetometer sensitivity can be assessed experimentally in several ways. In a CW magnetometer, one can measure a constant applied field continuously, and calculate the noise floor after taking the Fourier transform. In a pulsed magnetometer measuring discrete values, the standard deviation (normalized to 1 Hz bandwidth) of the measured magnetic field values is calculated. The typical sensitivities with an NV ensemble are $\approx$ 15 pT/$\sqrt{Hz}$ for DC magnetometer and $\approx$ 1 pT/$\sqrt{Hz}$ for AC magnetometer.

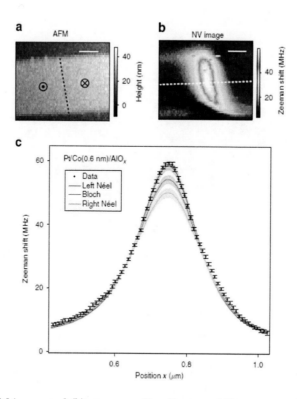

FIGURE 12.9 (a) AFM image and (b) corresponding Zeeman shift map recorded by scanning the NV magnetometer above a DW in a 500-nm-wide magnetic wire of Pt/Co (0.6 nm)/AlO$_x$. (c) Linecut extracted from (b) (markers), together with the theoretical prediction (solid lines) for a Bloch (solid line in the middle), a left-handed Néel (solid line at the top) and a right-handed Néel DW (solid line at the bottom). Black dots represent experimental data points. The hypothesis Néel-type DW structure with left-handed chirality is the one that matches best with the experimental data. (From reference [5] with permission from Springer Nature.)

12.5.5 Some experimental results

Fig. 12.8 shows the the structure of domain wall (DW) in a 1.5-μm-wide magnetic wire made of a Ta (5 nm)/Co$_{40}$Fe$_{40}$B$_{20}$ (1 nm)/MgO (2 nm) trilayer stack obtained with a scanning-NV magnetometer [5]. A single NV defect hosted in a diamond nanocrystal is grafted on the tip of an atomic force microscope (AFM), and a quantitative map of the stray field emanating from a nanostructured sample is provided by scanning across the sample (see Fig. 12.8(a)). The local magnetic field is obtained within a detection volume of atomic size by monitoring the Zeeman shift of the NV defect electron spin sub-levels through optical detection of the magnetic resonance. The left-hand panel of Fig. 12.8(b) shows the AFM image and the right-hand panel shows the corresponding Zeeman shift map recorded on a 1.5-μm wide stripe comprising a single DW. The stray magnetic field above the DW is computed for (i) right-handed Néel DW, (ii) left-handed Néel DW, and (iii) Bloch

DW after taking into account this DW spatial profile [5]. The computed magnetic field distributions were finally converted into Zeeman shift distribution, taking into account the quantization axis of the NV spin. Fig. 12.8(c) shows a linecut of the experimental data across the DW together with the predicted curves in the three above-mentioned cases. Excellent agreement with the experimental results is found if the DW is assumed to be a Bloch-wall.

Fig. 12.9 presents an image of the domain wall structure in a 500-nm-wide magnetic Pt (3 nm)/Co (0.6 nm)/AlO$_x$ (2 nm) wire, along with the linecuts across the DW compared with theoretical predictions [5]. The experimental results here clearly indicate a Néel-type DW structure with left-handed chirality.

12.6 Summary

In this chapter we have discussed nitrogen-vacancy (NV) magnetometry based on the electron spin associated with the nitrogen-vacancy (NV) defect in diamond. This is a new state-of-the-art technique, but even at the infant stage it displays a very impressive combination of capabilities. The principle of working of both DC and AC NV magnetometers have been discussed. The NV magnetometers are currently less sensitive than other technologies such as SQUID magnetometers, but their main strength lies in the high spatial resolution. With the use of single NVs, the typical sensitivities of magnetic field sensing range from tens of $\mu\text{TH}_z^{-1/2}$ for DC fields to tens of $\text{nTHz}^{-1/2}$ for AC fields.

Bibliography

[1] Casola, F., Van Der Sar, T., and Yacoby, A. (2018). Probing Condensed Matter Physics with Magnetometry based on Nitrogen-vacancy Centres in Diamond. *Nature Reviews Materials*, 3 (1): 1–13.

[2] Jensen, K., Kehayias, P., and Budker, D. (2017). Magnetometry with Nitrogen-Vacancy Centers in Diamond. In A. Grosz et al., eds., *High Sensitivity Magnetometers, Smart Sensors*, Cham: Springer International Publishing, pp. 553–576.

[3] Rondin, L., Tetienne, J. P., Hingant, T., et al. (2014). Magnetometry with Nitrogen-vacancy Defects in Diamond. *Reports on Progress in Physics*, 77 (5): 056503.

[4] Abe, E. and Sasaki, K. (2018). Tutorial: Magnetic Resonance with Nitrogen-vacancy Centers in Diamond – Microwave Engineering, Materials Science, and Magnetometry. *Journal of Applied Physics*, 123 (16): 161101.

[5] Tetienne, J. P., Hingant, T., Martinez, J., et al. (2014). The Nature of Domain Walls in Ultrathin Ferromagnets Revealed by Scanning Nanomagnetometry. *Nature Communications*, 6: 6733. DOI: 10.1038/ncomms7733.

Appendix A

Magnetic Fields and Their Generation

A magnetic field is called a steady field if the timescale of the increasing (or decreasing) magnetic field during an experiment is a few minutes or a few tens of minutes. On the other hand, if the magnetic field changes in a very short time, like milliseconds, it is called a pulsed-field [1]. Pulsed-fields with a duration of several hundred milliseconds up to several seconds are sometimes called long pulse fields [1].

A.1 Steady Field

Electromagnets are used if the magnetic field requirement is below 2 T. This is because of the saturation of iron cores used in conventional electromagnets. In 1933 the American physicist Francis Bitter introduced a design of an electromagnet to produce a higher magnetic field. This is known as the Bitter magnet where the current flows in a helical path through circular conducting metal plates stacked in a helical configuration with insulating spacers [2]. The schematic of a Bitter magnet is shown in Fig. A.1. The stacked metal plate design helps to withstand the Lorentz force due to the magnetic field acting on the moving electric charges in the plate. The Lorentz force causes an enormous outward mechanical pressure. The other important point in the design of Bitter magnets is about how to remove the huge amount of heat generated in this resistive magnet. This is achieved by flowing water along the axial direction through the holes incorporated in the stacked plates [1]. Bitter magnets are still used today to produce fields of up to 33 T at the National High Magnetic Field Laboratory (NHMFL) at Florida State University, Tallahassee, USA. Fig. A.2 shows a Florida Bitter disk from Tallahassee with highly elongated, staggered cooling holes. In more recent times a magnetic field of 37.5 T is produced

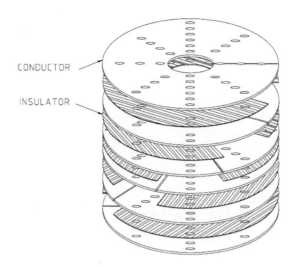

FIGURE A.1 Schematic of a Bitter magnet showing alternate conductors and insulators. (From reference [2] with permission from IOP Publishing.)

at room temperature by a Bitter electromagnet at the High Field Magnet Laboratory in Nijmegen, Netherlands.

Superconducting magnets are more popular nowadays in laboratories worldwide to produce fields up to 20 T. A magnetic field of 7 T using Nb_3Sn-based superconducting magnet was first generated in 1961 by Kunzler et al. [3], and the superconducting magnets that produced over 10 T became commercially available in the 1970s [1]. The quenching of the superconducting magnets, however, remained as a problem in the field region greater than 10 T, and it was only in the late 1980s that a 20 T superconducting magnet was finally built successfully.

Special techniques are needed for generating a magnetic field above 20 T. The maximum field limit before 2017 was around 23 T using a high T_C superconductor [4]. In late 2017 a 32 T superconducting magnet, consisting of a state-of-the-art 15 T low-temperature niobium tin and niobium titanium based superconductor outer magnet and a transformational 17 T high-temperature superconductor inner coil set using high T_C superconducting material YBCO coated conductor, has been successfully tested at the NHMFL of Florida State University [5]. The YBCO-coated conductor operates at the same temperature and in the same helium bath as their low temperature superconductor counterparts, but they remain in superconducting state far above the practical magnetic field limits inherent to niobium-based superconductors.

Superconducting magnets consume lots of liquid helium, and hence they are not quite user-friendly. In recent times cryogen-free superconducting magnets have been developed, but the available field for real experiments is still limited to around 15 T [1].

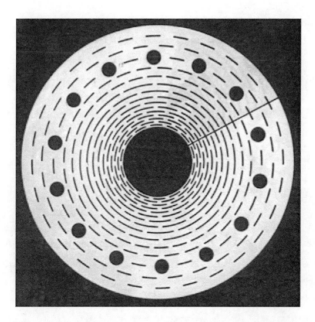

FIGURE A.2 Florida-Bitter disk from Tallahassee. (From reference [2] with permission from IOP Publishing.)

A hybrid magnet, which is a combination of a superconducting magnet and a Bitter magnet, is used to obtain a higher steady field than that obtained by a Bitter magnet. Such a 45 T hybrid magnet is now regularly used for experiments at NHMFL, Florida. This magnet combines a superconducting magnet of 11.5 T with a resistive Bitter magnet of 33.5 T. The Bitter magnet here requires 30 MW of power and the large bore superconducting magnet installed outside the Bitter magnet requires a huge amount of liquid helium for keeping it cold at 1.8 K.

A.2 Pulsed-Field

In a pulsed-field magnet the power consumption is much smaller than that of a steady field magnet, hence the technical problems in the generation of static and pulsed magnetic fields are of a different nature [1]. Kapitza at the Cavendish Laboratory was the first to build large pulsed magnets of field strength 35 T–50 T [6, 7], and used these magnets for many pioneering experimental studies on magnetoresistance, magnetization, magnetostriction, Zeeman effect, and the study of electron orbits in various solids. The pulsed-field is quite economical as compared with steady fields, and it is relatively easy to generate a pulsed-field of order 50 T for about 10 ms in a 20 mm bore [1]. The highest non-destructive pulsed-field obtained so far is 100.75 T and the highest field for a long-pulse magnet is 60 T [8].

In the methods to obtain magnetic fields higher than 100 T, destruction of the magnet is unavoidable. A metal will explode violently when it is exposed to such a high magnetic field pulse due to the combined effect of Joule heating and Maxwell stress [9]. This explosion starts due to the skin effect at the surface of the inner wall of the magnet coil and is accompanied by shock waves compressing the magnet coil material by pushing the surface away from the magnetic field. In the "single turn coil method", a large current is supplied to a small coil for a few microseconds and the destruction of the coil occurs after reaching the peak field due to the inertia of the coil material. However, since the coil expands during the experiment, the sample kept inside the coil survives. A magnetic field up to 300 T can be obtained using this method [1].

An extreme method of magnetic flux compression is employed to generate fields higher than 300 T. In this method a metallic cylinder called a "liner" with magnetic flux inside is imploded uniformly from outside, while maintaining the cylindrical symmetry. As a result, the magnetic flux gets compressed according to the ratio of the initial area and the final area of the cross section inside the cylinder. Using the method of electromagnetic flux compression a field up to 600 T can be generated. Even higher fields up to 1000 T can be achieved by compressing the magnetic flux by a high explosive [1]. Such experiments involve currents of the order of several mega amperes and the pulse duration is about 10–50 μs. However, both the magnet coil and the sample are destroyed during the experiments.

Bibliography

[1] Motokawa, M. (2004). Physics in High Magnetic Fields. *Reports on Progress in Physics*, 67 (11): 1995.

[2] Bird, M. D. (2004). Resistive Magnet Technology for Hybrid Inserts. *Superconductor Science and Technology*, 17 (8): R19.

[3] Kunzler, J. E., Buehler, E., Hsu, F. S. L. and Wernick, J. H. (1961). Superconductivity in Nb$_3$Sn at High Current Density in a Magnetic Field of 88 kgauss. *Physical Review Letters*, 6 (3): 89.

[4] Kiyoshi, T., Kosuge, M., Yuyama, M., et al. (2000). Generation of 23.4 T Using Two Bi-2212 Insert Coils. *IEEE Transactions on Applied Superconductivity*, 10 (1): 472–77.

[5] https://nationalmaglab.org/magnet-development/magnet-sciencetechnology/magnet-projects/32-tesla-scm.

[6] Kapitza, P. L. (1924). A Method of Producing Strong Magnetic Fields. *Proceedings of the Royal Society of London. Series A, Containing Papers of a Mathematical and Physical Character*, 105 (734): 691–710.

[7] Kapitza, P. (1927). Further Developments of the Method of Obtaining Strong Magnetic Fields. *Proceedings of the Royal Society of London. Series A, Containing Papers of a Mathematical and Physical Character*, 115 (772): 658–683.

[8] http://www.magnet.fsu.edu/mediacenter/news/pressreleases/2012/100tshot.html.

[9] Miura, N., and Herlach, F., eds. (2003). *Magnetic Fields: Science and Technology, 3-volumes*. Singapore: World Scientific.

Appendix B

Units in Magnetism

In 1960 the Systeme International d'Unites (SI) was recommended as the modern version of the metric system, which replaced the old CGS – centimetre, gram, and second – system with seven fundamental or "base" units: the metre, kilogram, second, ampere, kelvin, mole, and candela. Other "derived" units are constructed from these base units. Since then in almost every area of science and engineering, the SI units have been widely used unambiguously.

In the field of magnetism, however, a kind of confusing mixture of SI (in various versions) and CGS units are still being used sometimes. For example, "magnetic field" can mean "B-field" or "H-field". The SI units for these fields are tesla (T) or amperes per metre (Am^{-1}), whereas in CGS those are gauss (G) and oersted (Oe), all of which are currently in use. Another source of confusion arises due to the different expressions proposed for the magnetic induction $\vec{B}$ in a polarizable medium by Arthur Kennelly (1936) and Arnold Sommerfeld (1948) [1]. The Kennelly system is traditionally followed by electrical engineers, where $\vec{B}$ is expressed as:

$$\vec{B} = \mu_0 \vec{H} + \vec{J} = \vec{B}_0 + \vec{J} \tag{B.1}$$

Here μ_0 is the permeability of free space, $\vec{H}$ is the H field, $\vec{J}$ is the magnetic polarization, and $\vec{B}_0$ is the induction of free space that would remain in the absense of the medium. In the Sommerfeld convention, which has been adopted by the International Union of Pure and Applied Physics (IUPAP):

$$\vec{B} = \mu_0(\vec{H} + \vec{M}) = \vec{B}_0 + \mu_0 \vec{M} \tag{B.2}$$

Here $\vec{M}$ is the magnetization per unit volume. These equations are not in conflict once it is recognized that magnetization and magnetic polarization are different quantities. It is possible to use either the B-field or the H-field when one is dealing with the magnetic field. They can be distinguished by their behaviour at a boundary between media having different relative permeabilities (μ). Across the boundary of the media the normal component of the B-field and the tangential component of the H-field will be continuous. In both the Kennelly and Sommerfeld systems $\vec{B} = \mu_0 \vec{H}$ in the free space. In SI units $\mu_0 = 4\pi \times 10^{-7}$ henreys per metre (Hm^{-1}), hence B-field and H-field have different numerical values. On the other hand, in the CGS system $\mu_0 = 1$, hence B-field or the H-fields have identical numerical values. This causes some confusion in the nomenclature of "gauss" for the B-field and oersted for the H-field. It may be noted here that the relative permeability $\mu = \vec{B}/\vec{B}_0$ is dimensionless and has the same value in both the SI and the CGS systems.

The question here arises: what is then the difference then between the B-field and H-field? Traditionally B-field and H-field are used in the formulae dealing with force, energy, magnetic moment, etc. [1]. For example, the Lorentz force on a particle with charge e moving with velocity $\vec{v}$ in a magnetic field $\vec{B}$ is expressed as $\vec{F} = e\vec{v} \times \vec{B}$. The main reason for designating the field as B-field is possibly that the comparison of the results in SI and CGS systems is precise and simple, and that is why solid-state physicists working in fundamental magnetism tend to use B-field. For example, a 10 T superconducting magnet in SI units and 100 kOe or 100 kG in CGS units is not described as being a 7.958 MAm^{-1} magnet [1]. One needs to be cautious, however, to distinguish between the field (i.e., free-space induction) and the induction $\vec{B}$ within a polarized medium while working with B-field. On the other hand, H-field in kAm^{-1} is often used by electrical engineers and physicists working in applied magnetism. The advantage if any of using H-field, however, is not very obvious.

In summary, in SI, it is better to use the B-field. The free-space field is expressed with a special symbol such as $\vec{B}_0$ to distinguish it from the induction $\vec{B}$, and both $\vec{B}_0$ and $\vec{B}$ should be measured in tesla. With the decision to use the B-field and tesla, the definitions for various magnetic properties of interest – magnetization, susceptibility, hysteresis loops, etc., follow naturally. The main magnetic quantities are summarized in table B.1, and a brief explanation of those is provided below.

Magnetization

The Bohr magneton μ_B is the fundamental unit of the magnetic moment. This is the magnetic moment of one electron spin: $\mu_B = 9.2732 \times 10^{-24}$ JT^{-1}. The energy of a one Bohr magneton in a magnetic field of $\vec{B}_0$ is $\mu_B \vec{B}_0$ and is expressed in joules. In the CGS system $\mu_B = 9.2732 \times 10^{-21}$ erg Oe^{-1}.

In the case of a temperature variation study of magnetism in a sample, the mass of the sample is almost always better known than its volume. This is the reason for the more frequent use of the magnetization per unit mass $\vec{\sigma}$, which has units of JT^{-1}kg^{-1} in the SI system. The CGS unit for $\vec{\sigma}$ is ergOe^{-1}g^{-1}. It is numerically

TABLE B.1 Principal units used in magnetism.

Quantity	SI (Sommerfeld)	SI (Kennelly)	EMU (Gaussian)
Field ($\vec{H}$)	A m^{-1}	A m^{-1}	oersteds
Induction ($\vec{B}$)	tesla	tesla	gauss
Magnetization ($\vec{M}$)	A m^{-1}	–	emu/cm^3
Intensity of Magnetization (I)	-	tesla	-
Flux (Φ)	weber	weber	maxwell
Moment (m)	A m^2	weber metre	emu
Magnetic pole strength(p)	A m	weber	emu/cc
Field equation	$B = \mu_0(H + M)$	$B = \mu_0 H + J$	$B = H + 4\pi M$
Energy of moment (in free space)	$E = -\mu_0 m.H$	$E = -m.H$	$E = -m.H$
Torque of moment (in free space)	$\tau = \mu_0 m \times H$	$\tau = m \times H$	$\tau = m \times H$

equal to the SI unit, and often referred to as emu/gm. The unit for magnetization per unit volume $\vec{M}$ in SI is JT^{-1}m^{-3}. The CGS unit for $\vec{M}$ is ergOe^{-1}cm^{-3} (also called sometimes' emu/cm^3); it is equal to 10^{-4} of the SI unit.

Magnetic susceptibility

The susceptibility of a magnetic material is the ratio of the magnetization to the magnetic field which induces it. The mass susceptibility χ is defined as $\chi = \vec{\sigma}/\vec{B}_0$ and is measured in JT^{-2}kg^{-1}. The volume susceptibility κ is defined as $\kappa = \vec{M}/\vec{B}_0$ and has SI units of JT^{-2}m^{-3}. In the CGS system the corresponding units for susceptibility are ergOe^{-1}g^{-1} and ergOe^{-1}cm^{-3}, respectively. Sometimes emu/gm and emu/cm^3 are also used for these units.

It may be noted that the SI unit for the volume susceptibility κ is the inverse of that for μ_0, i.e., metres per henry (mH^{-1}). This is because $\mu = \vec{B}/\vec{B}_0 = 1 + \mu_0 \vec{M}/\vec{B}_0 = 1 + \mu_0 \kappa$ and μ is dimensionless.

Magnetic polarization

This is the part of the magnetic induction within a magnetized sample that is not the applied magnetic field. This is $\vec{J}$ in the Kennelly representation: $\vec{B} = \vec{B}_0 + \vec{J}$ [1]. $\vec{J}$ has the unit tesla in the SI system, the same units as $\vec{B}$ and $\vec{B}_0$. In CGS system the Kennelly equation is written as $\vec{B} = \vec{H} + 4\pi \vec{I}$, where $\vec{I}$ is the intensity of magnetization, a quantity which is not defined in the SI system [1].

Magnetization hysteresis

Magnetization hysteresis or hysteresis loop is an important characteristic of ferromagnetic materials. In relatively weak applied magnetic fields, the magnetization, and hence the induction and polarization, depend on the field history

TABLE B.2 Conversion table between CGS and SI units in magnetism.

Quantity	Gaussian	SI Conversion factor	SI
Magnetic Field ($\vec{H}$)	oersted	$10^3/4\pi$	A m^{-1}
Magnetic Induction ($\vec{B}$)	gauss	10^{-4}	tesla
Volume magnetization ($\vec{M}$)	emu/cm^3	10^3	A m^{-1}
Mass magnetization ($\vec{\sigma}$)	emu/gm	1	A m^{2}kg^{-1}
Flux (Φ)	maxwell	10^{-8}	weber
Moment (m)	emu	10^{-3}	A m^2
Demagnetization factor	dimensionless	$1/4\pi$	dimensionless

of the sample under study. The properties displayed by the hysteresis loop are often very important from the technical point of view. It is possible to plot hysteresis loops in several ways, all of which are equally valid [1].

1. The induction $\vec{B}$ (in tesla) is plotted on the y-axis against the field $\vec{B}_0$ (in tesla) on the x-axis.
2. The mass magnetization $\vec{\sigma}$ (measured in JT^{-1}kg^{-1}) or the volume magnetization $\vec{M}$ (in JT^{-1}m^{-3}) is drawn on the y-axis against the field $\vec{B}_0$ (in tesla) on the x-axis. In another way $\vec{M}$ is plotted against $\vec{B}_0/\mu_0$ with both axes given in Am^{-1}.
3. The magnetic polarization J (in tesla) is drawn on the y-axis against the field $\vec{B}_0$ (in tesla) on the x-axis.

The magnetic remanence and coercive field are often estimated from the measured hysteresis loops and their values will depend on how the y-axis is defined.

An important characteristic for permanent magnets is the maximum energy product or storage density. This is determined from the $\vec{B}$ vs $\vec{B}_0$ induction hysteresis loop. It is the maximum value of the product $(1/\mu_0)|\vec{B}_0||\vec{B}|$ and has SI units of kJm^{-3}.

Table B.2 summarizes the conversion factor between CGS and SI units of major magnetic parameters.

Bibliography

[1] Crnagle, J., and Gibbs, M. (1994). Units and Unity in Magnetism: A Call for Consistency. *Physics World*, November: 31.

Appendix C

Demagnetization Field and Demagnetization Factor

C.1 Phenomenology

Maxwell equations in free space are expressed as:

$$\nabla . B = 0 \qquad (C.1)$$

$$\nabla \times E + \frac{\partial B}{\partial t} = 0 \qquad (C.2)$$

$$\epsilon_0 \nabla . E = \rho \qquad (C.3)$$

$$\frac{1}{\mu_0} \nabla \times B - \epsilon_0 \frac{\partial E}{\partial t} = j \qquad (C.4)$$

The electric field E(r,t) and magnetic field B(r,t) are averaged over elementary volume ΔV centred around the position r. Similarly ρ and j represent electric charge density and current density, respectively. Equation C.1 indicates the absence of magnetic charge and Eqn. C.2 represents Faraday's law of indication in differential form. These two equations do not depend on the sources of an electric field or magnetic field, and they represent the intrinsic properties of the electromagnetic field. Eqns. C.3 and C.4 contain ρ and j, and they describe the coupling between the electromagnetic field and its sources.

Let us now consider a sample of ferromagnetic material through which no macroscopic conduction currents are flowing. A ferromagnet is characterized by the presence of spontaneous magnetization that can produce a magnetic field outside the sample. The microscopic current density j_{micro} producing such a magnetic field can be associated with the electronic motion inside the atoms and electron spins, or elementary magnetic moments of the ferromagnetic materials. Such microscopic currents present in an elementary volume ΔV centred about a position r gives rise to an average current [1]:

$$j_M(r) = \frac{1}{\Delta V} \int_{\Delta V} j_{micro}(r') d^3 r' \qquad (C.5)$$

j_M is termed as magnetization current and represents the current density in Maxwell Eqn. C.4 for a ferromagnetic material. This magnetization current j_M does not represent any macroscopic flow of charges across the sample. It can rather be crudely associated with current loops confined to atomic distances. This, in turn, implies that the surface integral j_M over any generic cross section S of this ferromagnetic sample must be zero:

$$\int_S j_M.ds = 0 \qquad (C.6)$$

This, in turn, tells that $j_M(r)$ can be expressed as the curl of another vector $M(r)$:

$$j_M(r) = \nabla \times M(r) \qquad (C.7)$$

Now inserting Eqn. C.6 into Eqn. C.7 and with the help of Stoke's theorem, one can convert Eqn. C.6 into a line itegral along some contour completely outside the ferromagnetic sample:

$$\int_S \nabla \times M(r).ds = \oint M.dl = 0 \qquad (C.8)$$

The Eqn. C.8 will be satisfied under all circumstances provided $M(r) = 0$ outside the sample. This latter condition is true if we take M as the magnetization or magnetic moment density of the ferromagnetic sample. It can be seen from Eqn. C.7 that the magnetic field created by the ferromagnetic sample is identical to the field that would be created by a current distribution $j_M(r) = \nabla \times M(r)$. If we now consider also the presence of conduction current $j(r)$ then the total current to be considered is the sum of $j(r) + \nabla \times M(r)$. It can also be shown that similar considerations lead to the introduction of electrical polarization P as a way to describe atomic charge distributions $\rho(r) = -\nabla.P(r)$ in dielectric materials. With the inclusion of the effects of magnetization and electrical polarization in Maxwell Eqns. C.3 and

C.4 it is possible to derive the following expressions [1]:

$$\nabla \cdot D = \rho \tag{C.9}$$

$$\nabla \times H - \frac{\nabla D}{\partial t} = j \tag{C.10}$$

Here ρ represents free charge only and j conduction currents only, and the derived fields D and H are expressed as:

$$D = \epsilon_0 E + P \tag{C.11}$$

$$H = \frac{1}{\mu_0} B - M \tag{C.12}$$

If one neglects any time dependence of the electric field E, then the Maxwell Eqn. C.4 in a magnetized medium like ferromagnetic sample takes the form:

$$\nabla \times H = j \tag{C.13}$$

Here the field H has taken the place of B/μ_0 in Eqn. C.4.

We shall now consider the solutions of Eqns. C.1 and C.13 in a region where the conduction current j is zero.

$$\nabla \cdot B_M = 0 \tag{C.14}$$

$$\nabla \times H_M = 0 \tag{C.15}$$

The subscript M in the above equations represent the consequence of the spontaneous magnetization M present in the ferromagnetic sample. The set of Eqns. C.14 and C.15 can be rewritten in two equivalent forms. In one such form using Eqn. C.12, one can write:

$$\nabla \cdot B_M = 0 \tag{C.16}$$

$$\nabla \times H_M = j_M \tag{C.17}$$

Here as defined earlier $j_M = \nabla \times M$. In this case, the vector potential A becomes the natural potential and the solution is obtained as [1]:

$$A_M(r) = \frac{\mu_0}{4\pi} \int \frac{\nabla \times M(r')}{|r - r'|} d^3 r' \tag{C.18}$$

In the other form Eqns. C.14 and C.15 can be rewritten as:

$$\nabla \cdot H_M = \rho_M \tag{C.19}$$

$$\nabla \times H_M = 0 \tag{C.20}$$

Eqns. C.19 and C.20 are equivalent to the set of electrostatic equations for electric field. Here the quantity $\rho_M = -\nabla \cdot M$ plays the role of magnetic charge density. It is also possible to introduce a magnetic scalar potential ϕ_M expressed as:

$$H_M = -\nabla \phi_M \tag{C.21}$$

This satisfies the relation $\nabla \times H_M = 0$.

From Eqns. C.19–20 it can be shown that ϕ_M obeys Poisson's equation:

$$\nabla^2 \phi_m = -\rho_M \tag{C.22}$$

This, in turn, gives rise to the solution:

$$\phi_M(r) = -\frac{1}{4\pi} \int \frac{\nabla.M(r')}{|r-r'|} d^3r' \tag{C.23}$$

These two approaches are quite parallel. In one, magnetostatics is based on Amperian current (j_M), where B is the fundamental field and A_M is fundamental potential. The other approach is based on magnetic charges ρ_M, where H and magnetic scalar potential ϕ_M play important roles. It may be noted here that magnetic charge is only a useful concept and it does not have any physical significance. Despite criticism from the pedagogical point of view, the concept of magnetic charge is widely used to calculate the demagnetization field. This is clear from the fact that $\nabla.B_M = 0$ (Eqn. C.16) represents a fundamental property of the magnetic field. On the other hand, the parallel equation $\nabla \times H_M = 0$ (Eqn. C.20) has only a limited validity in the regions of space free from conduction currents.

In a ferromagnetic sample of finite dimension, the magnetization M undergoes an abrupt change at the body surface, since $M = 0$ outside the sample. This leads to a quasi-singular behaviour at the surface of the sample, which introduces an additional contribution to the integrals in Eqns. C.18 and C.23. If one now considers the ideal situation where the magnetization M of a ferromagnetic sample goes to zero discontinuously at the sample surface, Eqns. C.18 and C.23 can be expressed as:

$$A_M(r) = \frac{\mu_0}{4\pi} \int_V \frac{\nabla \times M(r')}{|r-r'|} d^3r' - \frac{\mu_0}{4\pi} \oint_S \frac{n \times M(r')}{|r-r'|} ds' \tag{C.24}$$

$$\phi_M(r) = -\frac{1}{4\pi} \int_V \frac{\nabla.M(r')}{|r-r'|} d^3r' + \frac{1}{4\pi} \oint_S \frac{n.M(r')}{|r-r'|} ds' \tag{C.25}$$

In these equations, the first integral is over the sample volume V, and the second one is a surface integral over the boundary surface S of the sample. The term n defines the unit vector normal to the surface element ds' and points to the direction away from the sample. The quantities k_M and σ_M defined describe the effect of discontinuous change in magnetization at the ferromagnetic sample surface [1]:

$$k_M = -n \times M \tag{C.26}$$

$$\sigma_M = n.M \tag{C.27}$$

Here k_M plays the role of surface magnetization current and $(\sigma)_M$ plays the role of surface magnetization charge density (Fig. C.1), which in general form not only at

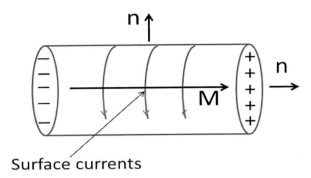

FIGURE C.1 A cylinder with uniform magnetization M, showing the formation of surface poles and surface currents.

the sample surface but wherever there is a discontinuous change in magnetization in the sample. Across the surface where magnetization M undergoes a discontinuous change from M_1 to M_2, ϕ_M and the component of H_M tangential to the surface varies continuously across the surface, but the component of H_M perpendicular to the surface changes discontinuously by an amount equal to $\sigma_M = n.(M_2 - M_1)$.

Let us now consider a ferromagnetic sample where the magnetization M is uniform inside the whole volume of the sample. In this situation $\nabla.M$ is zero everywhere within the sample volume, and only the surface integral in Eqn. C.25 will contribute to the magnetic scalar potential ϕ_M:

$$\phi_M(r) = \frac{1}{4\pi} \oint_S \frac{n.M(r')}{|r-r'|} ds' \qquad (C.28)$$

It is clear from the above equation that apart from magnetization M the magnetic scalar potential also depends on the shape or geometry of the sample. The field H_M calculated with the help of Eqns. C.21 and C.28 in the case of uniform magnetization M will also be uniform within the volume of the sample if the sample is of ellipsoidal shape. If the direction of magnetization M coincides with one of the principal axes of the ellipsoid, then the field H_M while having an intensity proportional to M will point to the direction opposite to M (Fig. C.2). This demagnetization field is expressed as:

$$H_M = -NM \qquad (C.29)$$

Here the field H_M inside the ferromagnetic sample is termed as demagnetization field since it opposes the magnetization. The coefficient N is termed as the demagnetization factor, and it depends on the ellipsoid axis the magnetization is aligned with. Thus, there are three demagnetization factors N_a, N_b, and N_c associated with the three principal axes a, b, and c of the ellipsoid. These demagnetization factors are constrained to follow the general relation [1]:

$$N_a + N_b + N_c = 1 \tag{C.30}$$

In the case of a sphere it can be seen immediately from the symmetry condition that $N_a + N_b + N_c = 1/3$. The results of more general interest are for the ellipsoids of revolution, where two of the principal axes are of equal lengths and the third is an axis for the rotational symmetry of the body. There are two relevant demagnetization factors, one in the direction of the symmetry axis and termed as $N_\parallel$ and the other $N_\perp$ perpendicular to the symmetry axis. They are subjected to the constraint relation $N_\parallel + N_\perp = 1$. The ratio r of the lengths of the symmetry and transverse axes then characterize the demagnetization properties of the ellipsoid shape ferromagnetic sample. When $r > 1$, one has a cigar-shaped or prolate spheroid sample, and $N_\parallel$ can be expressed as [1]:

$$N_\parallel = \frac{1}{r^2 - 1}\left[\frac{r}{\sqrt{r^2 - 1}} ln\left(r + \sqrt{r^2 - 1}\right) - 1\right] \tag{C.31}$$

For $r \gg 1$, $N_\parallel$ can be approximated as:

$$N_\parallel \approx \frac{ln2r - 1}{r^2} \tag{C.32}$$

When $r < 1$, one has a disc-shaped or oblate spheroid sample, and $N_\parallel$ can be expressed as [1]:

$$N_\parallel = \frac{1}{1 - r^2}\left[1 - \frac{r}{1 - \sqrt{r^2}} arcsin\left(\sqrt{1 - r^2}\right)\right] \tag{C.33}$$

The above results indicate that the demagnetization effect will be small in an elongated ferromagnetic sample with a high aspect ratio when the magnetization points along the long axis. It may be noted that the demagnetization factors expressed in Eqns. C.32 and C.33 depends on the axis ratio r but not on the actual lengths of the axes. The demagnetization field in the ferromagnetic sample will not change by changing the size of the sample by modifying all its linear dimensions by the same scale factor. Thus, the demagnetization field of a ferromagnetic sample depends only on the shape or geometry of the sample and not on its absolute dimensions.

It may appear by looking at Figs. C.1 and C.2 that the sources of the demagnetization field lie at the magnetic charges on the sample surface. Accordingly, it may be possible to nullify the effect of these magnetic charges by taking samples of progressively larger sizes to send the charges to infinity. This line of reasoning, however, is erroneous. As we have discussed above, the effect of magnetic charges does not vanish with the increase in body size. Even in the limit of infinite size, the

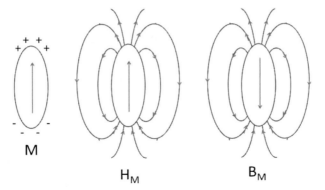

FIGURE C.2 Magnetization M, demagnetization field H_M, and magnetic induction B_M for an ellipsoid with uniform magnetization.

demagnetization effect will still depend on the body shape, i.e., on how the magnetic charges are sent to infinity.

In the general case when the direction of magnetization M in a ferromagnetic sample does not coincide with a principal axis of the sample, it is possible to utilize the linearity of the problem to decompose M along the principal axes, and then superpose the three resulting demagnetization fields. The three demagnetization factors N_a, N_b, N_c will then be the three eigenvalues of a demagnetization tensor, which is diagonal when the cartesian x-, y-, and z-axes coincide with the principal axes of the ellipsoid. In this case the relation between the magnetic field and magnetization is expressed in a matrix form [1]:

$$\begin{pmatrix} H_{Mx} \\ H_{My} \\ H_{Mz} \end{pmatrix} = - \begin{pmatrix} N_a & 0 & 0 \\ 0 & N_b & 0 \\ 0 & 0 & N_c \end{pmatrix} \begin{pmatrix} M_x \\ M_y \\ M_z \end{pmatrix} \qquad (C.34)$$

C.2 Experimental aspects

The ferromagnetic samples used in experiments are not necessarily ellipsoids, and for an arbitrarily shaped sample, H_M is generally nonuniform. If, for example, a uniform field is applied axially to a cylindrical shaped sample with length-to-diameter ratio, i.e., aspect ratio γ and uniform susceptibility χ, either or both the magnetic field and the magnetization inside the cylinder are nonuniform [2]. The demagnetization factor for such a cylinder is defined as the negative of the ratio of the average demagnetizing field to average magnetization. This is in contrast to the demagnetization factor of an ellipsoid, which is defined as the negative of the ratio of the demagnetizing field to the magnetization. In literature, the demagnetizing factor is distinguished between "magnetometric" and "fluxmetric or ballistic" demagnetizing factors N_m, and N_f, respectively, according to their application in two types of magnetic measurements [3]. N_m refers to an average of magnetization over the entire sample volume and is appropriate for small samples measured with magnetometers, whereas N_f refers to an average of magnetization at the mid-plane of the sample and is appropriate for fluxmetric measurements with short detector coils.

A long cylindrical sample with high aspect ratio, i.e., $\gamma \gg 1$ is appropriate for reducing experimental error due to demagnetization effects. In a saturated ferromagnet $\chi \approx 0$ and the following approximate formulae can be used for the demagnetization factors N_m and N_f [2]:

$$N_f(\chi=0,\gamma) = \frac{1}{2\gamma^2} \tag{C.35}$$

$$N_m(\chi=0,\gamma) = \frac{4}{3\pi\gamma} - \frac{1}{8\gamma^2} \tag{C.36}$$

The accuracy of both Eqns. C.35 and C.36 is better than 1.5% if $\gamma \geq 20$ for N_f and $\gamma \geq 1.4$ for N_m.

In the case of a cylindrical sample of soft ferromagnet at low fields, the susceptibility χ tends to infinity, and in such a case the approximate formulae for $N_{f,m}(\chi \to \infty, \gamma)$ for $10 \leq \gamma \leq 100$ can be expressed as [2, 3]:

$$N(\gamma) = \frac{1}{\gamma^2 - 1}\left[\gamma(\gamma^2-1)^{-1/2}\cosh^{-1}\gamma - 1\right]$$

$$\approx \frac{1}{\gamma^2}(ln2\gamma - 1) \tag{C.37}$$

Now the experimentally measured susceptibility χ_{exp} of magnetic sample in an externally applied field H is given by $\chi_{exp} = M/H$, whereas the actual susceptibility of the sample is $\chi_{int} = M/H_{int}$. Here H_{int} is the actual field seen by the sample, which is expressed as:

$$H_{int} = H - NM \tag{C.38}$$

Thus, due to the demagnetization effect the intrinsic susceptibility of the sample is not actually what is measured experimentally. Let us now see how the two quantities χ_{exp} and χ_{int} can be related, so that the intrisic property of the sample can be determined from the measured susceptibility. Eqn. C.38 can be rewritten as:

$$\frac{M}{\chi_{int}} = \frac{M}{\chi_{exp}} - NM \tag{C.39}$$

or

$$\chi_{exp} = \frac{1}{\frac{1}{\chi_{int}} + N} \tag{C.40}$$

Magnetic suceptibility of a ferromagnetic sample follows the Curie–Weiss law:

$$\chi_{int} = \frac{C}{T - T_C} \quad (C.41)$$

Here C is the Curie constant and T_C is the Curie temperature. In the temperature region $T \gg T_C$ χ_{int} is small, and hence $\frac{1}{\chi_{int}} \gg N$. In this temperature region we can see from Eqn. C.40 that $\chi_{exp} \approx \chi_{int}$. The susceptibility χ_{int} diverges at $T = T_C$, and the term $\frac{1}{\chi_{int}}$ in Eqn. C.40 becomes zero. Hence, for temperatures $T \leq T_C$ the experimentally measured susceptibility acquires a constant value:

$$\chi_{exp} = \frac{1}{N} \quad (C.42)$$

We say that in the ferromagnetic region the experimentally measured susceptibility instead of diverging becomes demagnetization limited. With the increase in the aspect ratio of a ferromagnetic sample, say, in a cylindrical shape, or rectangular bar shape there is a decrease in the demagnetization factor N, and one can observe progressive increase in the value of the measured susceptibility at $T = T_C$. Below T_C the temperature dependence of the susceptibility is, however, quite complicated and influenced by other intrinsic properties of a ferromagnet, namely magnetic anisotropies.

Bibliography

[1] Bertotti, G. (1998). *Hysteresis in Magnetism: For Physicists, Materials Scientists, and Engineers*. Houston: Gulf Professional Publishing.

[2] Chen, D. X. (2001). Demagnetizing Factors of Long Cylinders with Infinite Susceptibility. *Journal of Applied Physics*, 89 (6): 3413–3415.

[3] Chen, D. X., Brug, J. A., and Goldfarb, R. B. (1991). Demagnetizing Factors for Cylinders. *IEEE Transactions on Magnetics*, 27 (4): 3601–3619.

Index

AC susceptibility, 117, 118, 120, 121
alternating gradient magnetometer, 96, 97, 99
Alnico magnets, 10
Ampere law, 6
Ampere, Andre-Marie, 6, 7
angular momentum, 20, 26, 27, 29, 30, 32, 37, 42–44, 84, 126, 127, 253
antiferromagnetism, 9, 55, 56, 58, 68, 69, 71, 72

Bernoulli, Daniel, 6
Bloch equations, 130, 140
Bloch wall, 81, 82, 317
Bohr magneton, 9, 26, 29, 137, 138, 142, 326
Bohr–van Leeuwen theorem, 29, 34
Boltzmann distribution, 38, 40, 138
Brillouin function, 40, 65, 67, 69
Brillouin light scattering, 90, 182–184, 190

cantilever beam magnetometer, 99, 100
chemical shift, 129, 132, 164
coercivity, 10, 86, 134
compass, 4, 5
Coulomb, Charles-Augustin de, 6
Coulomb interaction, 47, 57, 58, 146, 165, 239
crystal field, 44–48, 58, 59, 61, 79, 126, 143, 144, 166, 223, 234
Curie constant 337
Curie temperature, 8, 65, 67, 79, 85, 117, 157, 337

Curie–Weiss law, 67, 69–71, 337
Curie, Pierre, 7, 8

demagnetization factor, 86, 328, 329, 331, 333–336
demagnetization field, 132, 133, 146–149, 157, 329, 332–335
diamagnetism, 7, 34, 37, 41, 43, 118
direct exchange, 55, 57, 73
domain walls, 81, 85, 264, 270, 272, 279, 283, 287, 298, 305
domains, 63, 79–81, 85, 86, 146, 147, 170, 178, 181, 189, 190, 231, 232, 250, 262, 264, 266, 270, 272–274, 281–283, 287, 294, 296
double exchange, 59, 60
Dzyaloshinski–Moriya interaction, 61, 83

electromagnet, 10, 94, 95, 98, 104, 112, 116, 126, 139, 149, 319, 320
electron holography, 273, 277, 279,
electron paramagnetic resonance, 90, 136
electron spin resonance, 136, 305, 306
exchange interaction, 9, 54, 55, 57–59, 61, 62, 64–68, 73–75, 80, 145, 161, 166, 248, 267, 284,
extraction magnetometer, 104, 116

Faraday balance, 94, 95–97
Faraday effect, 94, 169, 171
Faraday rotation, 169
Faraday, Michael, 6, 7, 169

ferrimagnetic, 145, 170, 208, 209, 238, 241
ferrimagnetism, 9, 71
ferromagnetic resonance, 90, 145
ferromagnetism, 7, 8, 58, 59, 61, 63, 65, 68, 72, 114
fine structure, 42–44, 255
force method, 94, 101, 122

g-factor, 33, 127, 137, 138, 142
gauss, 20, 21, 325–327
Gauss, Carl Friedrich, 6, 7
Gilbert, William, 5, 6
Goodenough–Kanamori-Anderson rules, 58
Goudsmit, Samuel, 8, 9
Gouy balance, 95
gyromagnetic ratio, 29–31, 76, 125, 150–152, 304, 305, 314

Hall effect, 94, 288
Heisenberg exchange, 64
Heisenberg Hamiltonian, 53
Heisenberg, Werner, 9
helical magnetic order, 71, 72
Hubbard model, 57
Hund's rules, 42, 43
hyperfine field, 126, 132–136, 143, 157, 166
hyperfine interaction, 126, 129, 133, 134, 143, 164, 166

Jahn–Teller effect, 48

Kerr effect, 90, 169, 171, 172, 173, 189
Kerr rotation, 174, 175, 177

Langevin function, 39
Langevin, Paul, 8, 38
Larmor precession, 20, 31, 152, 312
Lodestone, 3, 4
Lorentz force, 27, 260, 262, 271, 319, 326
Lorentz microscopy, 298
low angle diffraction, 275
LTEM, 270–274

magnetic anisotropy, 83–85, 101, 134, 177, 254, 255
magnetic circular dichroism, 91, 250, 255, 256, 290
magnetic dipole, 49, 90, 111, 125, 126, 142, 143, 146, 191, 198
magnetic excitations, 150, 223, 225, 227, 234, 305

magnetic flux, 31, 93, 102, 103, 109, 110, 114–116, 119, 219, 222, 262, 278, 280, 283, 289, 322
magnetic imaging, 91, 181, 259, 262, 263, 265, 281, 284–286, 289, 291, 294–297, 310
magnetic induction, 31, 75, 76, 93, 102, 104, 117, 230, 231, 270, 271, 274–280, 325, 327, 328, 335
magnetic linear dichroism, 250, 254, 290
magnetic resonance, 14, 90, 125–127, 129, 166, 301, 304, 305, 308–311, 313, 314, 316
magnetic scattering, 191, 204, 207, 212, 213, 217, 221, 231, 241, 242, 244–249
magnetic skyrmion, 82, 83
magnetic susceptibility, 7, 31, 32, 35–37, 39, 41, 42, 67, 69, 93, 97, 117, 118, 120, 327
magnetic x-ray scattering, 238–243, 248, 255
Magnetic x-ray Transmission Microscopy (M-XTM), 293, 294
magnetite, 3, 9
magnetocrystalline anisotropy, 145, 146, 147, 149, 166
magnon, 75, 78, 184, 187, 189, 223, 227, 228
Maxwell equations, 7, 329
Maxwell, James Clerk, 7, 8
Mermin–Wagner theorem, 79
MOKE, 90, 169–172, 174–177, 179
molecular field, 8, 9, 64–66, 68, 70, 71
Mössbauer effect, 161, 162, 166
Mott insulator, 11, 59,
muon spin rotation, 151–154, 166

Néel temperature, 69, 118
Néel wall, 80, 82, 282
Néel, Louis, 9
neutron diffraction, 90, 192, 202, 213–216, 218–220, 222, 232
nitrogen-vacancy (NV), 301–303, 306, 308, 317
NV magnetometry, 301, 305, 306, 308, 309, 317
nuclear magnetic resonance, 126–129

oersted, 20, 21, 23, 25, 26, 325–328
Oersted, Hans-Christian, 6, 7
orbital angular momentum, 29, 32, 33, 37, 42, 44, 64, 171

paramagnetism, 20, 37, 40–43
Pauli exclusion principle, 43, 51, 54, 56
Pauli, Wolfgang, 9
Poisson, Simeon Denis, 6
polarized neutron scattering, 218, 241

resonant X-ray scattering, 91, 244, 245, 247, 250
RKKY interaction, 62, 73
Russel–Saunders coupling, 43

scanning electron microscopy, 260, 263,
scanning Hall probe microscopy (SHPM), 287–290
spin polarized scanning tunneling microscopy (SP-STM), 284
scanning X-ray microscopy, 291, 292
SEMPA, 263, 264, 266, 298
SNOM 90, 177–182, 189
spallation neutron source, 192–196, 216, 227, 228
spin glass, 61, 72–74, 86, 107, 108, 120–122, 161, 162
spin wave, 74–79, 90, 145, 150, 182–190, 223, 225, 227, 228, 234, 237, 305
spin–lattice relaxation, 129, 134, 135, 139–141
spin–orbit coupling, 42, 83, 125, 146, 166, 171, 253, 254
spin-SEM, 263–266
SPLEEM, 266–268, 294, 298
SP-STM, 284, 285, 287, 296–298

SQUID magnetometer, 104, 108–115, 117, 122
SQUID-VSM, 115
superconducting magnet, 106, 108, 109, 112, 114, 139, 252, 320, 321, 326
superexchange, 55–61
Sushruta, 3, 4

tesla, 20, 325–328
Thales of Miletus, 3, 4

van Vleck paramagnetism, 41
Verdet constant, 169
vibrating sample magnetometer, 104, 106, 115

Weiss, Pierre, 8

X-PEEM, 295, 296
X-ray photoelectron microscopy, 294, 295
XMCD, 91, 250–252, 254, 256, 290, 291, 293–296, 298
XMLD, 254, 255, 290, 294, 296

Zeeman energy, 64, 143